2nd International Conference on Advanced Technologies in Design, Mechanical and Aeronautical Engineering (ATDMAE 2018)

IOP Conference Series: Materials Science and Engineering
Volume 408

Dalian, China
1 – 3 July 2018

ISBN: 978-1-5108-7258-5
ISSN: 1757-8981

Printed from e-media with permission by:

Curran Associates, Inc.
57 Morehouse Lane
Red Hook, NY 12571

Some format issues inherent in the e-media version may also appear in this print version.

Printed by Curran Associates, Inc. (2018)

For permission requests, please contact the Institute of Physics at the address below.

Institute of Physics
Dirac House, Temple Back
Bristol BS1 6BE UK

Phone: 44 1 17 929 7481
Fax: 44 1 17 920 0979

techtracking@iop.org

Additional copies of this publication are available from:

Curran Associates, Inc.
57 Morehouse Lane
Red Hook, NY 12571 USA
Phone: 845-758-0400
Fax: 845-758-2633
Email: curran@proceedings.com
Web: www.proceedings.com

Table of contents

Volume 408

2018 2nd International Conference on Advanced Technologies in Design, Mechanical and Aeronautical Engineering (ATDMAE 2018)

1-3 July 2018, Dalian, China

Published online: 1 October 2018

Preface

Papers

Mechanical Engineering

IOP Conf. Series: Materials Science and Engineering **408** (2018) 011001 doi:10.1088/1757-899X/408/1/011001

PREFACE

ATDMAE 2018 was held in Dalian, China during Jul.1-3, 2018. ATDMAE 2018 was co-organized by Dalian University of Technology, Nazarbayev University and Hong Kong Society of Mechanical Engineers (HKSME). The conference provides a useful and wide platform both for display the latest research and for exchange of research results and thoughts in Advanced Technologies in Design, Mechanical and Aeronautical Engineering. The participants of the conference were from almost every part of the world, with background of either academia or industry, even well-known enterprise. The success and prosperity of the conference is reflected high level of the papers received.

The proceedings are a compilation of the accepted papers and represent an interesting outcome of the conference. This book covers 3 chapters: Mechanical Engineering; Aeronautical Engineering; Design Engineering.

We would like to acknowledge all of those who supported ATDMAE 2018. Each individual and institutional help were very important for the success of this conference. Especially we would like to thank the organizing committee for their valuable advices in the organization and helpful peer review of the papers.

We sincerely hope that ATDMAE 2018 will be a forum for excellent discussions that will put forward new ideas and promote collaborative researches. We are sure that the proceedings will serve as an important research source of references and the knowledge, which will lead to not only scientific and engineering progress but also other new products and processes.

Prof. Yong Zhao, Nazarbayev University, Kazakhstan

COMMITTEES

Conference Chairman

Prof. Yong Zhao, Nazarbayev University, Kazakhstan

Conference Chair

Prof. Xiaohui Su, Dalian University of Technology, China

Conference Program Co-Chairs

Prof. Zijian Wang, University of Kansas, USA

Prof. Bin Wang, Brunel University London, UK

Prof. Tongming Zhou, The University of Western Australia, Australia

Dr. Yang Wenming, National University of Singapore, Singapore

International Technical Commmitte

Prof. Weihua Liu ,Nanjing University of Aeronautics and Astronautics, China (Chair)

Dr. Redouane Zitoune, Clément Ader Institute of Toulouse University, France

Prof. B. T. N. Sridhar, Anna University, India

Dr. Zhangming Wu, Cardiff University, UK

Dr. Pierre Kerfriden, Cardiff University, UK

Prof. Kuei-Yuan Chan, National Taiwan University, Taiwan

Prof. K.M.Parammasivam, Madras Institute of Technology, India

Prof. S.Thanigaiarasu, Madras Institute of Technology, India

Prof. TEH Soo Lee, Nazarbayev University, Kazakhstan

Dr. GUELAILIA Ahmed, Satellite Development Center, Algerian Space Agency, Algeria

Dr. Zhauyt Algazy, Satbayev University, Kazakhstan

IOP Conf. Series: Materials Science and Engineering **408** (2018) 011003 doi:10.1088/1757-899X/408/1/011003

Peer review statement

All papers published in this volume of *IOP Conference Series: Materials Science and Engineering* have been peer reviewed through processes administered by the proceedings Editors. Reviews were conducted by expert referees to the professional and scientific standards expected of a proceedings journal published by IOP Publishing.

IOP Conf. Series: Materials Science and Engineering **408** (2018) 012001 doi:10.1088/1757-899X/408/1/012001

One Micro-thrust Measurement Method Based on the Dynamic Photoelastic Theory

Yang Ou*, Jianjun Wu*, Yu Zhang, Jian Li and Xinru Du

College of Aerospace Science and Engineering, National University of Defense Technology, 410073, Changsha, P.R.China

*Corresponding author: naihe620@163.com; jjwu@nudt.edu.cn

Abstract. A new method is adopted to measure the micro-thrust of the plasma thruster. In this method, the dynamic photoelastic theory is used to record the stress loaded on the special structure. This technique is called dynamic photoelasticity method. The photoelastic element is the core of the whole system as it can transform the thrust changes into the variation of fringe images. The variation of thrust can be recorded by a CCD camera and then the thrust changes in real-time could be obtained, which is helpful for developing a higher precision propulsion system. This method built in this paper is not only non-contact and non-destructive but also easy to operate with simple structure and high precision.

1. Introduction

Normal thrust measurement methods for the plasma thruster, such as the pendulum, twist and cantilever beam often fail to recognize such values of the thrust which are lower than the mN magnitude. Accurate measurements of the visatergo are not only prerequisite for orbit raising, drag makeup, attitude controlling and constellation station keeping but also thus useful for aerospace applications [1] [2].

Many studies have been conduct on the accurate measurement of the thrust. As reported in the pioneering work by Kurt A. Polzin [3], an electric propulsion thrust stand capable of supporting testing of thrusters has been developed on the basis of the pendulum while such method cann't achieve a high precision. The most sensitive of these past devices appears to be the optical angular readout from Ref. [4] designed in Brian C D's papers, which can measure thrust as low as 7 nN with loads up to 12kg. In a word, limited attempts having been made to measure the thrust cann't show an absolute result, that is to say, there is no high accuracy measurement system at present. And so it's urgent to raise the measurement accuracy of the measurement as exact as possible for promoting the development of micro-satellites.

One way to overcome these problems is to develop the sensor with better performance but this technique is rather limited. Therefore, the thrust is converted into other signal which could be recognized in a high precision. With the mechanical knowledge, we know the overall stress of the thruster varies with the thrust and such variation meets one certain physical rule. What's more, it provides one good idea to measure the thrust converted from the stress. The thrust can't be changed naturally while we can change certain parameters to get the large stress. As a stress monitoring item, the photoelasticity method has been widely used as a reliable tool to record the variation of the stress.

This paper builds one new measurement method basing on thephotoelastic theory for approving the situation that the traditional way of micro-thrust measurement cannot get the transient thrust value with high precision and the measuring process of piezoelectric sensors is sensitive to the electromagnetic interference. The photoelastic element is the core of the whole system as it can transform the thrust changes into the variation of fringe images. Thus adopting a CCD to record these images so that the thrust changes in real-time could be obtained, which is helpful for studying the mechanism of micro thrusters. This method built in this paper is not only non-contact and non-destructive but also easy to operate with simple structure and high precision.

2. Photoelastic theory

The dynamic photoelastic method that has gained wide application in industry was first described by Tuzi in his fundamental 1928 study. The author proposed a new method for measuring the stresses in an isotropic material. The phenomenon that the method is based on is in the difference of the refractivity to the direction of stresses in the materials under load, and it is called the anisotropy of the stressed medium. Traditionally, a Cranz-Schardin system with array cameras was used to capture a series of photoelastic fringe patterns as showed in Cranz and Schardin's papers. This phenomenon has been termed the photoelastic effect and method for assessing the stress-strain state of structural elements is referred to as the photoelastic method. Nowadays, it is very easy to arrange a digital dynamic photoelastic system so that dynamic photoelastic fringe patterns can be taken using a digital video camera and analyzed later [5] [6].

Photoelasticity is a branch of mechanics that studies the interaction of structures, fluids, and lights. It involves ships, underwater weapons, aircraft, automobiles, buildings, chemicals, marine and water conservancy projects, and many other fields. It is an experimental technique for analyzing stress or strain fields in solids with birefringent properties. These normal photoelastic materials contain glass, epoxy and so on. In principle, when a photoelastic material is deformed, the electric vector of an incident polarized light beam can be decomposed into two components along the axes of the principal stresses in the material due to its temporary birefringence. Because of the difference in refractive indices along the axes of the principal stresses, the two components will be retarded differentially as they propagate through the material, leading to a relative phase shift. Consequently, an interference pattern will be formed as the two components are recombined outside the material by a downstream polarizer which is called the ''analyzer'' as well. Photoelasticity is, therefore, a classic common-path interferometric technique, with the surface of the photoelastic material acting as the beam splitter.

Based on the correlation between stress and the indices of refraction for temporary birefringent materials formulated by Maxwell, in the simplest case, the stress-optic law for a plate with only variable in-plane principal stresses can be written as [7]:

$$\sigma_1 - \sigma_2 = \frac{Nf_\sigma}{h} \tag{1}$$

where σ_1, σ_2 are the principal stresses in the temporary birefringent material, N is the relative retardation which is also known as the ''fringe order'', f_σ ($=\lambda/C$, with λ being the wavelength and C the stress-optic coefficient) is the material fringe value (in units of N/m fringe) corresponding to the minimum difference in principal stresses to produce a unit change in the fringe order in a material of unit thickness, and h is the distance traversed by the light in the stressed plate [7].

Equation (1) applies to both static and dynamic loading conditions, except that the material-fringe value under dynamic loading (f_σ *) is usually 10%–30% higher than that for the static case [5].

Dynamic photoelasticity has been widely used for transient stress analysis in fracture mechanics and geophysics. Dynamic photoelasticity was used by Jian Chang to visualize the fundamental behavior of polycarbonate gears and observe the stress under the impact loadings with diffenert pulse width [8]. Image-processing techniques and stroboscopic photoelasticity were combined to observe the wave stress field generated by commercial ultrasonic probes in Nam and Lee's papers. Xiaomao Li carried out the material surface strain measurement with photoelastic coatings for solving such problem that it is not easy to interpret distribution of the surface strain field [9]. However, seldom is such method applied to measuring the thrust, more application are in stress testing. And the goal of this study is to obtaining the thrust data on the spatial and temporal dynamics of the variations when the micro-thrusters work.

Such method is not only a non-destructive method for measuring stress, but also one new means to analysis the non-transparent material's stress as for its reliability, toughness, economical efficiency and easy-using. The fringe images change in accordance with the stress; therefore, we can get specific value of the stress by recording these images with one CCD device. What's more, if the exposure time is arranged to be short enough, the fringe images recorded by the CCD at different times can be close to the real time variation of the stress of the photoelastic element. And then, the thrust can be calculated according to the equation of mechanics of materials as follows:

$$\sigma = \frac{F}{S} \tag{2}$$

Where F is the internal force, S is the cross-section and σ is the principal stress which obtained from the fringe image.

Control the loading so that the thrust is equal to the internal force so that we can get the thrust by measuring the stress.

3. Experimental setup and theory

In this paper, the photoelastic element is regarded as the core of the system as it can reflect the variation of thrust through the change of fringe images. And such measurement method is built according to the characteristics of micro thruster and photoelastic theory. On the basis of the system, we can obtain the real-time change of the thrust when using high-speed signal collector to record the fringe images. This measurement is non-contact and non-destructive for a sensor or other measuring devices installed on the thrusters are not needful. Therefore, the system reduces the interference of mechanical vibration and power supply cables while solves problem like zero drift and unstable equilibrium position calibration, difficulty and low accuracy. General speaking, it has the advantages of simple structure, easy operation and high reliability while the most important thing is to provide a method to solve the problem of the traditional thrust measuring method which is difficult to measure the transient thrust.

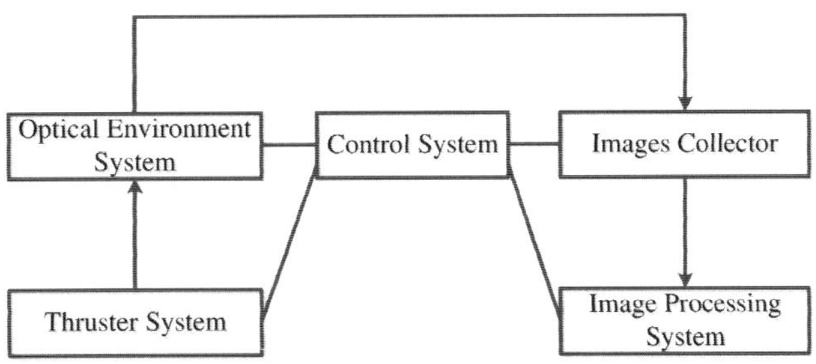

Figure 1. System composition.

The instantaneous thrust measurement system basing on the principle of photoelasticity consists of four parts, which are optical environment system, thruster system, control system, fringe images collector and image processing system. The overall composition is shown in Figure 1.

The main function of the optical environment system is to create an optical environment for the photoelastic element so that the stress change of the acoustic elastic element can be reflected by the fringe images. The thruster system generates thrust and to amplify and transfer the thrust to the photoelastic element so that it is in a stress state. The duration of the pulsed laser is set to be 30-50 ns which correspond to the shutter time of the high-speed camera at a speed of million images per second in the control system. With this arrangement, we can think that the deformation of the photoelastic element is approximate zero within the time interval of light passing from such element to the camera. The key of such measurement method by photoelasticity is the fringe pattern of the elastic element. Therefore, the quality of fringe pattern directly determines the accuracy of thrust measurement. Matlab is adopted in the image processing system. The control system and the fringe image collector ensure that the conversion thrust is approximate equal to the average thrust during the set time interval.

3.1. Optical environment system

As shown in Figure. 2, the optical environment system includes a light source 1, a collimation lens 2, a polarizer 3, 1/4 wave plates 4 and 6, a photoelastic element 5, an analyzer 7 and a condenser 8. According to different ranges of the thrust, the photoelastic element 5 can be replaced to meet the requirement of measurement. Generally, the material of 5 can choose epoxy resin or polycarbonate with high optical sensitivity and temporary birefringence effect after loading. The light source 1 uses white light. In order to make the fringes clear in high speed photography imaging, the distance between the light source 1 and the Collimation lens is set to be a proper value. The function of the collimation lens 2 is to turn the conical beam from the light source into parallel light so that the light can be vertically entered into the photoelastic element. The polarizer 3 and the analyzer 7 are the same type of optical instrument but play different roles. Natural light is turned into plane polarized light by the polarizer 3 while the analyzer 7 is arranged to check the polarization state of the light wave. The 1/4 wave plate 4 acts to transform the plane polarized light into circularly polarized light. On the contrary, 6 is designed to transform the circularly polarized light into plane polarized light. The condenser 8 focuses parallel lights to the CCD camera 9.

The photoelastic element is in a very stable optical environment for producing photoelastic effect with the combination of these elements. More importantly, this optical path has high stability which can effectively reduce the error in late loading and data acquisition.

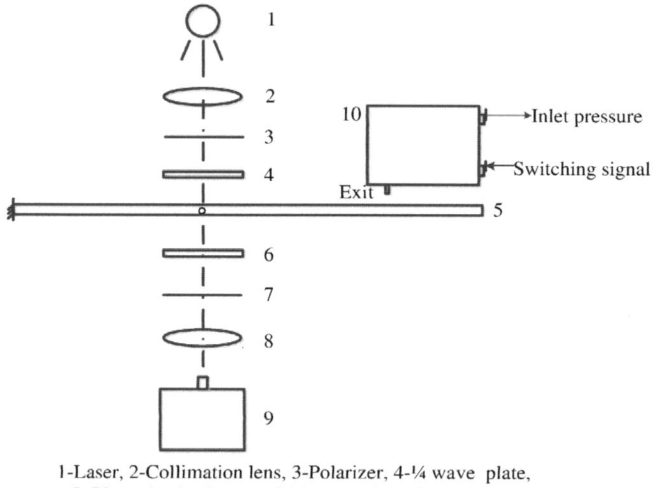

1-Laser, 2-Collimation lens, 3-Polarizer, 4-¼ wave plate,
5- Photoelastic element, 6-¼ wave plate, 7- Analyzer,
8- Condenser, 9- CCD camera, 10- Thruster

Figure 2. Optical Environment System

3.2. Thruster system

The purpose of this paper is to realize the force measurement. Therefore, different types of thrusters can be replaced according to the requirements of the force measurement without considering the structure and function of the thruster. Pulsed plasma thruster is taken as an example. The thruster sprayed the plume which impacted on the photoelastic element and made such element under stress.

According to the three laws of Newton, we can know that thrust is directly related to the plume. Meanwhile, the sound velocity variation of elastic element is directly related to the thrust's change. The whole thruster is integrated into one hull as shown in Figure 2. The pressure value at the nozzle is one important parameter of the thruster working condition. Therefore the wireless pressure conveyer measures and sends the pressure value to the computer for displaying the pressure value. In order to reduce the amount of calculation, the load can be equivalent to a concentrated load, so the plume ejection area is small.

The key of hull has a small outlet that causes the plume to be sprayed out from that outlet. The fringes variation of elastic element is directly related to the thrust's change. The outlet can amplify the thrust in an appropriate ratio, and then provides a larger thrust for recording the velocity variation more precisely. Finally, real data is obtained by scaling the value according to the outlet parameters.

3.3. Control system

The system structure diagram is shown in Figure 3. For getting more fringe images with higher quality, it is necessary to make the light triggering time, thrust loading time and transient images acquisition time accurate synchronization control. The time for the thruster start working is equivalent to the origin of the coordinates in the coordinate system as shown in Figure 3. When the thuster works, the plume is ejected from the nozzle. And then the pressure on the exit will be recorded and a sensor 13 is triggered at the same time. The pressure trigger 13 produces a signal PC as the first trigger time of the light source 1, thus the first pulse light is generated at t_{Q1}. When the photoelastic element 5 is subjected to imact, the electronmagnetic trigger will produce another signal CT as the second trigger time of the light source due to the plasma electronmagnetic shielding effect. The time t_{Q2} in which the light source produces second pulses of light is controlled by t_k. Δt is up to the electronic control system of light source, and Δt is the time when the photoelastic element needs to be collected fringe images.

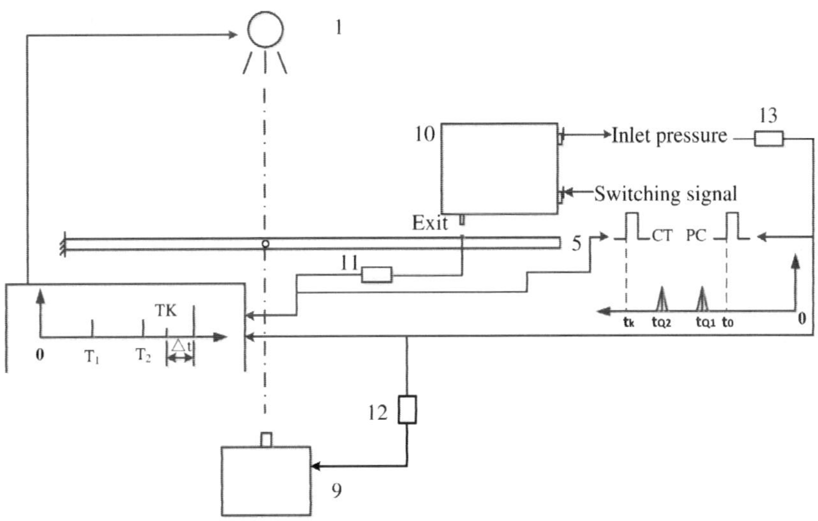

11-Electronmagnetic trigger, 12-Time delayer, 13-Pressure trigger

Figure 3. Control System

3.4. Images collector

The acquisition of image is mainly realized by CCD camera. It is important to note that the fringe image collected is not reflected the stress of the elastic element as the propagation of light between 5 and 9 also takes time. However, such time is small comparing to the light velocity, we can approximate the information of the 2 moment in the error range so as to simplify the calculation process.

3.5. Images processing system

The measurement results obtained by using the images collector are lots of stress fringe images. These pictures contain two important parameters: the stress direction and fringe order. The fringe order

IOP Conf. Series: Materials Science and Engineering **408** (2018) 012001 doi:10.1088/1757-899X/408/1/012001

represents the difference of the principal stress, but the principal stress can't be calculated by the difference and the direction angle of stress. Therefore, the important work of obtaining the fringe image is the separation of the stress.

In this paper we use boundary element method to separating main stress. It can simplify the problem and provide a practical method for separating main stress. The element 5 is bored as shown in Figure 2 which introduces one free boundary, thus the main stress magnitude can be measured directly by the tripe series. Two symmetry axis of the isochromatic fringes depends on the direction of the principal stress. Because the stress value obtained after drilling is deviated from the value before drilling, it is necessary to calculate the stress value before opening by the corresponding formula from the stress value after opening. These steps can be realized by MATLAB.

The picture of each strain on the instantaneous thrust should always, and the control system and the transient image acquisition system ensures high-speed camera to get the instantaneous thrust roughly the size of the thrust as the set time interval, based on a series of stress image processing to obtain the thrust moment has successfully connected to can approximately reflect the real-time change of thruster thrust.

4. Discussion

In this paper, the photoelectric method, which has the advantages of simple structure, easy operation and high reliability, is first applied to the field of instantaneous thrust measurement field. Moreover, this method provides a basic idea for measuring the thrust of plasma thrusters, can realize high precision and accurate measurement of instantaneous thrust and impulse. Thrust measuring methods of the non-contact measurement, and non-destructive measurement methods, does not require a sensor or other measuring device is directly installed on the structure, can reduce the interference of mechanical vibration and power supply cables for thrust measurement, solve the present commonly used micro thruster and micro impulse measuring device of zero drift, balance position stable, low accuracy and calibration difficult problem.

References

[1] Haag T W 1991 Thrust stand for high-power electric propulsion devices Review of Scientific Instruments vol 62 chapter 5 pp 1186-91

[2] Cubbin E, Ziemer J, Choueiri E and Jahn R 1997 Pulsed thrust measurements using laser interferometry Review of Scientific Instruments vol 68 chapter 6 pp 2339-46

[3] Polzin K A, Markusic T E, Stanojev B J, DeHoyos A 2006 Thrust stand for electric propulsion performance evaluation Review of Scientific Instruments vol 77

[4] D'Souza B C, Ketsdever A D 2005 Direct Impulse Measurements of Ablation Processes from Laser-Surface Interactions 36th AIAA Plasmadynamics and Lasers Conference

[5] Asundi A K , Marokkey S 1994 Low-cost digital polariscope for dynamic photoelasticity Optical Engineering vol 33 chapter 9 pp 3052-3055

[6] Wang W C, Chen T, Lin S H 1990 Digital photoelastic investigation of transient thermal stresses of two interacting defects Journal of Strain Analysis for Engineering Design vol 25 chapter 4 pp 215-228

[7] Xi X, Zhong P 2001 Dynamic photoelastic study of the transient stress fields in solids during shock wave lithotripsy Journal of the Acoustical Society of America vol 109 chapter 3 pp 1226.

[8] Chang J 2005 Static and dynamic numerical calculation and dynamics photoelasticity experiments (in Chinese)

[9] Li X 2014 The experimental research and measurement of material surface's strain field based on photoelastic coating (in Chinese)

IOP Conf. Series: Materials Science and Engineering **408** (2018) 012002 doi:10.1088/1757-899X/408/1/012002

Numerical Investigation on Work Performance of Tocardo-type Turbine Using BEM Method

Renqiang Liu[1], Xiaohui Su[1], Junyu Zhao[2] and Dong He[2]

[1] School of Hydraulic Engineering, Dalian University of Technology, Liaoning Province, China, 116023.
[2] ChongQing Gearbox Co. Ltd, Chongqing City, China, 402260.

Renqiang_Liu@mail.dlut.edu.cn

Abstract: In this paper, work performance of Tocardo-type turbine is numerically investigated by using BEM method. Tocardo-type turbine is significantly different from lift-type turbine in which it owns long chord and thick blade and it is of bidirectional characteristics. Working performance of Tocardo-type turbine is obtained by using BEM method and the results release that it is not of lift-type turbine because the parameter of Cl/Cd is not chosen as maximum value. Compared with the performance of lift-type turbine, working coefficient of Tocardo-type turbine is little lower at optimum working condition. The results also show that the optimized tip speed ratio of Tocardo-type turbine is about 4.5 and working power reaches about 183kW as coming velocity is 4m/s.

1. Introduction

Energy is an important driver of economy. However, energy demand pressure, now, is coming more and heavier, the researchers start to study renewable energy gradually. In the new energy sector, tidal current energy has achieved significantly economic interests. Now, many countries are all raising tidal current projects supported by governments and companies [1], In different kinds of turbines, horizontal current turbine is popular because of its high efficiency, big self-starting torque etc., compared to other kind turbines. [2]

BEM[3] is most classical theory among those methods and widely used to analyze dynamics of wind turbine and current turbine. BEM theory assumes fluid pressures are equal between front and rear of wind rotor, so it ignore blade pressure reduction generated by trailing vortex, which leads to relatively large differences between test results and calculated results of BEM theory about wind turbine. Joukowsky improved BEM method in the situations of low tip speed ratio (TSR) through considering trailing vortex [4]. Micallef et al. did CFD simulation for solving a problem that BEM theory ignores kinetic momentum generated by tangential velocity. That work provided a few of fundamental insights and experimentally and numerically confirmed that radial flows reach appreciable magnitudes especially in the tip region [5]. Fluid motion, in the tip of blade, is extremely complex because of not only tangential motion but also turbulence flow. In the BEM computations, airfoil data are corrected for three-dimensional and rotational effects. However, such corrections are of another nature than what is needed near the tip, where the flow is largely attached and rotational effects are negligible. In order to include three-dimensional tip loss effects, Shen et al. suggested to compute the resulting force coefficients in equations [6]. Khchine combines new tip loss correction and new axial flow induction factor to optimize blade geometry based on Shen's method[7].

In this paper, we extend scope of BEM applications to study a non-lift-type turbine based on the conservations of momentum and angle momentum. One company have invented a new turbine called Tocardo-type turbine owing long chord and thick blade, which the geometry of the turbine is significantly different from traditional lift-type turbine designed by BEM. In the beginning, this paper introduces BEM and come up with a new form of BEM analysis Tocardo-type turbine. Next, various characteristic of Tocardo-type turbine are analyzed compared to lift-type turbine.

2. Blade elements momentum theory

2.1 The computational process and characters of lift-type turbine

BEM split blade to several elements for simulating two dimensional fluid motion based on assumption that adjacent blade elements make no influence and two dimensional fluid homogeneity flowing through blade element. As far as the horizontal current turbine, fluid has two flow modes that one is flowing along the horizontal axis and another is circular motion opposite to the direction of blade rotation. Considering momentum loss in axial and tangential motion, axial flow induction factor (**a**) and tangential flow induction factor (**a'**) are introduced to compute axial and tangential velocity for analyzing blade element output power. Finally, whole rotor output states equal to total all elements power.

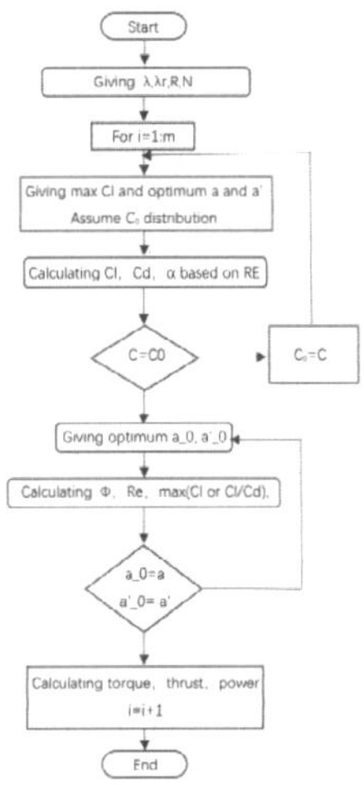

Figure 1. The flow chart of BEM for lift-type turbine

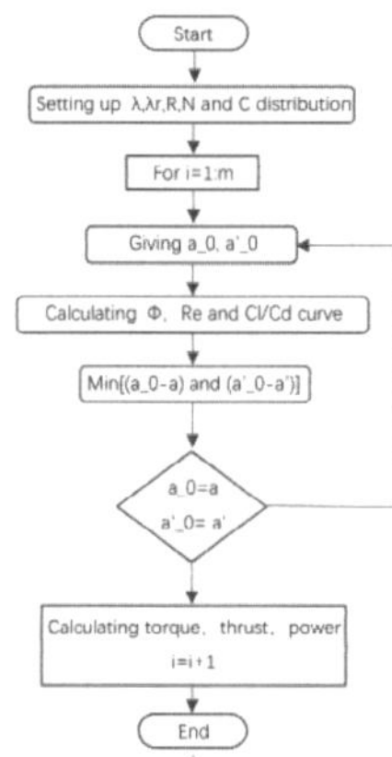

Figure 2. The flow chart of the new treatment to search for **a** and **a'**.

Essentially, the calculation of current turbine performance is continual convergent process of the axial and tangential flow induction factor. The process is plotted in figure 1. In the flow chart, λ is tip speed ratio, R is radius, N is numbers of blade, α is angle of attack, C is chord, Φ is angle from fluid velocity to the plane of rotation.

At the same time, it is necessary that one blade element's **a** and **a'** are iteratively calculated after chord is found based on the assumption of optimal **a** and **a'**. According to Reynolds number computed by assumed **a** and **a'**, a serious of Cl and Cd are computed in different angle of attack. Maximum Cl or Cl/Cd is selected from various Cl and Cd to obtain new **a** and **a'**. The process is repeating until convergent solutions are found. The lift-type turbine with small chords and refined blade, designed by using traditional form of BEM, can generate high lift and high power because maximum Cl or Cl/Cd is selected.

2.2 The computational process and characters of Tocardo-type turbine

Under the condition of giving chord distribution, the treatment with maximum Cl/Cd does not valid any more for the solution of **a** and **a'** in the BEM application of Tocardo-type turbine. The possible reason causing misconvergence of **a** and **a'** is seeking for then. Firstly, longer chord distribution of Tocardo-type turbine along with radial direction is discussed. Bigger chord may directly affect Reynolds number of each blade element, which may result in the misconvergence of **a** and **a'**. In this paper, changing original optimization objective of searching maximum Cl/Cd to minimum (**a-a0**) and (**a'-a'0**) is implemented. The process of this new treatment is shown in figure 2.

Comparing with traditional BEM, the new method proposed in this paper for Tocardo-type turbine still obeys the basic conservations of momentum and angle momentum, only optimization objective is different. General speaking, the treatment is valid for any chord length during the solution of **a** and **a'** in which the calculation of a and a' for chord distribution of lift-type turbine is one specified case.

3. Analyze of Tocardo-type turbine based on BEM

3.1 Geometry comparison between Tocardo-type turbine and lift-type turbine

The blade geometry of Tocardo-type turbine and lift-type turbine are obviously different shown in figures 3 and 4. Tocardo-type turbine blade is more bigger and fatter than that of lift-type turbine. The quantitative differences of chord and angle of attack are illustrated in figures 5, 6.

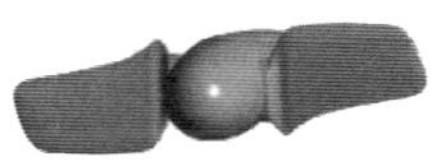

Figure 3. Geometry of Tocardo-type turbine

Figure 4. Geometry of lift-type turbine

Figure 5 shows that the chord of Tocardo-type turbine chord is significantly longer than that of lift-type turbine and the chord, in middle part of Tocardo-type turbine blade, is slightly changed while the chord changed rapidly in the location of tip and root. In addition, the distributions of twist angle for two kind of turbine are plotted in figure 6. It is clearly that all twist angle of blade of Tocardo-type turbine is bigger than that of lift-type turbine. Those data including chord distributions and twist angles is line drawn based on the picture from Tocardo company websites.

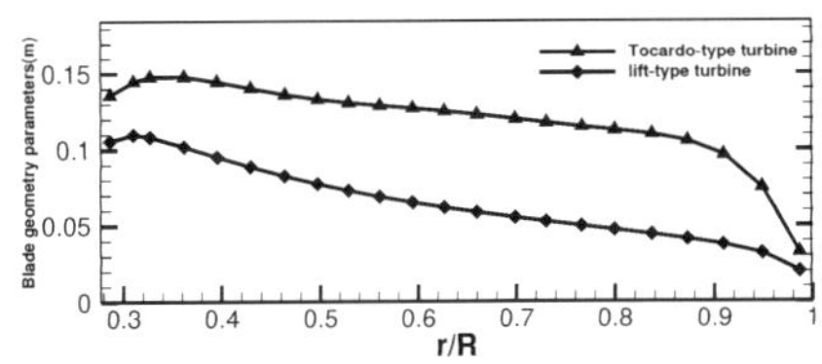

Figure 5. Blade geometry parameters of Tocardo-type turbine and lift-type turbine

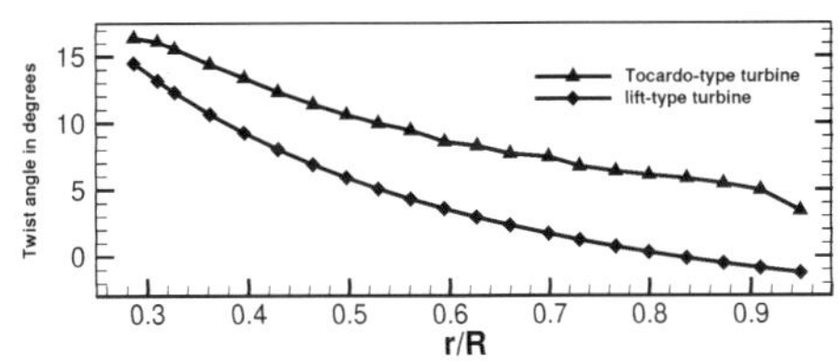

Figure 6. Twist Angle distributions of Tocardo-type turbine and lift-type turbine.

3.2 Selection of Airfoils

Different series airfoils are studied for analyzing its influence of Tocardo-type turbine before further investigation. Three low-Reynolds number airfoils, S805A, ClarkV and NACA63210, respectively are chosen shown in figure 7. Under same coming velocity, 4m/s, tip speed ratio, 6 and radius, 2.15m, the proposed BEM treatment is implemented to design turbines based on three different airfoils.

Figure 8 shows the solutions by using the proposed BEM treatment. The results illustrate that airfoils have little influence on turbine power but serious influence on turbines' twist angle. Among those turbines, S805A is fit for Tocardo-type turbine since twist angle of S805A is smaller and blade is thicker. Therefore, S805A is taken as basic airfoil in later work of this paper.

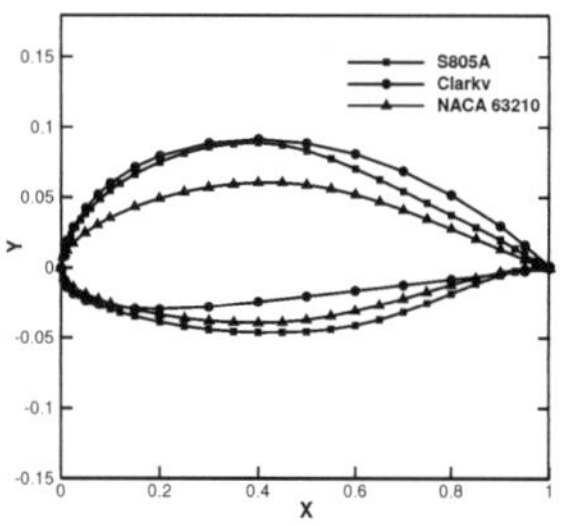

Figure 7. Three Low-Reynolds number Airfoils

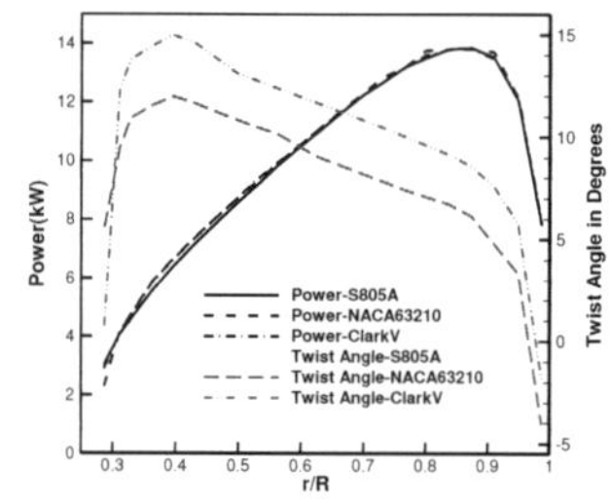

Figure 8. Power and twist angle distributions of three turbines

3.3 The Performance Analysis between Tocardo-type turbine and lift-type Turbine

Work performance of Tocardo-type turbine is computed by using the proposed BEM treatment. All the results have been obtained at the conditions of incoming velocity, 4 m/s, tip speed ratio, 6. Tip and root loss corrections are included in the process of analysis.

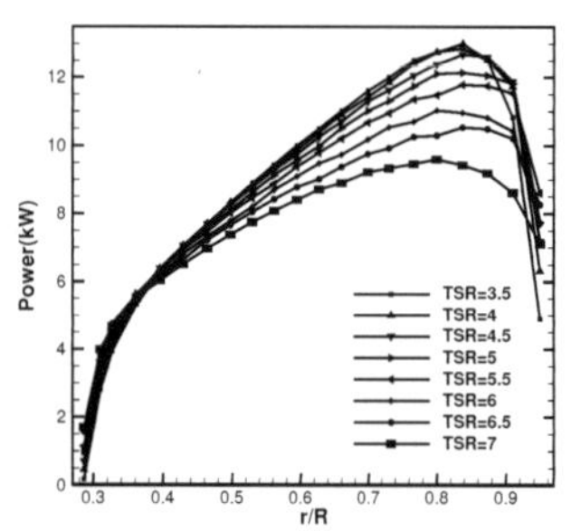

Figure 9. Tocardo-type turbine power distributions in different TSRs.

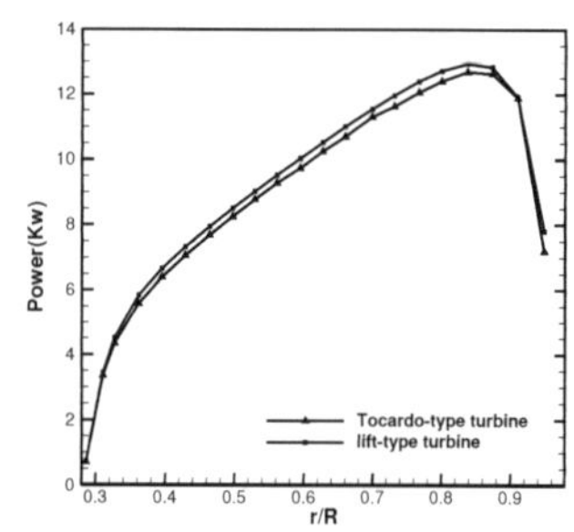

Figure10. Comparison of power between Tocardo-type turbine and lift-type turbine in TSR=4.5.

The calculated working power results of Tocardo-type turbine are plotted in figure 9. Figure 9 shows that the power of Tocardo-type turbine declines as TSR raising from 5 to 7. Moreover, the power of Tocardo-type turbine gradually increases with TSR from 3.5 to 4.5 and reaches to maximum efficiency as TSR is equal to 4.5.

Figures 10-12 shows the power comparisons between Tocardo-type turbine and lift-type turbine as TSR is 3.5, 4.5 and 5.5. The results indicate that the output power of lift-type turbine is larger than that of Tocardo-type turbine. The more power results are summarized in the figure 13. Among those TSRs, 4.5 is found optimized TSR and the power of Tocardo-type turbine at that TSR is about 183kW and the corresponding working coefficient is about 39.5%. The power of lift-type turbine is higher than that of Tocardo-type turbine except for the case that TSR equals to 3.5. In addition, the difference becomes more and more larger as TSR raising from 4 to 7. The reason for the difference between two kinds turbines is explained that the blade of Tocardo-type turbine with longer chord losses the optimum working condition of airfoil and then causes the reduction of efficiency.

IOP Conf. Series: Materials Science and Engineering **408** (2018) 012002 doi:10.1088/1757-899X/408/1/012002

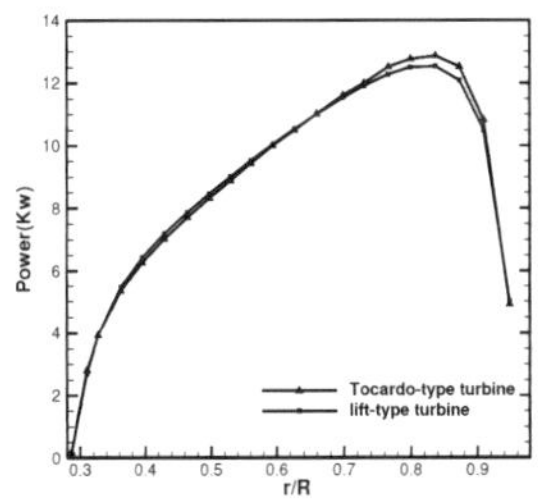

Figure 11. Comparison of Power between Tocardo-type turbine and lift-type turbine in TSR=3.5

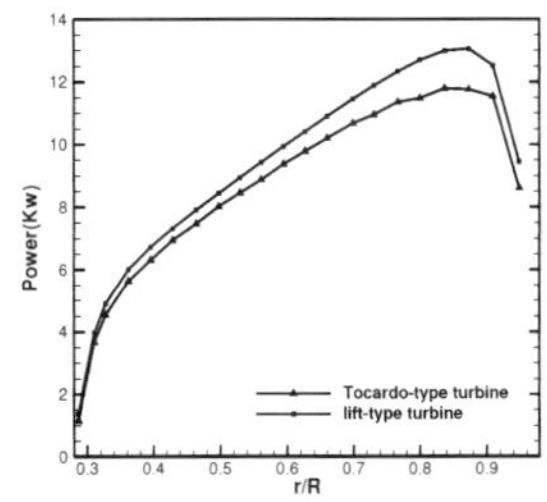

Figure12. Comparison of Power between Tocardo-type turbine and lift-type turbine in TSR=5.5

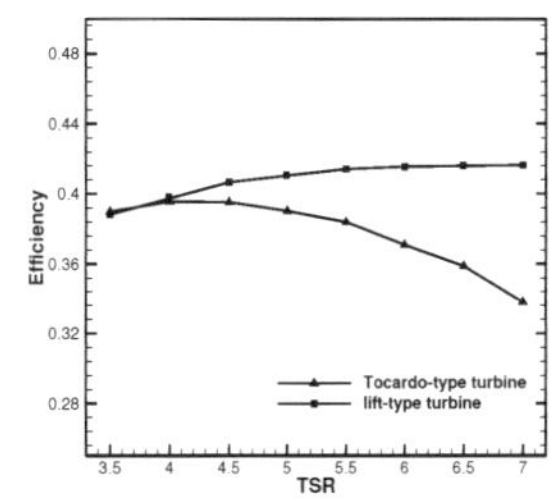

Figure13. Comparison of Efficiency between Tocardo-type turbine and lift-type turbine

3.4 The difference of Cl/Cd between Tocardo-type turbine and lift-type Turbine

In order to verify difference of Cl/Cd between Tocardo-type turbine and lift-type Turbine, Cl and Cd, corresponding convergent **a** and **a'**, are selected to further study based on same conditions: incoming velocity is 4m/s, TSR is 4.5, radius is 2.15m. The distributions of Cl/Cd shown in figure 14 and 15.

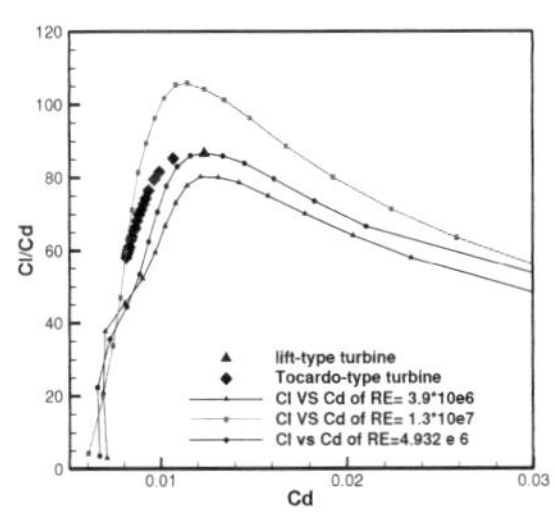

Figure 14. Comparisons of Cl/Cd versus Cd between Tocardo-type turbine and lift-type turbine

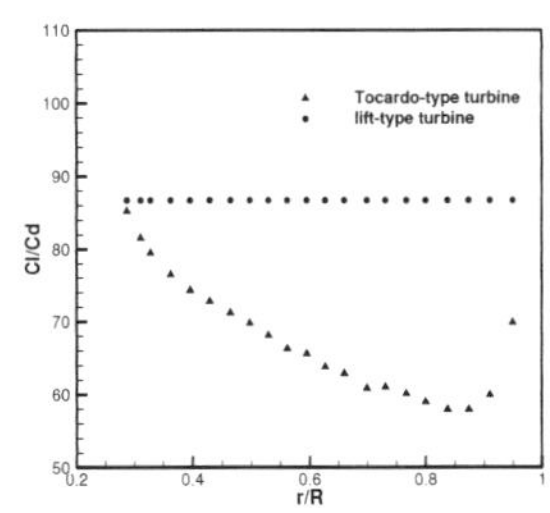

Figure 15. Comparison of Cl/Cd versus r/R between Tocardo-type turbine and lift-type turbine

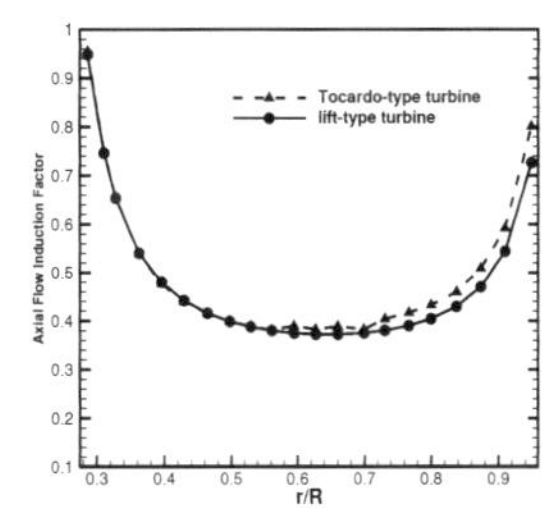

Figure 16. Axial flow induction factor between Tocardo-type turbine and lift-type turbine

Figure14 and 15 show that Cl/Cd of different blade elements of lift-type turbine are equal and located in maximum value when **a** and **a'** are convergent. In contrast, the values of Cl/Cd are not equal to the maximum value in the design of Tocardo-type turbine. The values of Cl/Cd for Tocardo-type turbine gradually decrease as r/R raising from 0.28 to 0.87. Cl/Cd of lift-type turbine always higher than Tocardo-type turbine which is believed the reason that lift-type turbine gives better output power than Tocardo-type turbine. At the same time, the conclusion could be drawn that Tocardo-type turbine is not lift-type turbine.

At last, the reason how the values of Cl/Cd influent turbines' power is analyzed. The distributions of **a** between Tocardo-type turbine and lift-type turbine are drawn in figure 16. The values of **a** for Tocardo-type turbine is more and more higher than that of lift-type turbine as r/R raising. The phenomenon is believed due to the difference Cl/Cd between two type turbines. Since larger value of **a** implies smaller incoming kinetic energy to be converted by turbine, the power of Tocardo-type turbine decreases.

4. Conclusion

In this paper, the Tocardo-type turbine has been calculated and analyzed by using BEM. The results are compared with that of lift-type turbine and are analyzed in detailed. The following remarks are addressed:

1. The Tocardo-type turbine is not lift-type turbine.
2. Different airfoils have a little influence on output power of Tocardo-type turbine.

3. Working efficiency of Tocardo-type turbine is about 39.5% as TSR equal to 4.5 and output power is about 183kW. The output power is little lower than that value of lift-type turbine at optimum working condition.

Reference

[1] The World Offshore Renewable Energy Report 2004-2008. *R.*

[2] Zhang Liang, Li Xin-zhong, Geng Jing and Zhang Xuewei. Tidal Current Energy Update 2013. 2013. *J. Advances in New and Renewable Energy*. **1** 53-68.

[3] Zhang Huiying. Design and Analyze of Horizontal Tidal Current turbine Blade of DUTTCT_H120. 2016. *D.Dalian University of Technology.*

[4] Sun Zhenye, Chen Jin. Improved blade element momentum theory for wind turbine aerodynamic computations. 2016. *J. Renewable Energy.* **96** 824-31.

[5] Micallef D, Bussel G, Ferreira C S. An investigation of radial velocities for a horizontal axis wind turbine in axial and yawed flows. *2013. J. Wind Energy*. **16(4)** 529-44.

[6] Shen Wen Zhong, Mikkelsen Robert and Nørkær Sørensen Jens. Tip Loss Corrections for Wind Turbine Computations. *2005. J. Wind Energy.* **8** 457-75.

[7] khchine Y El, Sriti M. Tip Loss Factor Effects on Aerodynamic Performances of Horizontal Axis Wind Turbine. 2017 *Germany April. 2nd International Conference on Advances on Clean Energy Research, ICACER 2017.*

IOP Conf. Series: Materials Science and Engineering **408** (2018) 012003 doi:10.1088/1757-899X/408/1/012003

A Computational Fluid Dynamic Study for Temperature Uniformity at a Wind Tunnel Outlet with Effects of Heat Transfer along Tunnel Walls

Li Song[1,2,a], Chen XiaoBing[3], Huang ZhiLong[1,2] and Liao DaXiong[1]

[1] State Key Laboratory of Aerodynamics, China Aerodynamics Research and Development Center, Mianyang Sichuan 621000, P R China
[2] Facility Design and Instrumentation Institute, China Aerodynamics Research and Development Center, Mianyang Sichuan 621000, P R China
[3] Advanced Analysis Group, Advisian Singapore (WorleyParsons Group), #13-03 TripleOne Somerset, 111 Somerset Road, Singapore, 238164

[a] Corresponding author: lisonic@foxmail.com

Abstract. A Computational Fluid Dynamics (CFD) Modelling was carried out to investigate temperature uniformity at a wind tunnel outlet region to benefit its design and construction. Low temperate (110K) nitrogen entered the wind tunnel at inlet, passed through two turning area with 18 turning vanes and went into expansion session, stable session and congestion session, finally towards outlet discharge nozzle session. The model considered static nitrogen isolation layer in physical model and three more isolation layer outside the tunnel as well as possible heat transfer from ambient environment. Two case scenario for supersonic and subsonic at outlet (Mach Number 0.9 and 1.3) were conducted with a uniform temperature imposed at inlet, as well a 3rd case scenario (Mach Number 0.9) with a non-uniform temperature distribution at inlet. Existence of porous zone (to model honeycomb session) and porous interfaces (to model damping interfaces) at the stable session were also included. Contours of temperature and velocity on representative cross sections were presented, and effects of Mach number, total pressure and initial temperature distribution at inlet on temperature uniformity at outlet were discussed. Based on the numerical results, improvement of tunnel geometry for better flow uniformity at outlet regions were also suggested.

1. Introduction

Wind tunnel has been a well-established discipline and it is a mature tool used in aerodynamic research [1-2]. Large-scale cryogenic wind tunnels, such as ETW and NTF, are the best groud testing facilities to obtain aircraft flow characteristics in the real flight conditions.[3] Design and testing of such wind tunnels become an important task before construction of these tunnels, in which CFD plays an increasingly important role[4-6]. With the development of modern technology, it is possible to do full-scale CFD modelling with high performance computers, which will reduce demand and cost for wind tunnel physical testing [7-9].

To assess temperature uniformity at a wind tunnel outlet region and finally to benefit its design and construction, a Computational Fluid Dynamics Modelling was carried out to investigate temperature uniformity at a wind tunnel outlet region. The target was to find effects of incoming flow at inlet (low and high Mach number, total pressure, and uniform or non-uniform temperature distributions) as well as four layer isolation materials outside the tunnel wall. Possible heat transfer from ambient environment was also considered.

CFD simulation results were presented with contours of temperature and velocity on representative cross sections. Based on the numerical results, improvement of tunnel geometry for better flow uniformity at outlet regions were also suggested.

The High-performance computing for CFD model construction, Meshing, set-up and running was finished with Ansys WorkBench.

2. Model and Boundary Condition

3D geometry was imported into Ansys SpaceClaim and simplified into a half due to its symmetry property at the center interface (Figure 1). The structure was defeatured or re-organized to fit requirements for meshing. There was slight geometry difference for subsonic (Ma=0.9) and supersonic (Ma=1.3) at the discharge nozzle outlet region: one was flat extended (Ma=0.9), and another one was enlarged towards exit (Ma=1.3).

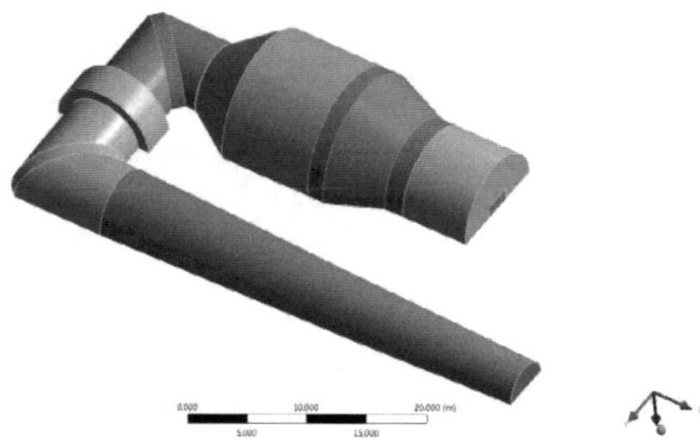

Figure 1. Simplified model for tunnel fluid domain and outer isolation material domain.

Another three layer of isolation materials was imposed as shell boundary with their thickness values to consider heat transfer in radial and axial directions (Table 1).

Table 1. Isolation layer material properties.

Material	Location	Conductivity(w/m/k)	Remark
Aluminium	1st layer touching with tunnel fluid	120.0	Shell boundary
Static Nitrogen	2nd layer	0.0242	Physical model
Isolation material	3rd layer	0.037	Shell boundary
Steel	Last layer (4th) to environment	16.0	Shell boundary

2.1. Meshing

A mixed type of mesh was used for both fluid and solid regions, totally 21.5 million cells (Figure 2). The mesh was composed of hexahedral and tetrahedral cell types in different tunnel sessions, and 10-15 prismatic cell along wall boundary as well as turning vane walls at the two turning corners. A minimum three-layer cells were imposed for the solid domain to consider heat transfer with 2nd order accuracy.

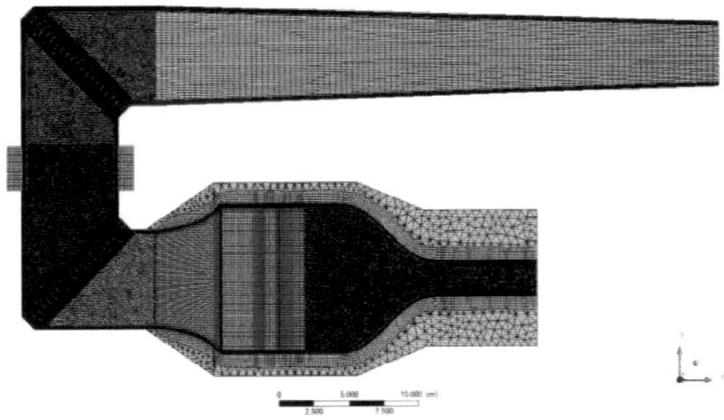

Figure 2. Overall mesh illustration for both solid and fluid domain.

2.2. Boundary condition

An overview of boundary conditions was illustrated in Figure 3. For the three case scenario, inlet conditions for total temperature and pressure were adjusted to meet their requirements at outlet (Table 2). For case 1 and case 2, a uniform temperature field was imposed at inlet; for case 3, a non-uniform temperature field was imposed:

$$T = 2.5 \times \sin(2\pi \times y) + 110 \tag{1}$$

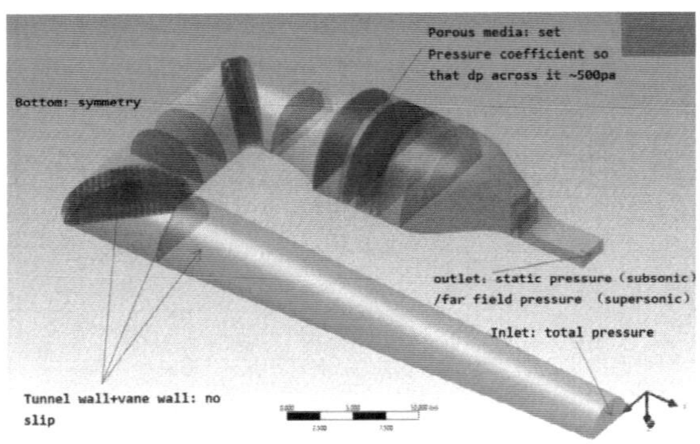

Figure 3. Set-up for boundary conditions.

To consider existence of honeycomb structure at the stable session, one porous media zone was imposed with a pressure coefficient set up for a pressure jump across it around 500 pascal. To consider effects of damping nets there, 7 layers of porous resistance interface were imposed as porous-jump interface condition.

Table 2. Three case outlet scenario.

	Total temperature (K)	Total pressure (Pa)	Mach number	Temperature at inlet
Case 1	110	115	1.3	uniform
Case 2	110	450	0.9	uniform
Case 3	110	450	0.9	sinusoidal

Heat transfer from ambient environment through tunnel walls and isolation materials was considered with a heat transfer coefficient h:

$$Q = h \times (T - T_{ambient}) \tag{2}$$

where the environmental temperature $T_{ambient} = 25\ °C$.

An ideal-gas law physical model was applied for the fluid inside and a k-epsilon turbulent fluid model with standard wall function treatment was implemented. The numerical model was set-up and the simulations were running using Ansys Fluent 18.2. Due to the model size, parallel simulations over 8 cores were implemented and each case scenario took around 24-48 computer hours to get converged solutions.

3. Results and Discussion

Figure 4 to Figure 6 show velocity streamlines in the tunnel, velocity and temperature contours along the symmetry bottom surface for case 2 Ma = 0.9. Velocity and its related Mach number follow what expected: velocity develops and become uniform after passing by porous media zone and porous interfaces at the stable tunnel session. After entering the discharge area, due to decrease of tunnel region, the velocity increases and temperature decreases until outlet location (Figure 6).

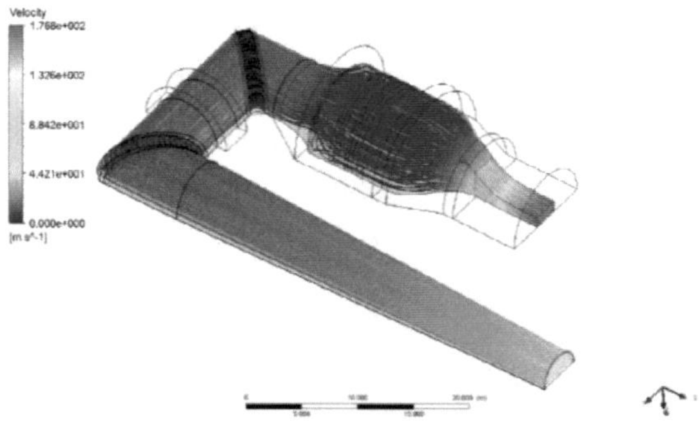

Figure 4. Overall streamlines for case 2 (Ma = 0.9).

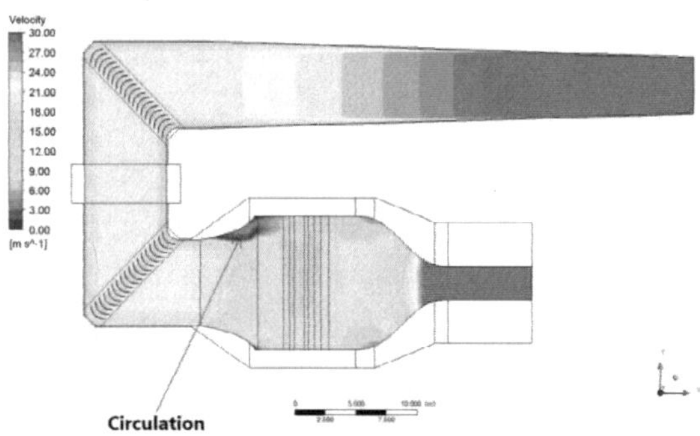

Figure 5. Overall velocity contour for case 2 (Ma = 0.9).

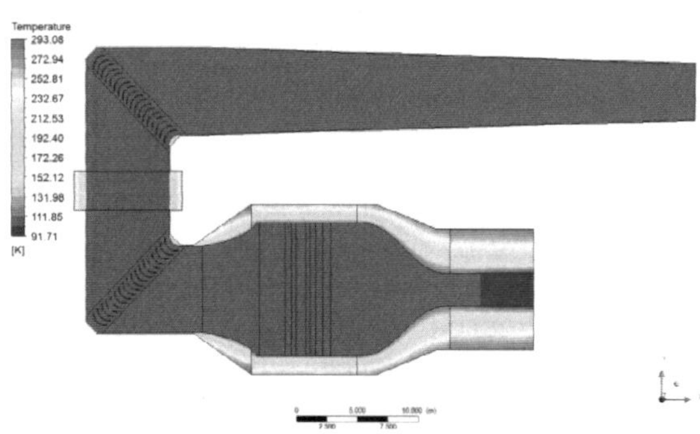

Figure 6. Overall temperature contour for case 2 (Ma = 0.9).

IOP Conf. Series: Materials Science and Engineering **408** (2018) 012003 doi:10.1088/1757-899X/408/1/012003

It is discovered that due to the existence of 2nd turning corner, there are fluid separation and a big circulation in the downstream (Figure 5), before the porous honeycomb zone. It is also noted there is a slightly small circulation at the 1st turning corner (circled in red in Figure 5). These two circulation may have negative impact on the flow and temperature uniformity in the downstream.

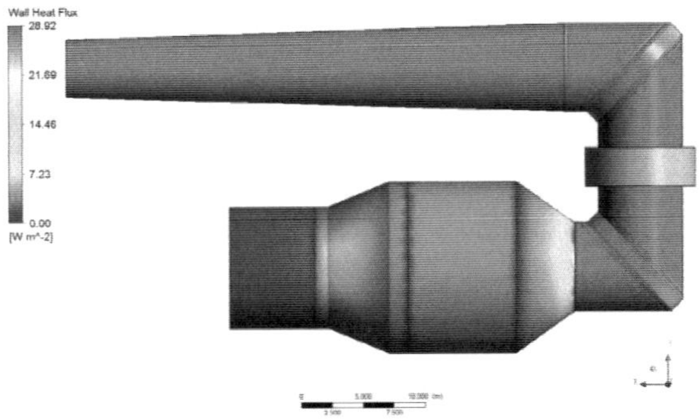

Figure 7. Overall heat flux contour for case 2 (Ma = 0.9).

Figure 7 presents wall heat flux distribution for case 2. It is noted that normally heat flux with thicker isolation material layers close to the outlet region is smaller compared to those with thinner isolation material layers in the upper stream. A maximum heat flux of 28.92 w/m^2 is found, which is within the design criteria requirement.

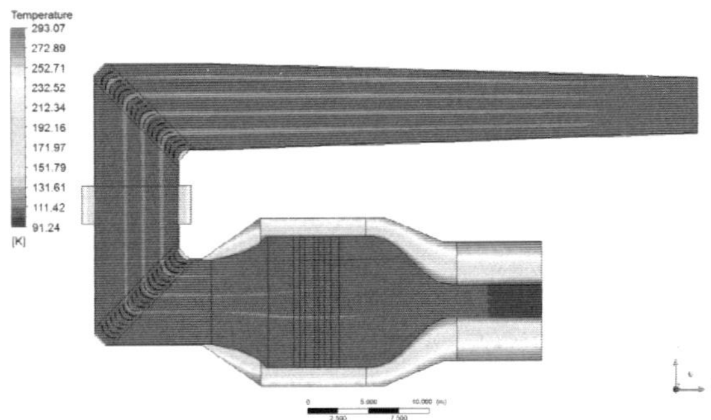

Figure 8. Overall temperature contour for case 3 (non-uniform temperature at inlet, Ma = 0.9).

Figure 8 shows the temperature contour for Case 3 with non-uniform temperature imposed at inlet. It is noted that temperature variation exists along the flow downstream, and after passing the porous media (honeycomb) and porous-jump interface (damping interface), the temperature field becomes more uniform. This shows that existences of honeycomb region and porous interface are able to help to improve flow and temperature uniformity.

Figure 9 show temperature difference contour (T-T_{avg}) at outlet surface (80% of the outlet surface, without boundary layer) for the 3 cases. It is found that when Mach number is smaller (Ma=0.9, case 2), the temperature variation at outlet is smallest ($\delta t = 0.1168$ K); when Mach number is larger (Ma=1.3, case 1), the temperature variation at outlet also increases ($\delta t = 0.5219$ K). However, a non-uniform temperature at inlet (Ma=0.9, case 3) gives a highest temperature variation at outlet ($\delta t = 0.8766$ K). This variation value exceeds the maximum allowable temperature non-uniformity of 0.5 K. This can be improved by changes of structure at the 1st and 2nd turning corner to have smaller circulations, or adjusting the locations of porous honeycomb and porous damping interfaces.

IOP Conf. Series: Materials Science and Engineering **408** (2018) 012003 doi:10.1088/1757-899X/408/1/012003

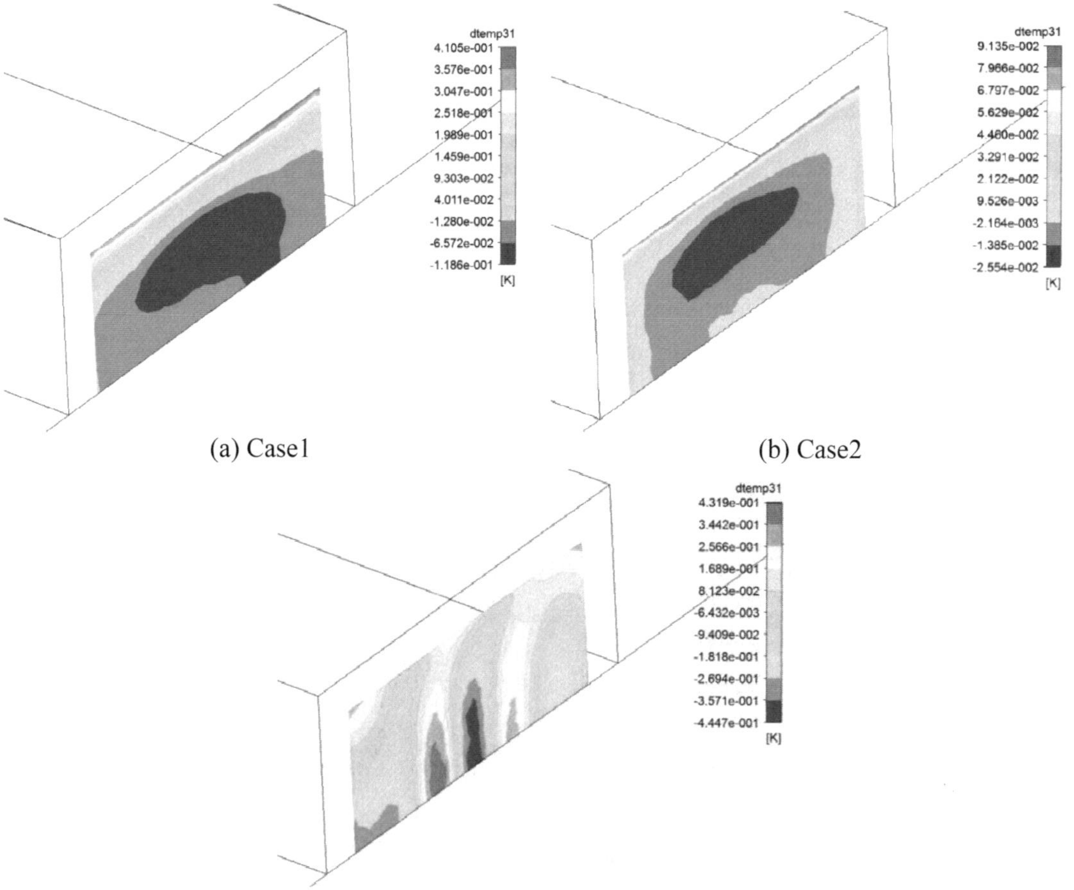

(a) Case1 (b) Case2

(c) Case3

Figure 9. Temperature contour (T-T_{avg}) at outlet location (80% of the outlet surface, without boundary layer).

4 Conclusion

CFD Modelling analysis was carried out to investigate temperature uniformity at a wind tunnel outlet region to benefit its design and construction. Low temperate (110K) nitrogen entered the wind tunnel at inlet, passed through two turning area with 18 turning vanes and went into expansion session, stable session and congestion session, finally towards outlet discharge nozzle session. The model considered static nitrogen isolation layer in physical model and three more isolation layer outside the tunnel as well as possible heat transfer from ambient environment. Two case scenario for supersonic and subsonic at outlet (Mach Number 0.9 and 1.3) were conducted with a uniform temperature imposed at inlet, as well a 3rd case scenario (Mach Number 0.9) with a non-uniform temperature distribution at inlet. Existence of porous zone (to model honeycomb session) and porous interfaces (to model damping interfaces) at the stable session were also included in the model.

The results showed that two circulation zones were found at the 1st and 2nd turning corner, which may impact flow and temperature uniformity downstream. It was also proved that existences of porous honeycomb and porous damping interface were capable to improve temperature uniformity at the outlet regions. However, it was suggested that improvement of structure of turning vanes at the 1st turning corner, and 2nd turning corners, location changes of honeycomb and damping interfaces might be necessary to have smaller circulation zones so that the impact on downstream is minimum.

References

[1] Wu Fenglin, Wang Zhenyu. The principles of wind tunnel design[M]. Beijing: Beijing Publishing House, 1985.
[2] J.E. Cermak. Wind-tunnel development and trends in applications to civil engineering[J]. Journal of Wind Engineering and Industrial .Aerodynamics, 2003, 91(3):355-370.
[3] Liao Daxiong, Huang Zhilong, Chen Zhenhua, Tang Gengsheng. Review on large-scale cryogenic wind tunnel and key technologies[J]. Journal of Experiments in Fluids Mechanics, 2014, 28(2):1-6.
[4] Cong Chenghua, Liao Daxiong, Chen Jiming, Qin Honggang. Numerical investigation on performance of second throat in transonic wind tunnel[J]. Journal of Aerospace Power, 2010, 25(9):2050-2056.
[5] Li Hongzhe, Liao Daxiong, Cong Chenghua. Numerical simulation of flow conditioning device design in wide angle diffuser of continuous transonic wind tunnels[J]. ACTA Aerodynamics Sinica, 2015, 33(2): 198-203.
[6] Cong Chenghua, Ren Zebin, Yang Gaoqiang, Wang Ning. Numerical simulation for optimal design of wide angle diffuser in limited condition[J]. Journal of Aerospace Power, 2016, 31(4): 910-917.
[7] Moonen P, Blocken B, Carmeliet J. Numerical modelling of the flow conditions in a closed-circuit low-speed wind tunnel[J]. Journal of Wind Engineering and Industrial Aerodynamics, 2006, 94:699-723.
[8] Li Qichen, Yang Zhigang. Application of CFD for the desing of aero-acoustic wind tunnel[J]. ACTA Aerodynamics Sinica, 2009, 27(3):373-377.
[9] Dai Yi, Chen Zuogang, Ma Ning, Ren Zebin. Numerical simulation of flow field inside the low-speed wind tunnel[J]. ACTA Aerodynamics Sinica, 2014, 32(2):203-208.

IOP Conf. Series: Materials Science and Engineering **408** (2018) 012004 doi:10.1088/1757-899X/408/1/012004

Numerical Investigation on Flow Making System in Multi-purpose Basin

Zheng Xu* and Xiaohui Su

School of Hydraulic Engineering, Dalian University of Technology, Dalian, China, 116023

*Xuzheng123@mail.dlut.edu.cn

Abstract. In this paper, numerical investigation on flow field of flow making system is carried out in the 1:10 scaled model of the multi-purpose basin for the deep sea engineering research center of Dalian University of Technology. According to the requirement of obtaining specified flow velocity in the working area, two kinds of numerical methods simulating pump are used. The calculation results show that the pump simulating method with controlling boundary conditions at inlet and outlet is not valid in the presence of free water surface. The problem is solved by the second pump simulating method in which the pump rotation is simulated in the multiple frame of reference (MRF) system. It is believed that the establishment of current model will provide an effective calculation tool for the simulation of flow characteristics of the deep water flow system.

1. Introduction

In order to upgrade the offshore engineering test technology [1,2], Dalian University of Technology (DUT) is planning to build an oceanic deep water multi-purpose basin with functions of flow making, waves making and wind making. The flow making system is the most important part of the project [3-6].

The pool of flow making system in multi-purpose basin is designed as 50m long, 30m wide and 12.9m deep. The designed free water surface height is 12m. Based on the initial design, 14 water pumps are equipped. The depth of the floating bottom can move in the range of 0~10m in the water depth direction. In order to make the flow of water into the multi-purpose basin and achieve a stable flow, a few of perforated walls and deflectors are set in the system. A working area with $10\times10m^2$ of the middle basin is set for testing.

In this paper, the 1:10 scaled model of the making flow system is numerically investigated. Since those 14 pumps and channels of inflow system are identical, only one of the inflow channels is numerically simulated.

2. Numerical Methods

Reynolds-averaged multiphase incompressible fluid Navier-Stokes equations and continuity equations are implemented to describe flow motion in the flow domain. VOF (Volume of Fluid) method is used for free surface capturing [7,8].

The standard k-ε model is used for dealing with turbulence happening in the flow field. The standard k-ε model is mainly based on the turbulent kinetic energy and turbulent dissipation rate of two parameters to describe the turbulence.

IOP Conf. Series: Materials Science and Engineering **408** (2018) 012004 doi:10.1088/1757-899X/408/1/012004

When studying the calculation domain of the pump impeller, the multiple frame of reference (MRF) is adopted. The pump and the main body of flow making system belong to rotating part and stationary part separately.

3. Pump Simulation I

3.1 Model setup

3.1.1 Geometry description. A single inflow channel with 5m length, 0.215m width, 1.29m height and free water surface height 1.2m is modelled. The height of floating bottom is set to 0.8m. Flow making system model diagram is shown in Fig. 1. The perforated wall model is shown in Fig. 2.

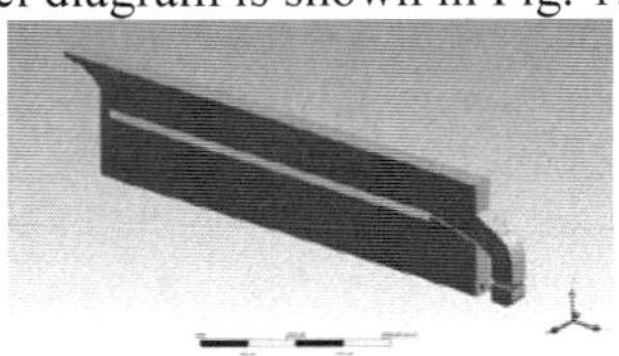

Figure 1. Geometry of flow making system model.

Figure 2. Zoomed in Perforated wall.

3.1.2 Mesh information. The number of grid cells in the entire model is about 890,000 after the mesh convergent test. The overall grid diagram of flow making system and the grid of perforated wall are shown in Fig. 3 and Fig. 4, respectively.

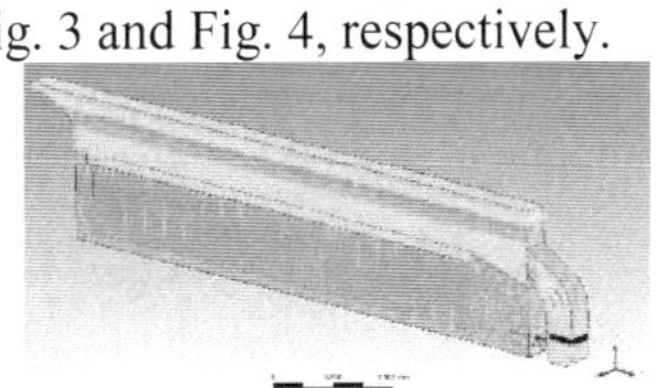

Figure 3. Grid of flow making system.

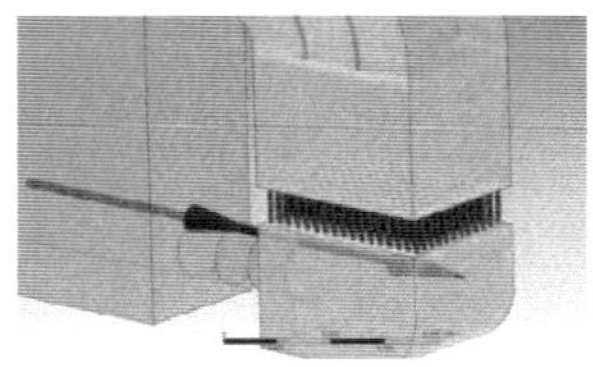

Figure 4. Grid of perforated wall.

3.2 Boundary conditions and calculation conditions

Entrance boundary is set as total pressure (11800pa). Export boundary is set as mass flow rate (2.752kg / s). Wall condition is set as no slip wall. The two side walls of the main body of the flow making system are set as the symmetrical boundary. Water and air interface is set as free boundary. The boundary conditions are denoted in Fig. 5.

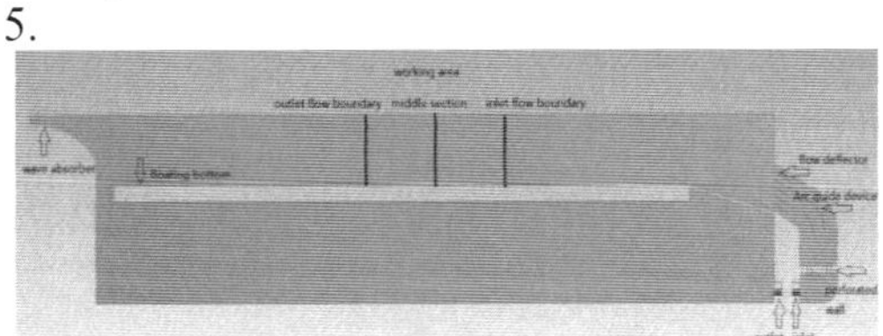

Figure 5. Boundary conditions of the flow making system.

3.3 Calculation results and analysis

3.3.1 Streamlines of flow making system and free water surface. The position of free water surface captured by the VOF method at 60s is plotted as green plane in Fig. 6. The height of free water surface at that time is about 1.25m. Turbulent flow of water entering into making flow system through inlet channel is obvious, and there is a large whirlpool.

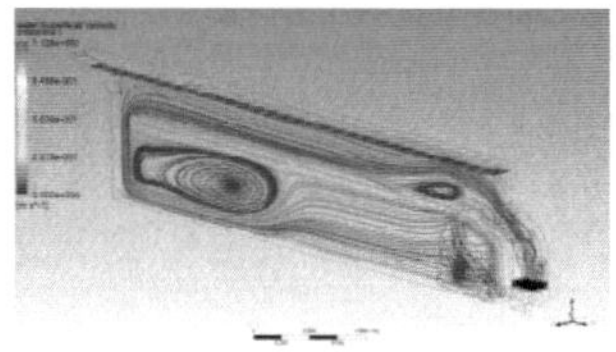

Figure 6. Free water surface and streamlines at 60s.

3.3.2 Vertical velocity profiles in working area boundaries. In this paper, focus is made on flow field characteristics in the working area. The sketch of working area and investigating lines are plotted in Fig. 7. It states that the inlet flow boundary (X = 3.0m) is set at the right side of the working area, the middle section (X = 2.5m) is set in the center of working area, and the outlet flow boundary is set at the left side of the working area (X = 2.0 m).

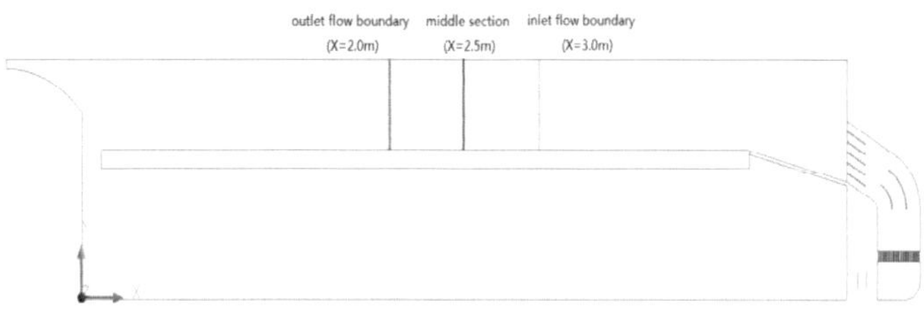

Figure 7. Boundaries of working area.

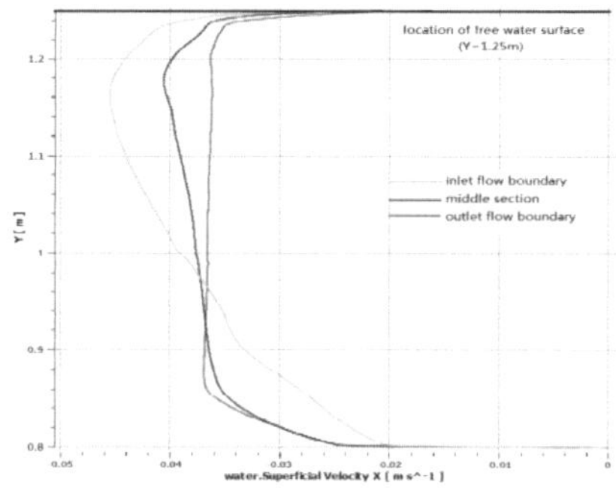

Figure 8. Vertical velocity profiles.

In Fig. 8, the green line represents the inlet flow boundary, the blue line represents the middle section, and the red line represents the outlet flow boundary, respectively. The average velocity at the inlet flow boundary is larger than those values at the middle section and the outlet flow boundary.

The instantaneous results at 60s were chosen to check mass convergence. It found that the mass flow at inlet is 3.28 kg /s while the value at outlet is 2.70 kg/s, indicating that this simplified pump simulation increases the non-conservative amount of water mass in the system.

4. Pump Simulation II

4.1 Model setup

4.1.1 Geometry description. The model is 5m long, 0.215m wide, 1.29m high, the height of the free water surface is still 1.2m and the height of the floating bottom is set to 0.63m. A four-blade water pump model is inserted into the rotation domain. Model diagram is shown in Fig. 9, the pump model is shown in Fig. 10.

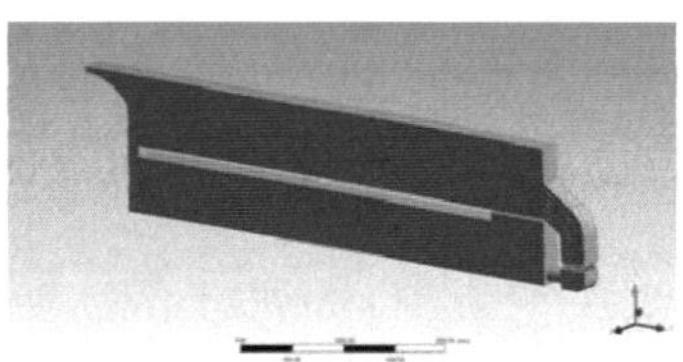

Figure 9. Geometry of flow making system.

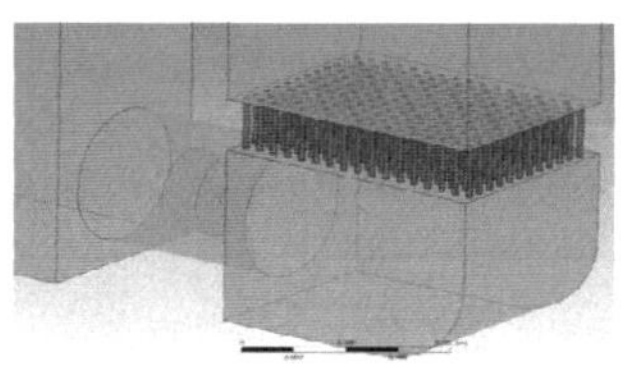

Figure 10. Water pump model.

4.1.2 Mesh information. The amount of total grid is about 5,370,000, in which the amount of rotating domain mesh reaches 3,620,000, the amount of stationary domain mesh reaches 1,750,000. Schematic diagram of the overall model of the flow making system is shown in Fig. 11, the pump model diagram is shown in Fig. 12.

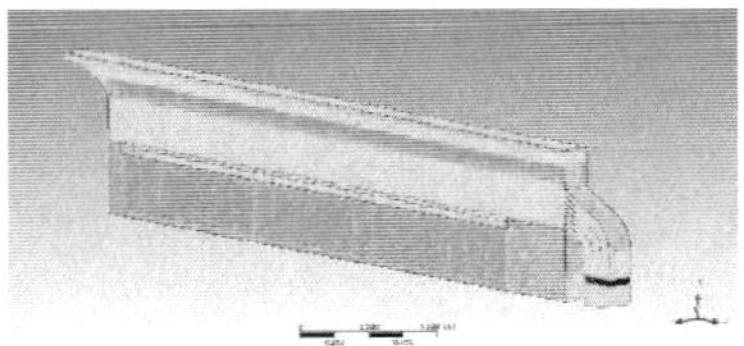

Figure 11. Overall grid of flow making system.

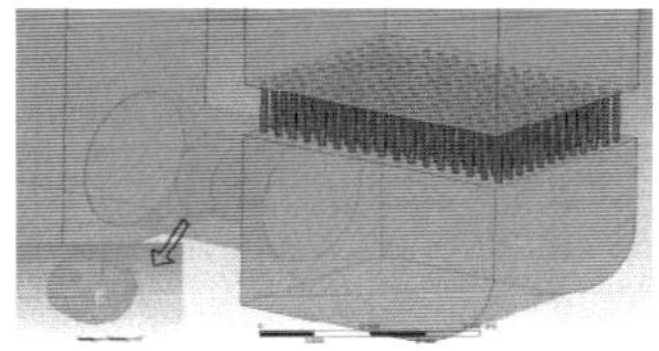

Figure 12. Grid of water pump.

4.2 Boundary conditions and calculation conditions

The calculated boundary includes the wall (no slip), the symmetrical boundary and the interface between the rotating area of the pump and the stationary area of flow making system, the sketch of boundaries is plot in Fig. 13.

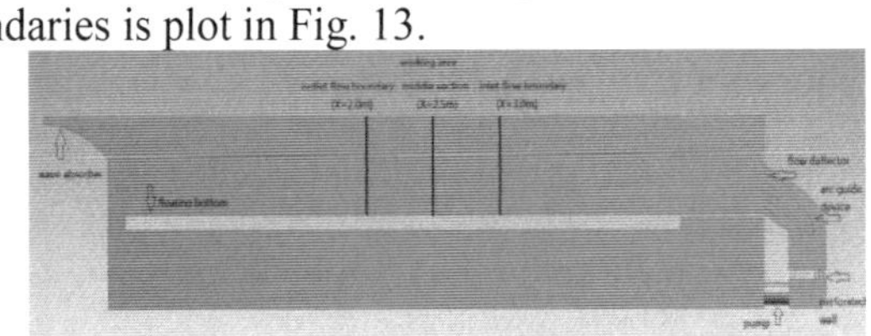

Figure 13. Boundary conditions of flow making system.

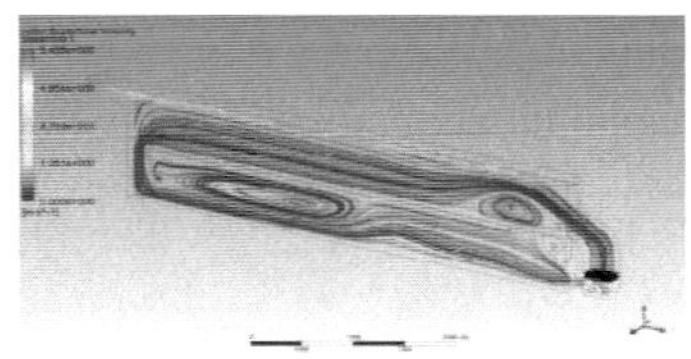

Figure 14. Streamlines in flow field at 55s.

In order to obtain specified velocity, 0.032m/s, in the working area in which the corresponding flow velocity of prototype is 0.1m/s according to the gravity similarity criterion, a few of rotating speeds of pump, range from 800rpm to 1600rpm with 200rpm interval, have been tried. Finally, rotation rate, 1200rpm, is found to be proper one.

4.3 Calculation results and analysis

4.3.1 Streamlines of overall flow field. There exists a vortex in front of working area, which is believed due to flow deflector. After the vortex, flow passes through channel smoothly and there is no vortex any more before flow reaches the end of the channel. Streamlines in overall flow field at 55s is shown in Fig. 14.

4.3.2 Vertical velocity profiles in working area. Same as the treatment of vertical velocity profile investigation in Section 3, velocity profiles in the working area are obtained through the velocity in the vertical direction in the working area set in Fig.15.

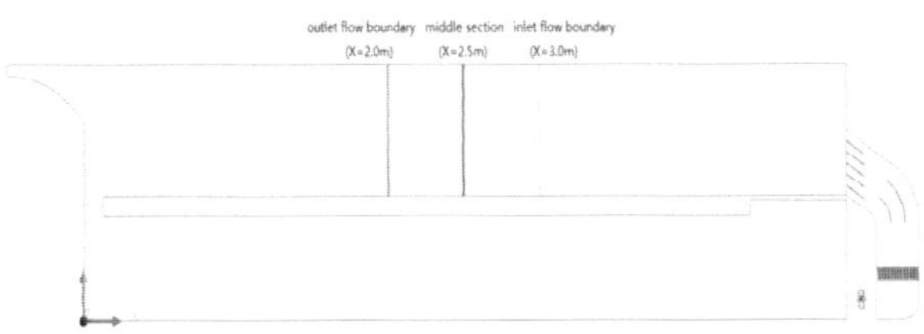

Figure 15. The sketch of working area.

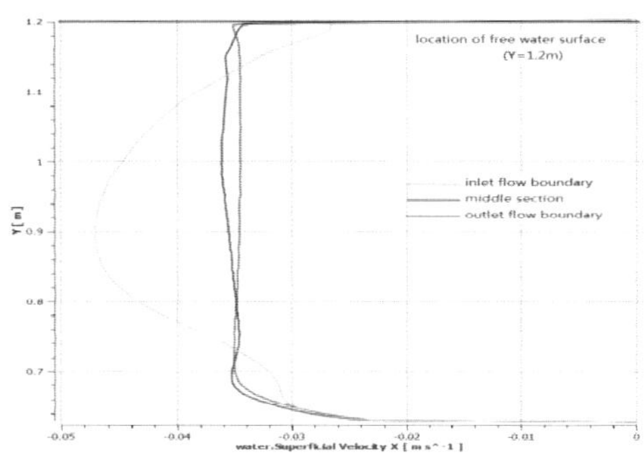

Figure 16. Velocity profiles in the working area boundary at 55s.

Fig.16.shows vertical velocity profiles at different positions in the working area according to the results at 55s. The profiles show that velocity varies greatly along vertical direction and the averaged value is larger at inlet flow boundary (green line), compared with the other two colorful lines. It can be explained that the energy of water flow is affected by the vortex generated at the position of front of working area. At the middle section (blue line) of the working area, the variation of water flow

IOP Conf. Series: Materials Science and Engineering **408** (2018) 012004 doi:10.1088/1757-899X/408/1/012004

velocity in vertical direction decreases. The velocity at the outlet flow boundary of the working area (red line) tends to be uniform in the vertical direction. Therefore, it is reasonable to carry out model tests near the outlet flow boundary in the working area.

4.3.3 Velocity histories in specified points. There are totally nine points specified here for the output of velocity histories. Point 1 (Y = 0.7m), point 2 (Y = 0.9m) and point 3 (Y = 1.1m) are located at the inlet flow boundary of working area (X = 3.0m). Point 4 (Y = 0.7m), point 5 (Y = 0.9m) and point 6 (Y = 1.1m) are located at the middle of the working area (X = 2.5m). Point 7 (Y = 0.7m), point 8 (Y = 0.9m) and point 9 (Y = 1.1m) are at the outlet flow boundary of the working area (X = 2.0m). Fig. 17 shows location of the points. Fig. 18 show velocity histories of points 1-3, points 4-6 and points 7-9, respectively.

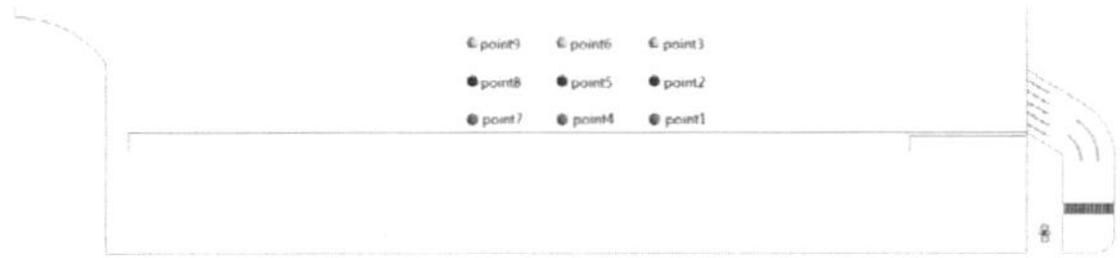

Figure 17. The layout of specified points.

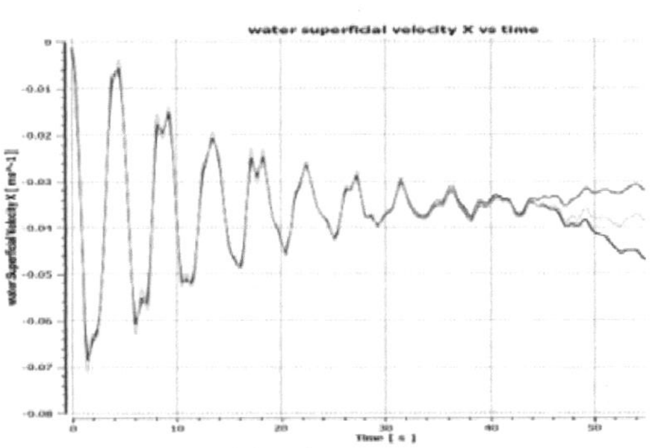

(a)Points 1-3

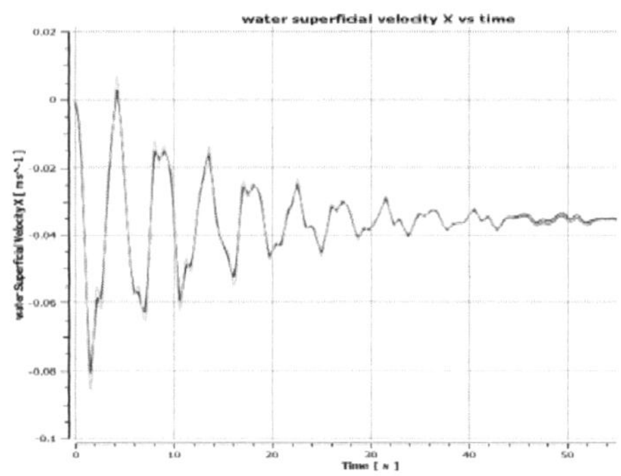

(b) Ponts 4-6

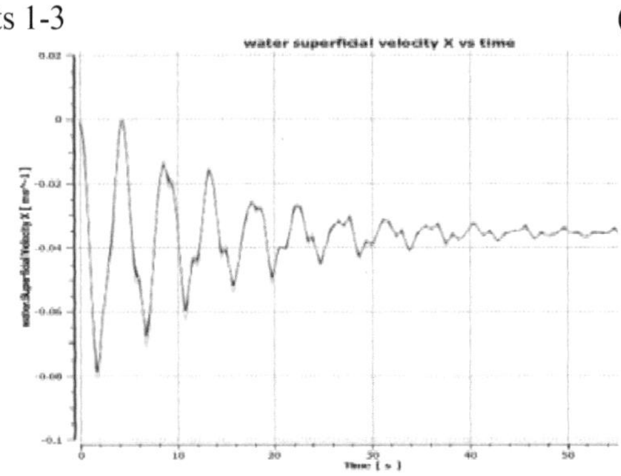

(c) Points 7-9

Figure 18. Water velocity histories of points.

Fig. 18 shows that flow field in working area becomes stable during 45 seconds running. The reason that vertical velocities at the inlet flow boundary begin different is mainly due to the effect of wall boundary of floating bottom. The profiles at of points 7-9 at the outlet flow boundary matches very well.

4.3.4 Pump hydraulic performance calculation. According to the calculation results in the case of rotating speed 1200rpm, those hydraulic characteristic parameters of the calculated pump are listed in Table 1.

Table 1. Pump hydraulic characteristics

Rotation speed N (rpm)	Inlet total pressure (Pa)	Outlet total pressure (Pa)	Mass flow rate (kg/s)	Torque on blade M (N.m)	lift (m)	Shaft power (KW)	effective power (KW)	efficiency (%)
1200	10426.24	13524.45	4.28	0.39	0.32	0.049	0.0134	27.3

5. Conclusions

In this paper, numerical investigation on flow field of flow making system is carried out in the 1:10 scaled model of the multi-purpose basin for the deep sea engineering research center of Dalian University of Technology. A few of conclusions are made as follows:

(1) The pump simulating method with controlling boundary conditions at inlet and outlet is not valid in the presence of free water surface. Non-conservation flow rate is produced. The problem could be solved by the second pump simulating method in which the pump rotation is simulated in the multiple frame of reference (MRF) system.

(2) According to the results of the MRF method, the velocities in the vertical direction are almost uniform at the outlet boundary in the working area. Therefore, it is suggested to conduct experiments near the outflow boundary of the working area.

(3) The effective power of this pump is predicted about 0.0134kW and the efficiency is about 27.3% at 1200rpm. The averaged flow velocity in working area is about 0.035m/s, which is slightly larger than the ideal flow rate of 0.032m/s.

Acknowledgement

This research is supported by the Liaoning Natural Science Foundation, project number 2015020629. The financial support is gratefully acknowledged.

References

[1] Yucheng Li 1998 The New Development of Offshore Engineering Technology. *J. China Offshore Platform,* **13**, pp9-12

[2] Runpei Li, Yonghe Xie and Zhi Shu 2003 A Review on the Technical Development of Deep Water Offshore Platform. *J. China Ocean Platform,* **18**, pp 1-5.

[3] Chenghong Mao, Jianmin Yang, Tao Peng and Xuefeng Wang 2007 The numerical simulation and analysis of water commutating equipment in current generation system of deepwater offshore basin. *Journal Of Hydrodynamics,* **22**, pp 483-490.

[4] Fen Li and Zaojian Zou 2003 Research Condition and Development Tendency of the Floating Structures. *Journal of Wuhan University of Technology (Transportation Science & Engineering).* **27**, pp 682-686.

[5] Guoyu Gao, Yongchun Xu and Haibei Lin 2015 Design of water flow engineering system for a deepwater offshore basin. *J. Water & Wastewater Engineering.* **51**, pp 56-58.

[6] Haining Lv, Jianmin Yang and Tao Peng 2006 Research on parameters of perforated wall in current generation system of deepwater offshore basin. *Journal of Hydrodynamics,* **21**, pp 198-204.

[7] Bin Liao and Shanqun Chen 2011 Simulation of water Oscillation Flow Field by VOF Method *J. Journal of Yangtze River Scientific Research Institute,* **28**, pp 27-30.

[8] Ling Li, Yongcan Chen and Yonghong Li 2007 Three-dimensional VOF model and its application to the water flow calculation in the spillway. *Journal of Hydroelectric Engineering* **26**, pp 83-87.

IOP Conf. Series: Materials Science and Engineering **408** (2018) 012005 doi:10.1088/1757-899X/408/1/012005

A Topology-alterative Algorithm for Transient Dynamic Hydraulic Response Investigation

Yao Lu[1,a], **Xiaohui Su**[1] **and Junjie Li**[1, 2]

[1]School of Hydraulic Engineering, Dalian University of Technology, Dalian, 116024, China
[2]School of Engineering, Tibet University, Lhasa, 850012, China

[a]luyao.88225@mail.dlut.edu.cn

Abstract. In this paper, a hexahedra-based topology-alterative algorithm is proposed to simulate dynamic boundary issues of the gate opening motions in water conservancy projects. The algorithm, which cooperates with the two-phase, incompressible solver of OpenFOAM platform, enables researchers to investigate transient hydraulic characteristic variations comprehensively. Two benchmark experiments were performed to validate the numerical results under the working conditions of moving sluice gates and unsteady weir head. The time history response of water level and pressure on specified monitoring points, flow pattern scenario near moving gate, furthermore, the development of antisymmetric-revolving vertical vortices around gate piers were recorded to assess the accuracy of the proposed model. The cross-contrasting results show that the relative error of weir flow time is less than 5.00 %. Moreover, good agreement has been obtained about time history curves of free surface elevation and pressure between numerical and experiment solutions. The vivid flow pattern shows the simulating capability of the current model and indicates an inspiring way to numerically investigate generation mechanism of vertical vortex inside gate grooves. The proposed algorithm is believed to be a useful utility on the further investigation on flow variation in real hydraulic engineering applications, such as improvement of semi-empirical discharge formulas, reduction of cavitation damage, elimination of pipe hammer pressure even spectral analysis of sluice gates with a fluid structure interaction interface.

1. Introduction

Hydraulic sluice gates are basically responsible for controlling water level and ejecting sediment, which enable hydraulic engineers to achieve various requirements, such as flood regulation, hydroelectric generation, agricultural irrigation and inland navigation. But it is more significant to understand the relationship between flow variations and gate opening before these engineering applications. Thus, a worldwide used semi-empirical technical architecture of hydraulic discharging response has been established, validated and practiced to date, based on the priori condition of a fixed rate of gate opening and a steady state assumption.

However, engineering problems and even industrial accidents have occurred because the hydrodynamic behaviour of water flow due to gate movement is not considered in the predictions. There are three primary issues that lead to economic losses for both the operation management department of the hydropower station and the water-using enterprises, i.e., accurate flux control, cavitation damage and vortex-induced vibration. Each issue may result in negative effects on hydraulic structures and endanger citizens as well as their property.

Unfortunately, few articles of theoretical analysis and prototype or model experiments focused on transient hydraulic response investigation until computational fluid dynamics (CFD) was involved recently. Dargahi [1] captured the transient pressurized and free-surface flow features of a bottom outlet associated with a plain opening gate using volume of fluid (VOF) method implemented in Flow3D. Xiao et al. [2] conducted prototype experiments and transient simulations to predict two ski-jump jets on the flip bucket while a radial gate was closing. The motion curve of the moving boundaries surrounded by an unstructured dynamic domain was governed by a user-defined function (UDF) of ANSYS FLUENT. An identical methodology was employed to inquiry the spatial and temporal distribution law of velocity and hydrodynamic pressure, relations between weir head or angular speed of the opening radial gate and unsteady discharge coefficient [3-4].

In this paper, a novel topological-alterative two-phase flow solver embedded with a dynamic mesh class named movingDamGateTopoFvMesh is carried out based on interDyMFoam solver of OpenFOAM platform, aiming to work out problems that plain and radial gates open on settled pathway [5-6]. In comparison with commercial or proprietary CFD codes, the proposed model performs preponderantly in three aspects: (1) providing new technological support for transient working conditions analyzing; (2) adhering to the open-source general public license (GPL); (3) introducing hexahedral structured cells to improve quality and to reduce quantity of mesh lying on the dynamic domain.

2. Methodology of the topography-alterative algorithm

The topography-alterative algorithm is designed to imitate sliding gates by adding hexahedral cells (cells 10, 11, 12 and 13 in figure 1) into the computational domain. A new layer of cells is generated above the specified gate lip patch at the scheduled time step with boundary conditions and fluid flux updated in real time. The following workflow of mesh motion will be triggered if the logical switch of the algorithm is toggled to yes, otherwise, the process will be suspended.

- All the information of mesh and fluid field are stored temporarily at the current time step t.
- The points lay on the gate lip patch, marked as 10, 11, 29 and 35, are about to match their inborn partners, points 13, 14, 26 and 39. New points, such as 44 and 45, will be generated to cooperate with points 20 and 33 using a self-adaptive offset matrix based on layer height h and gate thickness δ.
- A one-to-one mapping relationship is introduced to link the layer of new hexahedral elements (shadowed in figure 1a) with the original boundary cells of gate lip patch, for instance, cells 1 and 9.
- The problem of geometric closure is considered as the primary objective afterwards. Faces attached on new boundary patch, adjoined by two new cells and connected with a new cell and an original one will be produced, represented by faces (41 17 46 47), (47 46 44 45) and (33 20 11 29) in figure 1b.
- Updating mesh and zero-initializing new cells are indispensable processes before solving field information at time t + dt.

3. Benchmark experiments and numerical cases

3.1. Experimental setup

Table 1 briefs the two benchmark experiments conducted to validate the proposed model using the platform shown in figure 2. A self-tapping leading screw and a reduction gearbox were introduced to manually control the opening motion of the middle plain and radial gates. Meanwhile, a digital image and pressure acquisition system was equipped to panoramically collect the water surface elevation and pressure signal in real time. The working procedures are as follows:

- Fill the upstream reservoir until the pre-determined initial weir head H_0 is reached. Then, shut down the pump.
- Initialize the digital image and pressure acquisition system.
- Open the middle gate linearly with inlet volume flow rate $Q_{in} = 0$ m^{3}s^{-1} until the gate opening e reaches the maximum status e_{max}, and collect data within the total time t_{tot}.
- Extract water level and pressure data from cameras and sensors, and identify the transition time t_m at which the flow pattern changes from orifice flow to weir-type flow, the maximum opening time t_{max}.

3.2. Numerical setup
Two numerical cases are established to validate the accuracy of capturing the free surface elevation, pressure history and discharge capacity using the same configurations shown in table 1. The transport properties applied in all cases are: air kinematic viscosity $\nu_a = 1.0 \times 10^{-5}$ m^{2}s^{-1}, air density $\rho_a = 1.0$ kgm^{-3}, water kinematic viscosity $\nu_w = 1.0 \times 10^{-6}$ m^{2}s^{-1}, water density $\rho_w = 1000$ kgm^{-3}, and water surface tension $\sigma = 0.07$ kgs^{-2}.

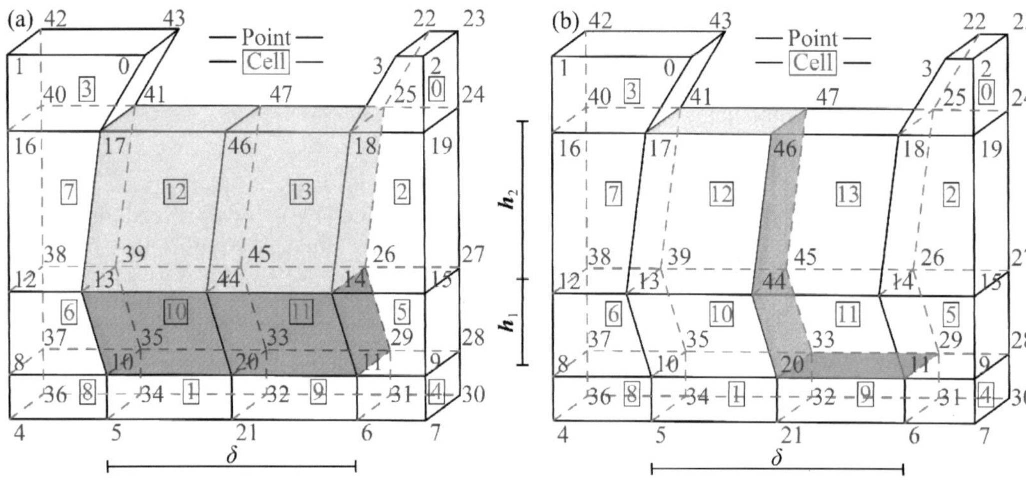

Figure 1. Sketch of a sliding gate lip patch: (a) cells and (b) faces.

Table 1. Parameters of the two benchmark experiments.

Exp.	Gate	Q_{in}	H_0	t_{max}	e_{max}	t_{tot}
		m^{3}s^{-1}	m	s	m	s
1	Plain	0	0.144	42.14	0.100	60
2	Radial	0	0.144	14.56	0.114	60

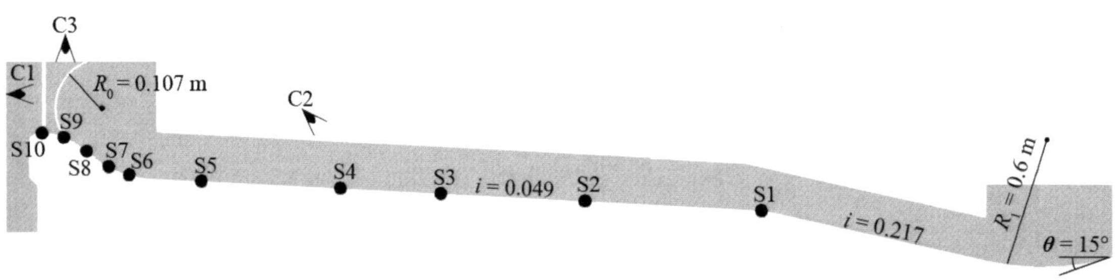

Figure 2. Sketch of the digital image and pressure acquisition system containing cameras (C1) ~ (C3) and pressure sensors (S1) ~ (S10).

4. Results of experimental and numerical cases

The process at t_m, provided in table 2, when the gate lip patch separates entirely from the free water surface, is recorded to assess the reliability of the developed code. The relative errors of t_m for both the plain and radial gate cases are less than 5.00 %, which validates the performance of the proposed model.

Table 2. Comparison of transition time t_m.

Gate	t_m (s)		Relative Error (%)
	Exp.	Case	
Plain	35.47	34.88	-1.66
Radial	12.77	12.34	-3.37

Figures 3 and 4 provide the time history of pressure and water level, which are used to quantitatively validate the accuracy of the proposed model via a comparison with the experimental data obtained from sensors S2, S7 and S9 and cameras C1 and C2. The left column presents the plain gate case, and the right column presents the radial gate case. The gate opening processes over time are described in the first row. In general, both figures 3 and 4 show perfect agreement between the experimental data and numerical results. This result suggests that the proposed model is accurate.

Flow patterns near the opening gate area are compared based on experimental data and numerical model results to qualitatively assess the proposed model. Figure 5 presents photographs of the free water surface in front of the gate from the view of camera C3 during the opening process for the middle plain and radial gates. Screenshots from the acquired experimental videos are shown on the left side, and the blue arrows are drawn to represent the main flow, which comes from the upstream reservoir. The locations of secondary flows are denoted by red arrowed helices. The contours of free water surfaces obtained from the numerical results are plotted on the right side for parallel comparison. Again, the experimental data and numerical model results match for both the surge in front of the gate and the secondary vortices produced.

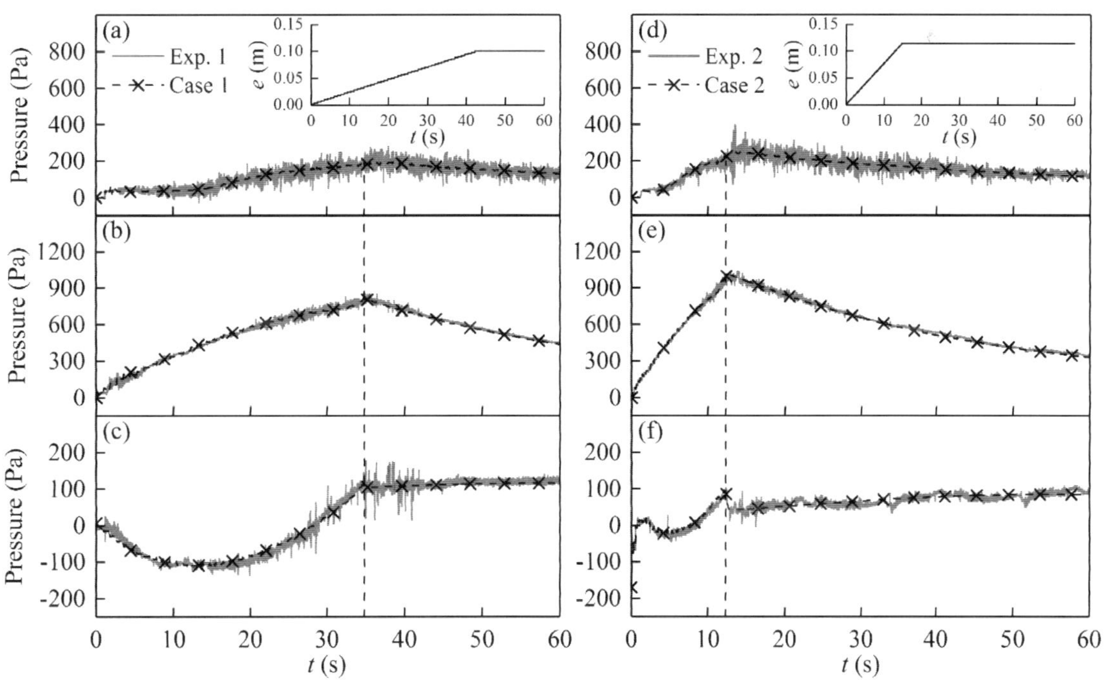

Figure 3. The historical pressure of plain gates at the positions of sensors (a) S2, (b) S7, (c) S9, and radial gates at the positions of sensors (d) S2, (e) S7, (f) S9.

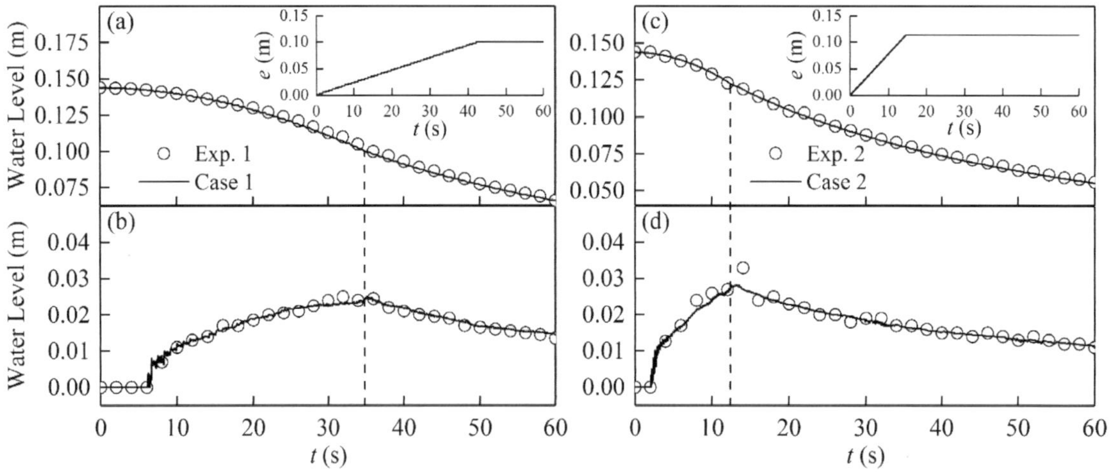

Figure 4. The historical water level under the conditions of plain gates from the view of cameras (a) C1, (b) C2, and radial gates from the view of cameras (c) C1, (d) C2.

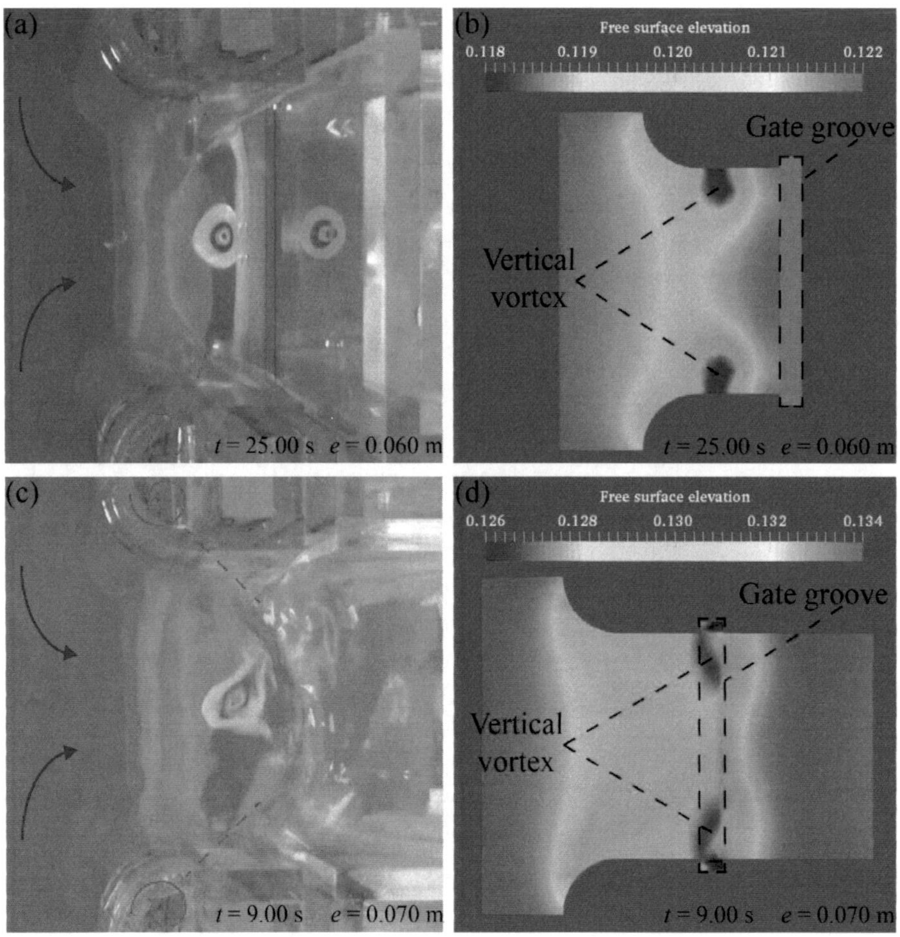

Figure 5. Comparison of experimental and numerical results from the view of camera C3 at the opening middle plain gate when t = 25.00 s for (a) experiment 1, (b) case 1, and at the opening middle radial gate when t = 9.00 s for (c) experiment 2, (d) case 2.

5. Discussion of potential applications

The accuracy of the proposed model was tested and validated in the previous sections. The proposed model can potentially be applied in hydrodynamic studies of water flow and water conservancy projects. Two examples are illustrated in this section. The first potential application is flow velocity simulation. Traditional methods of digital imaging and monitoring merely provide coarse-scale fluid field data. Velocity measurement methods, such as laser doppler velocimetry (LDV) and particle image velocimetry (PIV), provide high resolution for a single pole or fluid domain and can be efficiently applied in bench-scale experiments. However, these measuring approaches are inferior in advising large-scale hydro-complexes for dispatching management because of their high costs and rigorous imaging conditions. However, the proposed model provides a new solution to scale and cost issues, as well as issues associated with global transient variables. The second potential application of the proposed model is flow discharge prediction. Because the numerical results discussed above are quantitatively and qualitatively consistent with the experimental data, they can be used to predict flow discharge, particularly for the case of gate opening and the associated motion.

6. Conclusions

In this paper, a domain topology-alterative, two-phase incompressible flow model is applied to simulate flow variations associated with gate opening motions in water conservancy projects. The domain topology-alterative model is formulated to simulate domain changes due to gate movement based on the OpenFOAM platform. Then, using the interDyMFoam solver, the current model is applied to simulate flow patterns of plain and radial gate opening motions at a hydropower station.

In addition, the proposed model is assessed and validated quantitatively and qualitatively based on comparisons with data obtained from two experiments in the laboratory. Good agreement was observed among the free surface elevations, pressure histories and flow patterns of the numerical results and experimental data for different scenarios. Thus, the accuracy and capability of the proposed model are illustrated.

Finally, the potential applications of the proposed model, flow velocity simulation and discharge prediction are discussed for future research. Based on the results of this study, the proposed model is considered a useful tool for investigating flow variations in real hydraulic engineering applications as well as an efficient supplement for current standards of spillway design.

References

[1] Dargahi B 2010 *Journal of Hydraulic Research* ***48***(4) 476-82
[2] Xiao Y X, Wang Z W, Zeng J D, Zheng J T, Lin J Y and Zhang L J 2015 *Engineering Computations* ***32***(2) 289-307
[3] Qiu C and Liu C L 2014 *Applied Mechanics and Materials* ***580-583*** 1971-74
[4] Qiu C and Yue S B 2013 *Applied Mechanics and Materials* ***405-408*** 717-20
[5] Lu Y, Su X H and Li J J 2017 *Water Resources and Power* ***35***(8) 107-110
[6] Lu Y, Su X H and Li J J 2017 *2nd Int. Conf. on New Energy and Renewable Resources* (Lancaster, PA: DEStech Publications) pp 231-38

IOP Conf. Series: Materials Science and Engineering **408** (2018) 012006 doi:10.1088/1757-899X/408/1/012006

Experimental Research on a Kind of High-pressure Hot Gas Servo Valve Based on PWM Principle

Zhong Yang [1,3] and Xin Zhang[2]

[1] Collage of Transportation, Ludong University, Yantai, 264025, China
[2] School of Aerospace, Beijing Institute of Technology, Beijing, 100081, China

[3] ldu_yangzhong@163.com

Abstract. High-pressure hot gas servo valve based on PWM principle is widely used in the solid attitude control system of kinetic kill vehicle and other aircrafts. In this paper, a kind of hot gas servo valve is proposed and its mathematic model is established. Also, the pressure in the chamber is tested and compared with simulation results to prove the model valid. In order to test the dynamic performance, an indirectly method of testing the piston's position is put forward and the delay time of the electromagnet and valve is tested. The results reveal that the hot gas servo valve designed in this paper has good dynamic performance and can satisfy the requirements of solid attitude control system.

1. Introduction

The hot gas servo valve based on PWM (pulse width modulation) principle works under the hot gas which is supported by the gas generator. It received PWM signal and regulates average thrust by controlling the ratio of two nozzles' switch time, which can control the gas flux. The thrust system is capable of producing continuously variable control thrust according to task comparing with the conventional thruster. This can realize aircraft's high-speed and accurate maneuverability[1-3], so they are widely used in solid divert and attitude control system (SDACS) and trajectory correction system of kinds of aircrafts, such as kinetic kill vehicle (KKV), trajectory correction rocket projectile, etc[4-6].

The dynamic performance of the hot gas servo valve is very import which decides the precision of the control system[7,8]. The import indicator of the dynamic performance is dynamic delay time, which is the time interval from sending out the control signal to producing thrust. The increase of the dynamic delay time will influence the dynamic performance of the control system and increase the miss distance, which results in reducing guidance accuracy.

In this paper, we propose a kind of hot gas servo valve based on PWM principle and establish its mathematic model. Then the pressure in the control chamber is tested and compared with the simulation result to verify the validity of the established model. Experimental research is done on its dynamic performance and an indirect method of testing the delay time by testing the piston's position is put forward. The experiment method and result are expected to provide reference for the structural design and experimental research of the hot gas servo valve.

2. Mathematic model

The hot gas servo valve designed in this paper is shown in Figure 1. The hot gas flows into the valve from the inlet and a small part goes into the control chamber through the fixed orifices f_1 and f_3 as control media. The adjustable orifices f_2 and f_4 are enclosed by the electromagnet and the outlet of the control chamber. As shown in Figure 1, when the adjustable orifice f_2 is closed and f_4 is open, the pressure is increased in control chamber 1 and decreased in control chamber 2. Then the pressure difference is formed and increased gradually. When it's large enough, the piston is driven to move right. The rocker arm is pulled to revolve around its axis and block the Laval nozzle in the left side. Eventually, the hot gas flows out through the right Laval nozzle to produce thrust.

The orifice model of the valve is shown in Figure 2. The symbols p_0 and p_b are the gas supply pressure and atmospheric pressure separately. The orifice f_i (i =1,2,3,4) are the same with those shown in Figure 1. The symbol q_{mi} (i =1,2,3,4) means the mass flow of the four orifices and q_{ma}, q_{mb} are the mass flow of the control chamber 1, 2.

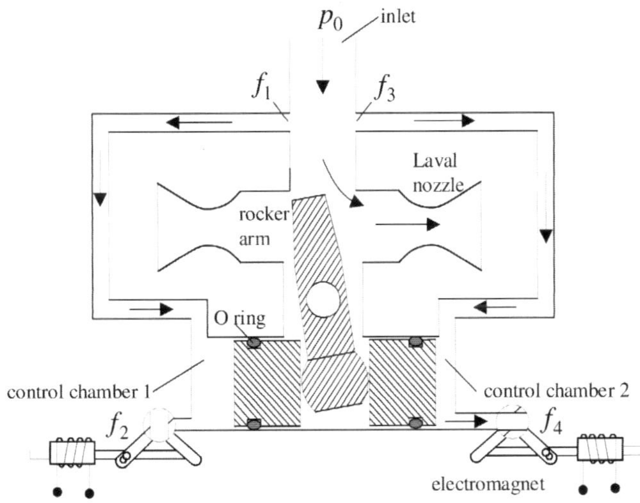

Figure 1. Schematic of the hot gas servo valve

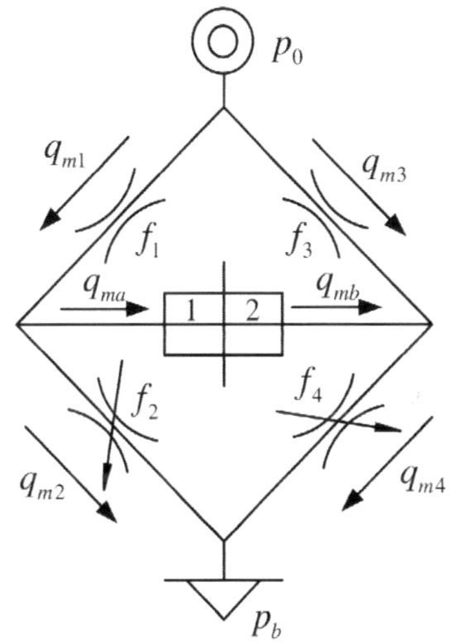

Figure 2. Orifice model diagram

In order to obtain the mathematic model of the valve, we assume that:

- The gas is perfect.
- The hypothesis of an adiabatic process is reasonable.
- The pressure and temperature within the control chambers are homogeneous.

Then the motion equation of the piston can be written as:

$$M\ddot{x} = A(p_1 - p_2) - F_L \tag{1}$$

Where M, x, A, p_1, p_2, F_L mean the mass, displacement, end-face area of the piston, the pressure of the two control chambers and the equivalent resistance of the piston respectively.

The pressure and temperature of the two control chambers are calculated by the following equations:

$$\begin{cases} \dot{p}_1 = \dfrac{k}{V_{01}+Ax}\left(q_1^+ RT_0 - p_1\dot{x}A\right) \\ \dot{p}_2 = \dfrac{k}{V_{02}-Ax}\left(q_2^+ RT_0 - q_2^- RT_2 + p_2\dot{x}A\right) \\ \dot{T}_1 = \dfrac{T_1\dot{x}A}{V_{01}+Ax} + \dfrac{T_1\dot{p}_1}{p_1} - \dfrac{RT_1^2 q_1^+}{p_1\left(V_{01}+Ax\right)} \\ \dot{T}_2 = \dfrac{T_2\dot{x}A}{V_{02}-Ax} + \dfrac{T_2\dot{p}_2}{p_2} - \dfrac{RT_2^2\left(q_2^+ - q_2^-\right)}{p_2\left(V_{02}-Ax\right)} \end{cases} \tag{2}$$

IOP Conf. Series: Materials Science and Engineering **408** (2018) 012006 doi:10.1088/1757-899X/408/1/012006

Where k, R, T mean the specific heat ratio, gas constant and the temperature of the working media. V_{01}, V_{02} are the initial volumes of the two control chambers respectively. q_1, q_2 mean the mass flow of the two control chambers. The subscript 0 means the parameters of the gas source and the superscript +, - represent the input and output, which are calculated by the equations as follows:

$$q_m = q_m\left(f, p, p_e, T\right) = \begin{cases} \dfrac{\mu f p B}{\sqrt{T}} \varphi\left(\dfrac{p_e}{p}\right) & \dfrac{p_e}{p} > C_{t1} \\ \dfrac{\mu f p C}{\sqrt{T}} & \dfrac{p_e}{p} \le C_{t1} \end{cases} \tag{3}$$

Where μ, f mean the flow coefficient and the area of the orifice. B, C, C_{t1} and $\varphi(p_e/p)$ are coefficients related with the gas property and given as:

$$B = \sqrt{\frac{2k}{R(k-1)}}, C = \left(\frac{2}{1+k}\right)^{\frac{1}{k-1}} \sqrt{\frac{2k}{R(k+1)}}, \varphi\left(\frac{p_e}{p}\right) = \sqrt{\left(\frac{p_e}{p}\right)^{\frac{2}{k}} - \left(\frac{p_e}{p}\right)^{\frac{k+1}{k}}}, C_{t1} = \left(\frac{2}{1+k}\right)^{\frac{k}{k-1}} \tag{4}$$

Then the mass flow q_{mi} (i =1,2,3,4) can be calculated as:

$$\begin{aligned} q_{m1} &= q_m\left(f_1, p_0, p_1, T_0\right); q_{m2} = q_m\left(f_2, p_1, p_b, T_0\right); \\ q_{m3} &= q_m\left(f_3, p_0, p_2, T_1\right); q_{m4} = q_m\left(f_4, p_2, p_b, T_2\right). \end{aligned} \tag{5}$$

3. Experiment research

3.1. Delay time of the electromagnet

The electromagnet receives the PWM control signal and overcome the load to drive the armature to plug the orifice f_2 or f_4. It's necessary to analyse the motion of armature as the delay time of the electromagnet is part of hot gas servo valve's delay time.

We connected a small resistor with the coil and test its voltage value to obtain the delay time of the electromagnet. The experimental results are shown in Figure 3 (without load) and Figure 4 (with load, the pressure is 7MPa).

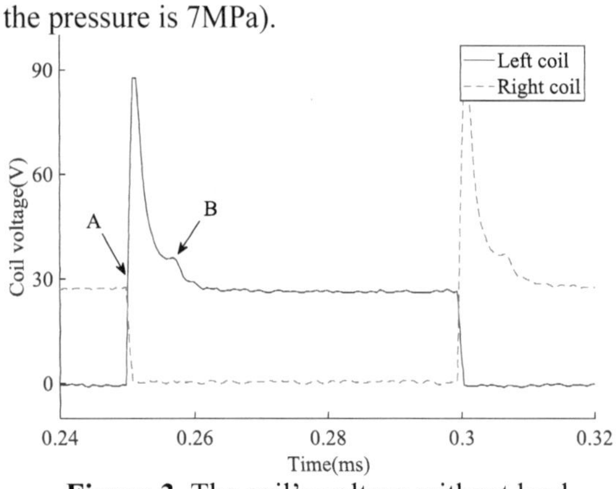

Figure 3. The coil's voltage without load

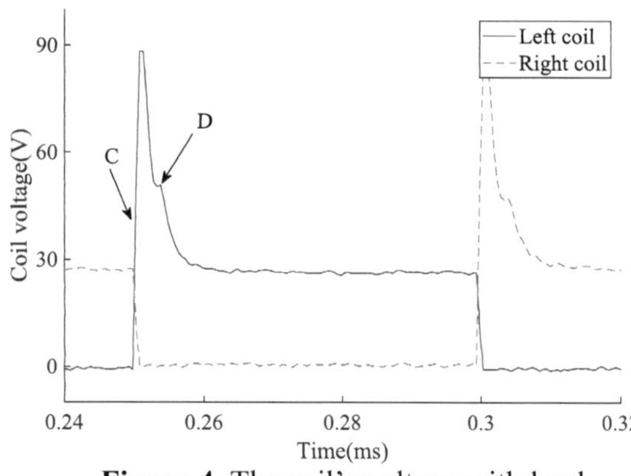

Figure 4. The coil's voltage with load

It's observed that the delay time of the electromagnet without load is 7ms (e.g. t_{AB}) in Figure 3 and 3ms (e.g. t_{CD}) with load in Figure 4. The delay time with load is shorter than without load because the direction of the force generated by the gas flowing out from the orifice is the same with the armature's motion.

3.2. Delay time of the valve

In order to test the delay time of the valve, two pressure sensors were applied to test the pressure in the two control chambers. Also, we assembled magnetic sheets on both sides of the piston and Hall sensors on the body correspondingly. The output signals of the Hall sensors are compared to obtain the piston's position signal which is contrast with the control signal to calculate the delay time indirectly. The principle of the test system is shown in Figure 5 and the experiment devices in Figure 6.

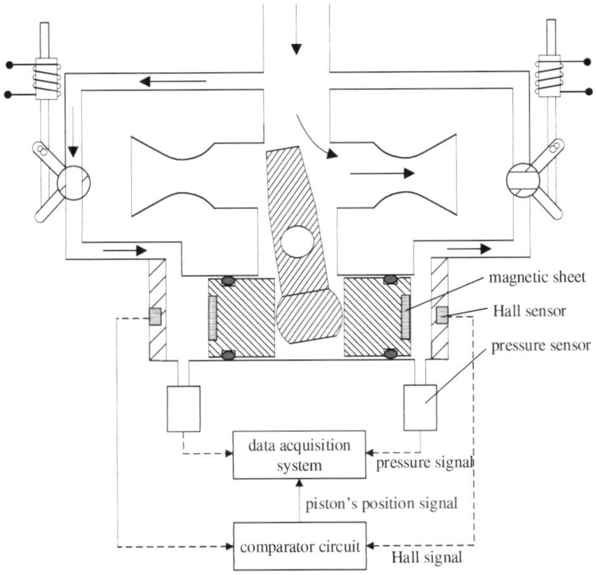

Figure 5. Schematic of the test system

Figure 6. Diagram of the experiment

3.2.1. Experiment of the pressure in the control chambers The pressure in the control chambers was tested and compared with the simulation result calculated by the model in section 2. When the control signal is send out at time 0, the experimental and simulation results in one cycle are shown in Figure 7.

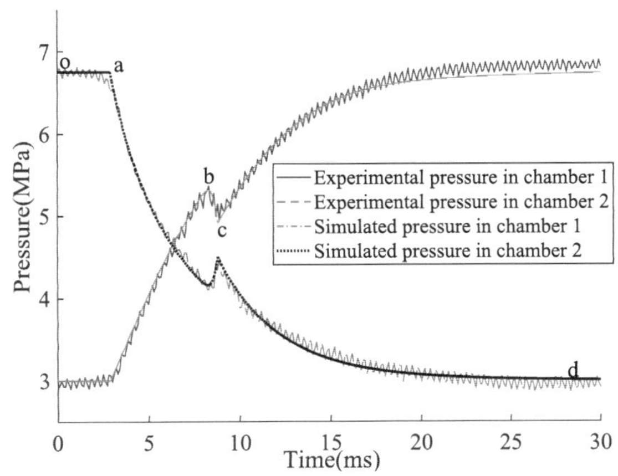

Figure 7. Simulation and experiment result of the pressure in the control chamber

In Figure 7, we can see that the experiment and simulation pressures are consistent, so the validity of the mathematic model in section 2 can be verified. Meanwhile, the time interval between the control signal and the pressure changing point (a in Figure 7.) is 0.3 millisecond, which is the electromagnet's delay time. This is consistent with the experimental result t_{CD} in Figure 4.

Then we divide the remaining working process into 3 parts. The first part is from point a to point b, which is called prepare phase. In this phase, the pressure in the two control are changing, but the pressure difference isn't large enough to drive the piston. The prepare phase's time t_{ab} is 5ms. The second part is from point b to point c, which is called motion phase. The piston moves to the other side under the pressure difference at this stage. The last part is from point c to point d, which is called stabilizing phase. The pressure becomes stable gradually until the next control signal comes.

IOP Conf. Series: Materials Science and Engineering **408** (2018) 012006 doi:10.1088/1757-899X/408/1/012006

3.2.2. Experiment of the piston's position The parts of testing the piston's position are shown in Figure 8. The Hall sensor is Honeywell's SS495A1 which is installed in the plug by high temperature glue and latching mechanism. SS495A1 is a kind of linear sensor and the relationship between its output voltage and its distance from the magnetic field is linear. But the magnetic field of the magnetic sheet is non-uniform, it's necessary to calibrate the sensor's output voltage and its distance from the magnetic sheet. The calibration result is shown in Figure 9. It can be seen that the output voltage has a linear relationship with the distance when it's between 2mm and 4mm. The piston's stroke is 1.6mm, so we set the distance in this range to realize the linear output.

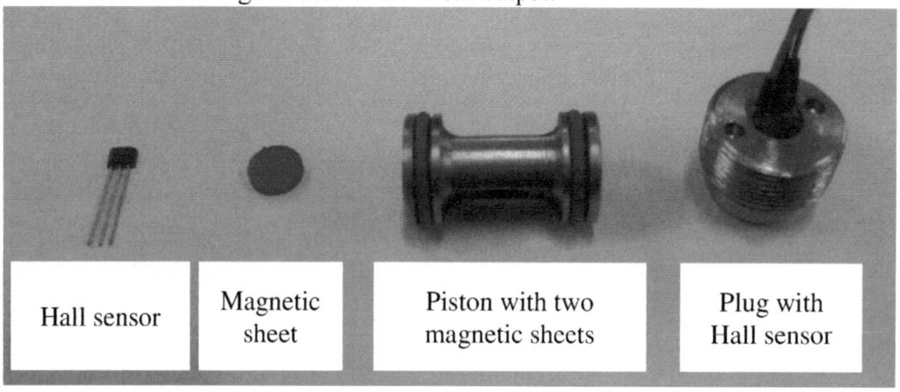

Figure 8. Parts of the testing system

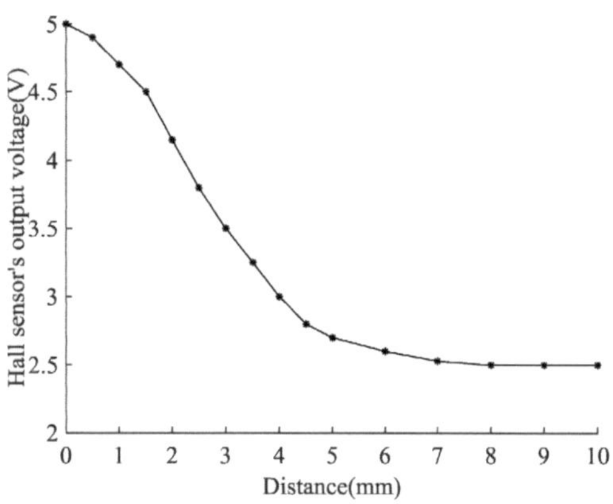

Figure 9. Curve between the Hall sensor's output and the distance

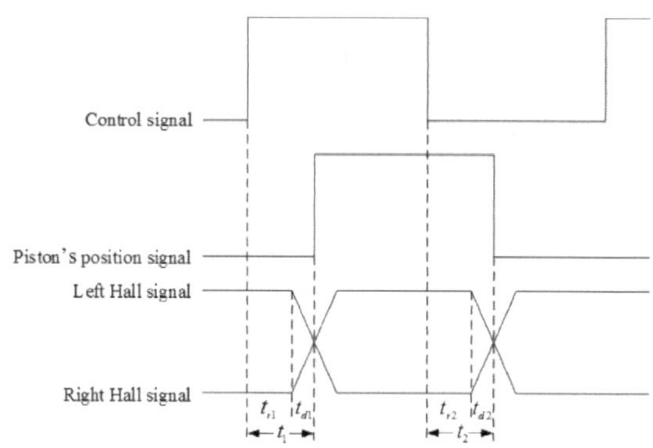

Figure 10. The sequence diagram of control signal and Hall sensor signal

The test principle of comparing the control signal and the Hall sensor's signal is shown in Figure 10. The electromagnet receives the control signal and the pressure differential drives the piston's motion. When the piston goes through the middle position, the piston's position signal switches by comparing the two Hall sensors' signal. In Figure 10, t_1 and t_2 mean the delay time of the valve at the rising and falling edge. The symbol t_r is the delay time of the electromagnet and the prepare phase while t_d is the delay time of the piston's motion to the middle position. So the delay time can be expressed as follows by comparing Figure 10 and Figure 7:

$$t = t_r + t_d = t_{oa} + t_{ab} + \left(t_{bc}\right), \quad t_r = t_{oa} + t_{ab} \tag{6}$$

Where (t_{bc}) means the time from point b to the piston's motion to the middle position, which is part of the time t_{bc}.

The experiment result is shown in Figure 11 and Figure 12. In Figure 11, we compare the control signal, the coil's voltage signal and the piston's position signal. It can be seen that the delay time of the hot gas servo valve is 8.5ms. In Figure 12, the piston's position signal and the pressure signal in the control chamber are compared. It can be seen that the time of the piston's motion to the middle phase (t_{bc}) is 0.5ms.

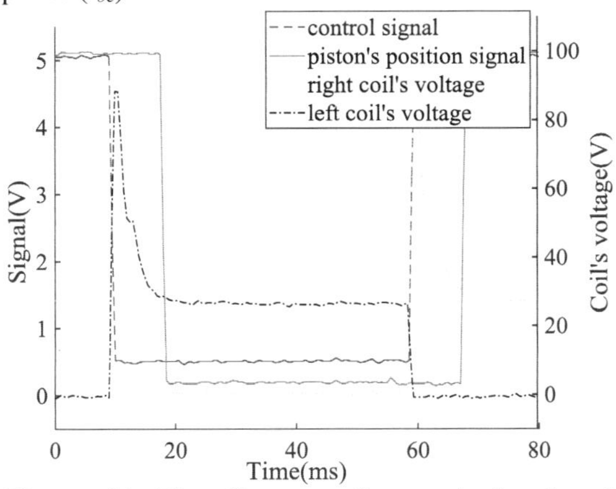

Figure 11. The diagram of control signal and piston's position signal

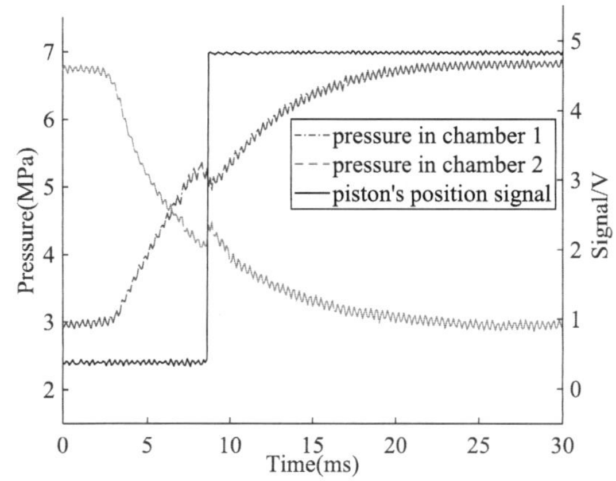

Figure 12. The diagram of pressure in control chambers and piston's position signal

4. Conclusion

The hot gas servo valve based on PWM principle can be used in solid attitude control system to produce attitude control thrust. A kind of hot gas servo valve is proposed in this paper and its mathematic model is established from the perspective of pneumatic theory. The pressure in the control chamber is tested and compared with simulation results to prove the model valid. In order to test the dynamic performance, the delay time of the electromagnet is tested and the piston's position is acquired by testing the Hall signal. So the delay time of the hot gas servo valve is tested to be 8.5ms. The results reveals that the gas servo valve designed in this paper has good performance and can be used in solid control attitude system.

Acknowledgments

This work was supported by the Training Fund of Shandong Natural Science Foundation (Grant NO. ZR2017PEE016) and the Talent Introduction Project of Ludong University (Grant NO. LB2016002).

References

[1] CHANG Heng, Wang Yi-bai and LIU Yu 2013 Experimental investigation on solid attitude control system based on PWM principle *J. Journal of Solid Rocket Technology* **36** 608-612
[2] Li Shipeng and Zhang Ping 1999 Analysis of solid propellant control thruster schemes for small KKV *J. Journal of Propulsion Technology* **20** 100-103
[3] HOU Xiao, FU Peng and WU Yuan 2017 Energy management technology of SRM and its development *J. Journal of Solid Rocket Technology* **40** 1-6+23
[4] P. Baker 1992 Army lightweight exo-atmospheric projectile (LEAP) *Aerospace Design Conf.* (Irvine, USA: AIAA) AIAA-92-1212
[5] J. G. Campbell 1993 Divert propulsion system for army LEAP kill vehicle *2nd Annual AIAA SDIO Interceptor Technology Conf.* (Albuquerque, USA: AIAA) AIAA-93-2637
[6] P. Baker and B. Kelley 1993 Lightweight exo-atmospheric projectile (LEAP) space flight test *2nd Annual AIAA SDIO Interceptor Technology Conf.* (Albuquerque: NM, USA/AIAA) AIAA-93-2696
[7] LIU Qing-hong, CHEN De-yuan and WANG Zi-cai 2003 Effect of system delay on guidance accuracy of interceptor *J. Journal of Astronautics* **24** 642-645
[8] DUAN Meijun and ZHOU Di 2017 Guidance law with finite time convergence considering thrust constraint and dynamics characteristic *J. Systems Engineering and Electronics* **39** 2068-207

IOP Conf. Series: Materials Science and Engineering **408** (2018) 012007 doi:10.1088/1757-899X/408/1/012007

Effect of Cold Working on the Residual Stress around the Notch front of Stress Corrosion Cracking

Zhao Lingyan[1,*], Cui Yinghao[2] and Yang Fuqiang[1]

[1] School of Science, Xi'an University of Science and Technology, Xi'an 710054, China

[2] School of Mechanical Engineering, Xi'an University of Science and Technology, Xi'an 710054, China

*gloomy2@foxmail.com

Abstract. Failures show that the austenitic stainless steel (SS) pipes with produced high weld residual stresses are more susceptible to stress corrosion cracking in the simulated high temperature water environments of pressurized water reactor. To understand the effect of cold working on the residual stress and strain of stress corrosion cracking in 304SS, the stress-strain field near the crack front of 304SS under different degree of cold working was simulated by ABAQUS. Results show that the higher tensile residual stress and the larger compressive residual stress simultaneously occurs in the lower cold working material, and cold working has greater effect on the residual stress than on the residual strain. The residual compressive stress and residual compressive strain both decrease rapidly in a small range around the notch fronts and tend to be consistent in a more larger distance. But cold working has little effect on the residual strain.

1. Introduction

Stress corrosion cracking (SCC) is a failure mechanism that is caused by environment, susceptible material and tensile stress of nuclear power structural materials under high temperature and high pressure water environments [1]. Failures show that the austenitic stainless steel (SS) pipes with produced high weld residual stresses is more susceptible to SCC in the simulated high temperature water environments of pressurized water reactor [2-4]. Generally, weld residual stresses (RS) in pipe joints are up to or even over the material yield stresses at service temperature. Tensile residual stress is one of dominant factors resulting in SCC of these joints [5]. RS may have a great contribution to the total stress field when pipe surface SCC in nuclear power plants is assessed. Therefore, before evaluating SCC growth at flaws in actual pipe joints of PWR plants, accurate RS distribution needs to be performed.

During the manufacturing and assembly process under the nuclear power equipment, a certain degree of cold working (CW) will occur [6]. Many research shows that cold working has a great effect on the local mechanical field around the crack tip [7-8]. The amount of CW will also affect the state of RS field and the SCC growth rate of the nuclear power equipment [9]. Therefore, it is very important to obtain the effect of cold working on local residual stress and strain around the crack tip for quantitative prediction nuclear structure life.

Pressure was introduced in a pre-crack CT specimen to produce RS around the crack front. Finite element models of 304SS CT specimens are established under different degree of CW. The mechanical properties of crack tip region were simulated in the presence of residual stress. The effects of cold working on local residual stress and strain around the crack front are also investigated.

2. Finite element modelling

2.1. Specimen model

A notched compact tension (CT) specimen (specimen width W=20 mm) was performed in the finite element analysis, and its geometry and dimensions are shown in figure 1. To eliminate the stress singularity generated at the crack tip, the crack notch radius was taken as 0.25 mm in the finite element simulation. To avoid the contact between upper and lower surfaces in the process of residual stress induced by pressure, a groove (width is 2.5 mm and length is 11 mm) was machined at the front of the specimen gap [10].

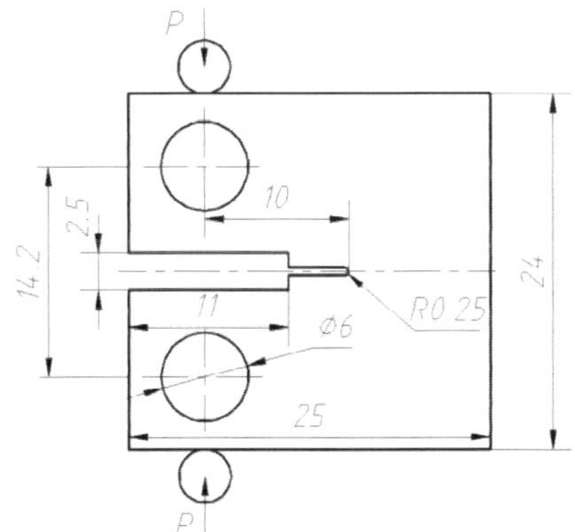

Figure 1. Geometry and dimensions of CT specimen (W=20 mm)

Due to symmetry, only one half of the specimen geometry was modeled, and the typical finite element model and meshes are illustrated in figure 2(a), which contains 20156 8-node biquadratic plane strain elements. To obtain more detailed and accurate stress-strain field, the mesh around notch front was refined as shown in figure 2(b).

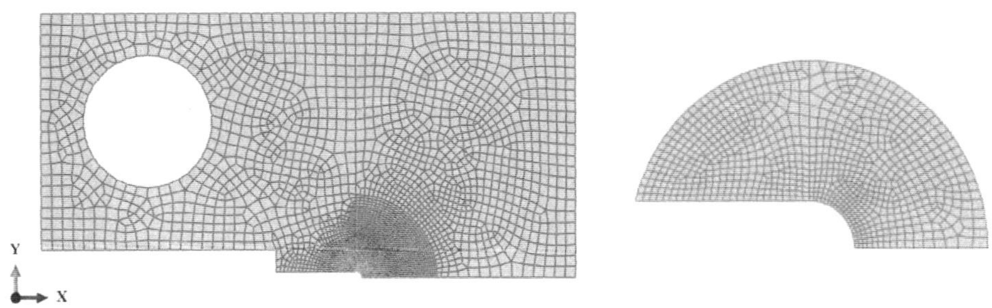

(a) Global mesh model (b) Mesh around the notch

Figure 2. Finite element mesh model of CT specimen

2.2. Material model

The specimen adopted in this study is 304SS and its chemical chemical composition includes C-0.043%, Si-0.4%, Mn-1.18%, P-0.033%, S-0.002%, Ni-8.04%, Cr-18.115%, N -0.046% etc. The geometry and dimensions of the specimen are shown in figure 3 below. Tensile experiment was carried out in fatigue stretcher machine under different cold working degrees, engineering stress and strain curve was derived and converted to true stress-strain curves. The material true mechanical curves was shown in figure 4 under the degree of cold working is 0%, 10% and 20%, respectively.

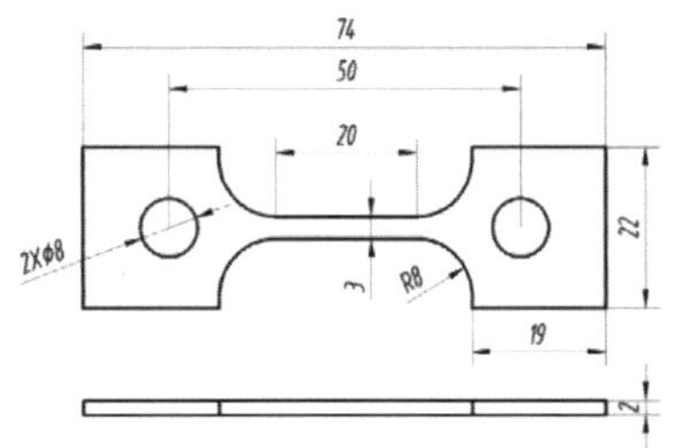

Figure 3. Plate tensile specimen

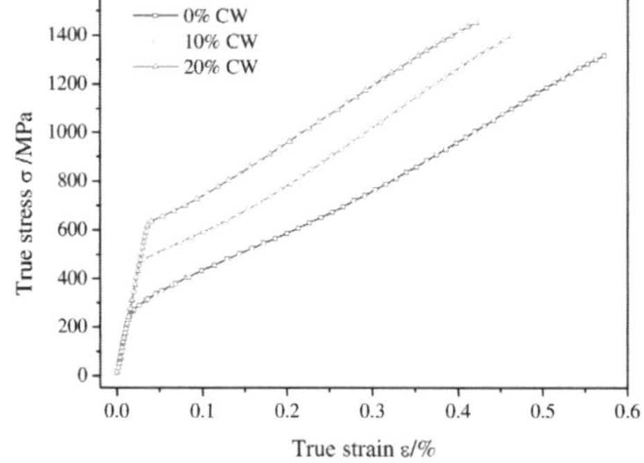

Figure 4. True stress-strain curves under different CW

2.3. Loading and boundary conditions

Two analysis steps were set in this simulation. The first one is the loading process, and displacement loading is applied to the reference point of rigid body and its value was set as 0.5 mm in the compressive direction. The other is the process of unloading, which sets the displacement load free. To eliminate the rigid body displacement during the period of loading and unloading, two points in the right side of the CT specimen was constrained in rotation and movement of X direction.

3. Results and Discussion

3.1. Effects of CW on residual stress and strain around the notch fronts

The residual stress distribution around the notch fronts under different degree of CW is shown in figure 5. The tensile residual stress decreases with the increasing CW. And the tensile residual stress mainly distributes in the near notch region, and gradually becomes compressive residual stress with the increasing distance away from the notch front. Cold working has great effect on the compressive residual stress.

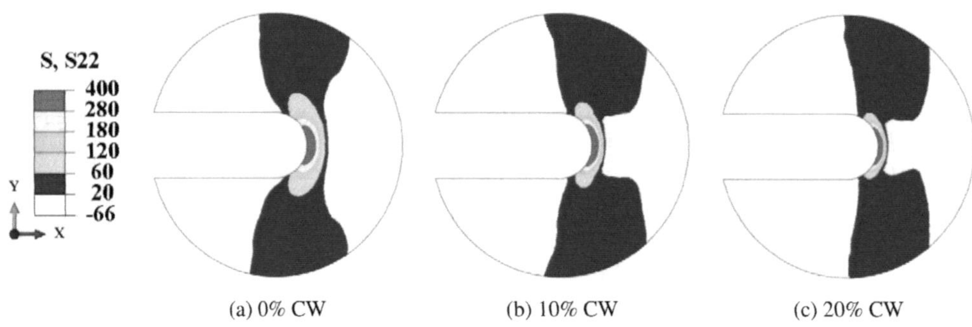

Figure 5 Residual stress around the notch fronts under different degree of CW (MPa)

The residual strain distribution around the notch fronts under different degree of CW is shown in figure 6. The residual tensile strain is mainly distributed in the direction of perpendicular to the notch and the compressive strain is mainly distributed in the front of the notch. The tensile residual strain increases with the increasing CW. Moreover, CW has little effect on the compressive residual strain.

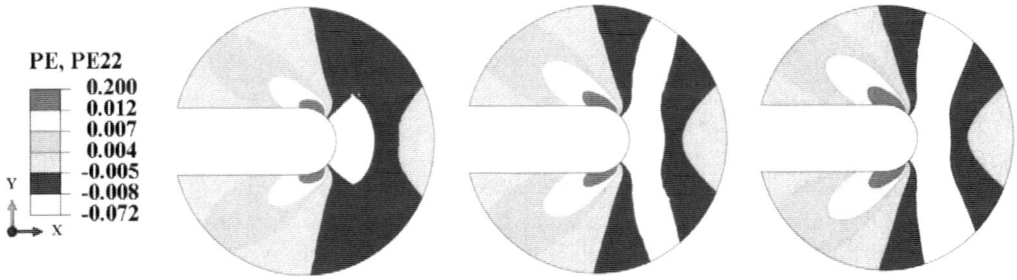

(a) 0% CW (b) 10% CW (c) 20% CW

Figure 6 Tensile residual strain around the notch fronts under different degree of CW

3.2. Residual stress around the notch fronts under the influence of CW

The tensile residual stress distribution in front of notch front under different degree of CW is shown in figure 7. The residual stress decreases rapidly from tensile to compressive in a small range of 0.1 mm in the notch zone and tends to be consistent in a more larger distance from notch front. The higher tensile residual stress and the larger compressive residual stress also occurs in the lower cold working material.

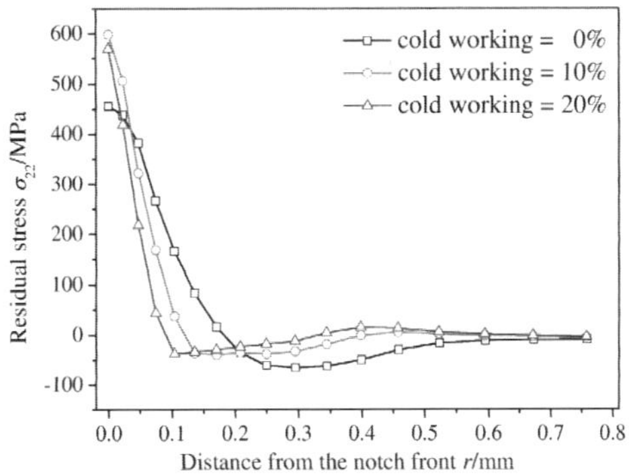

Figure 7 Residual stress around the notch fronts (MPa)

3.3. Residual strain around the notch fronts under the influence of CW

The tensile residual strain distributions around notch fronts under different degree of CW is shown in figure 8. The compressive residual strain decreases rapidly in a small range around the notch fronts and tends to be consistent in a more larger distance. The cold working has little effect on the residual strain.

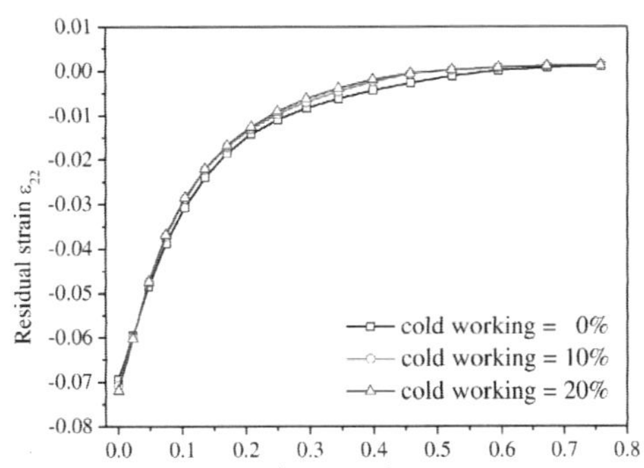

Figure 8 Residual strain around the notch fronts

4. Conclusions

(1) Near the notch region, residual stress gradually becomes compressive stress with the increasing distance from the notch front. Residual stress decreases rapidly from tensile to compressive in a small range around the notch fronts, and tends to be consistent in a more larger distance from notch front. The higher tensile residual stress and the larger compressive residual stress also occurs in the lower cold working material. CW has great effect on the residual stress.

(2) Tensile residual strain is mainly distributed in the direction of perpendicular to the notch and the residual compressive strain is mainly distributed in the front of the notch. The residual compressive strain decreases rapidly in a small range around the notch fronts, and tends to be consistent in a more larger distance. CW has little effect on the residual strain.

Acknowledgments

This work is financially supported by the National Natural Science Foundation of China (Grant No. is 11502195).

References

[1] Xue H, Sato Y, Shoji T. (2009) Quantitative estimation of the growth of environmentally assisted cracks at flaws in light water reactor components. *J. Press. Vess.-T. ASME,* 131, pp61-70

[2] Bamford W, Newton B, Seeger D. (2006) Recent experience with weld overlay repair of indications in alloy 182 butt welds in two operating PWRs. *ASME Pressure Vessels & Piping Conference (PVP)2006*, pp427-434

[3] Li G F, Congleton J. (2000) Stress corrosion cracking of a low alloy steel to stainless steel transition weld in PWR primary waters at 292°C. *Corros. Sci.*, 42, pp1005-1021.

[4] Li G F, Li G J, Fang K and Peng J. (2011) Stress corrosion cracking behavior of dissimilar metal weld A508/52M/316L in high temperature water environment. *Acta. Metall. Sin.*, 47, pp797-803

[5] Deng D, Kiyoshima S. (2010) FEM prediction of welding residual stresses in a SUS304 girth-welded pipe with emphasis on stress distribution near weld start/end location. *Comp. Mater. Sci.*, 50, pp612-621

[6] Hou J, Peng Q J, Shoji T, et al. (2010) Study of microstructure and stress corrosion cracking behaviour in welding transition zone of Ni-based alloys. *Acta. Metall. Sin.,* 46, pp1258-1266

[7] Du D H, Kai C, Hui L, et al. (2016) Effects of chloride and oxygen on stress corrosion cracking of cold worked 316/316L austenitic stainless steel in high temperature water. *Corros. Sci.*, 110, pp134-142

[8] Hou J, Shoji T, Lu Z P, et al. (2010) Residual Strain Measurement and Grain Boundary Characterization in the Heat Affected Zone of a Weld Joint between Alloy 690TT and Alloy 52. *Journal of Nuclear Materials*, 397, pp109-115

[9] Zhang L, Chen K, Du D, et al. (2017) Characterizing the effect of creep on stress corrosion cracking of cold worked Alloy 690 in supercritical water environment. *J. Nucl. Mater.*, 492, pp32-40

[10] Song X M, Wang G Z, Xuan F Z, et al. (2015) Investigation of residual stress effects on creep crack initiation and growth using local out-of-plane compression. *Eng. Fract. Mech.*,149, pp45-57

IOP Conf. Series: Materials Science and Engineering **408** (2018) 012008 doi:10.1088/1757-899X/408/1/012008

The Influence of the Assembly Gaps and Tightening Operations on the Stress Distribution during the Assembly Process

JunHao Chang[1], XueXhu Liu[2*], YuXing Yang[1], YiQi Wang[1] and Hang Gao[1,*]

[1] Key Lab. for Precision and Non-traditional Machining Technology of Ministry of Education, School of Mechanical Engineering, Dalian University of Technology, Dalian 116024, China;
[2] School of Automotive Engineering, Dalian University of Technology, Dalian 116024, China;

[*]Corresponding author's e-mail: liuxs@dlut.edu.cn;

Abstract. In the manufacturing process of the large-scale wing boxes of the airplane, the assembly gaps of the bolts play a significant role in the safety assessment of the whole plane. Moreover, unreasonable tightening operations can also affect the assembly quality. In this paper, some experiments about the different tightening sequences and torques applied on the gap-existing bolts are executed for the influence on stress distribution. The recording of the surface profile is based on the trilinear coordinates measuring instrument, while the deformations of the surface can be measured by the strain gauges. Both the tightening sequence and torques can cause the local stress concentrations, especially in the places where the gaps appear firstly. The residual gaps of this place is of nearly 30% of the maximum clearance, which cannot be eliminated without shimming.

1. Introduction

The carbon fibre reinforcement plastics materials has been widely admitted in the field of aerospace industry. Among the wing boxes manufacturing process, the assembly parts have played a more significant part in the quality assessment. However, the form deviations of the parts in the manufacturing stage such as component warpages might influence the assembly quality, which cannot be ignored randomly. The most obvious result of the mismatching of the components is the assembly gaps between the parts. The distribution of the gaps is casual and the size is hard to control, influencing the stability of the whole structure. The assembly gaps usually exist in the joint parts, represented with the bolts and nuts. Many researches have proved the gaps can affect the quality of the bolts connection. McCarthy researched the effects of clearance of the bolts in the load distribution [1]. Besides, McCarthy did other researches to find out the strength criteria in single-bolt, single-lap joints [2]. They also did the other related studies to verify the influence of the assembly gaps [3]. Moreover, the joints are also worth being noticed. The tightening operations of the bolts are the main dominant factors that dampen the performance of the joint parts. The assembly gaps cannot be eliminated randomly [4]. Kumakura researched the tightening sequences on the flange joints [5]; Alsahan did many experiments to figure out the effects of the different tightening sequences [6]; Ioannis studied the effects of different torques on failure or fatigue life of laminated bolts joints [7].

The influence of the assembly gaps in the joining parts and the tightening operations is studied in this paper. The research mainly contains the comparison of shape deformations under different tightening sequences and torques. While the strains of the testing components are listed to verifies the results.

2. Experiments

The experiments for the assembly gaps influencing the shape of the surface were set up as follows. The whole wing box is made of carbon fibre composite materials, while the bolts, gaskets and nuts used here are made of steel and the quality of them can meet the demands of the practical working conditions, seen in figure 1. The gaps exist between the ribs and upper skins, where the bolts exist. The gaps between the upper surface and the ribs are uneven. The producing place of the gaps is combined of a curving part of the rib and the upper surface. The maximum of the gaps happens in the middle part.

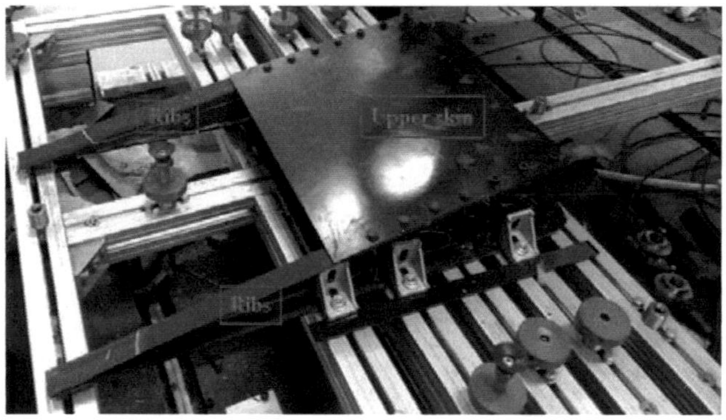

Figure 1. The structure of the CFRP wing box

The different tightening sequence can be mainly divided into two types: the single direction sequence and the multi direction sequence. The detailed tightening information are shown in Table 1. The outermost bolts were named O, the nearest bolts were named I and the middle bolts were named M. Besides, the applied torques mainly contains three stages: 3.5Nm 7.0 Nm and 10.5Nm finished by torque wrench. The testing of the surface shape was based on the trilinear coordinate measuring instrument, while fastening the bolts above the gaps under different torques and consequences.

Table 1. Tightening operations

Tightening sequence	Mark	Tightening torque(Nm)
Outer → inner side	OI	1.0
Inner → outer side	IO	3.5
Middle → inner →outer side	MIO	7.0
Middle →outer →inner side	MIO	10.5

To validate the experimental results of the shape deformations, the strain of the upper surface was also measured at the same tightening sequence and torques. The practical measurement set up can be seen in figure 2. The strain gauges were mainly put on the upper surface, where the middle line is the basement and the other gauges were put beside at a proper distance. Strains of the part whether gaps exist are both recorded under the different tightening sequence and torques. The comparison of the strains between both parts are also listed in next chapter.

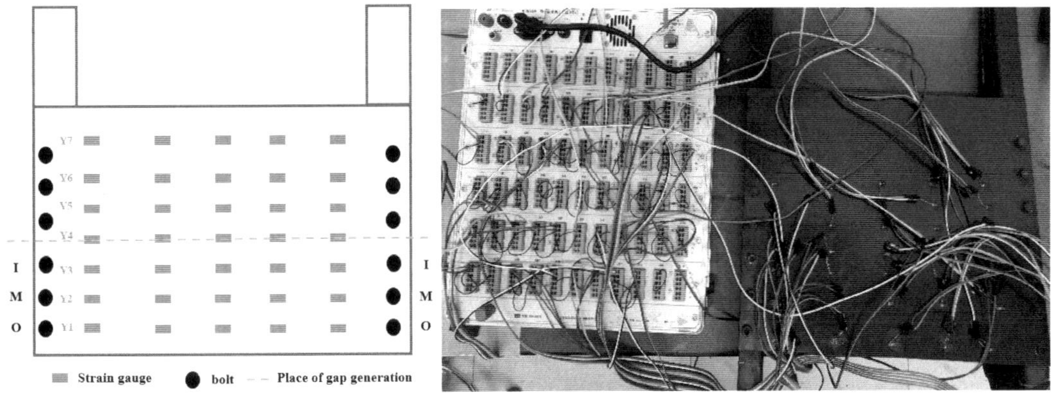

Figure 2. Practical experiment set up of strain gauges

3. Results and discussion

The surface deformations recorded by trilinear coordinates measuring instrument, under the torque 7Nm and the tightening sequence from O to I, are illustrated in the next picture. From the picture, the deformations of the district containing the gaps is more obvious than that without the assembly gaps. The max deformation of the districts containing the gaps happened in the border of the surface while the biggest deformation of the districts containing nothing happens in the central parts of the upper surface. A remarkable thing is that deformations of the places of gaps generation are quite even, compared with other places. Moreover, the trend of deformation is a between the adjacent districts, which are with and without assembly gaps, respectively. The deformation laws of the surface under different tightening operations perform the same trends as figure 3(a). The measuring results of surface strain are illustrated in figure 3(b). The specific chosen positions contains the outermost part (Y1) and the innermost part (Y7) and the place of gaps generation (Y4). It can be seen that the maximum strains of the parts with assembly gaps are positive, while the maximum strains of the parts without assembly gaps are negative representing the part is compressed. The strains beside the middle line of the parts whether containing the assembly gaps become smaller gradually which means the deformations are not so obvious, matching well with the measuring results of the upper surface. The strains of parts in the places of generation are quite even, which means the deformations of these places can be most dominated by the gaps.

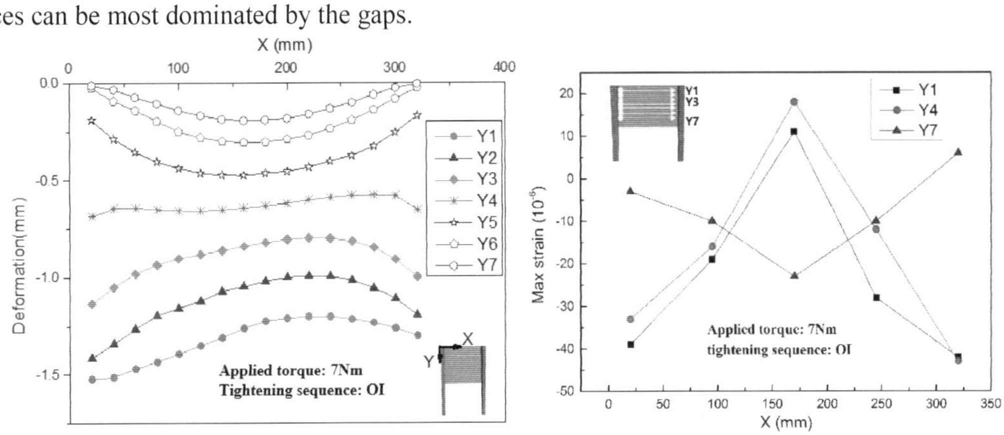

(a) (b)

Figure 3. Measuring results of the shape of surface: (a) deformation; (b) max strain

IOP Conf. Series: Materials Science and Engineering **408** (2018) 012008 doi:10.1088/1757-899X/408/1/012008

The tightening operations applied on the joints are researched. From the figure 4(a), it can be seen that the multi direction (MIO/MOI) tightening can cause less deformations compared to that with single direction tightening, where the maximum strain can be decreased by nearly 10%. After different tightening sequence of the bolts are carried out, the tightening torques are also studied. The torques are applied as designed while keeping the same tightening sequence. The measuring position is still at the places containing the max gaps. The deformations of the surface can be greatly restrained by adding torque as 1Nm, with 90% gap eliminated, seen in figure 4(b). The residual gaps are kept in a certain range as the applied torques increases. However, the residual gaps in the places of gap generation should be carefully noticed, where the gaps can hardly be removed even the torques are over the limited values (8.8Nm). Some specific compensating method should be executed so that the structure can meet the demands of actual use, shown as figure 4(c).

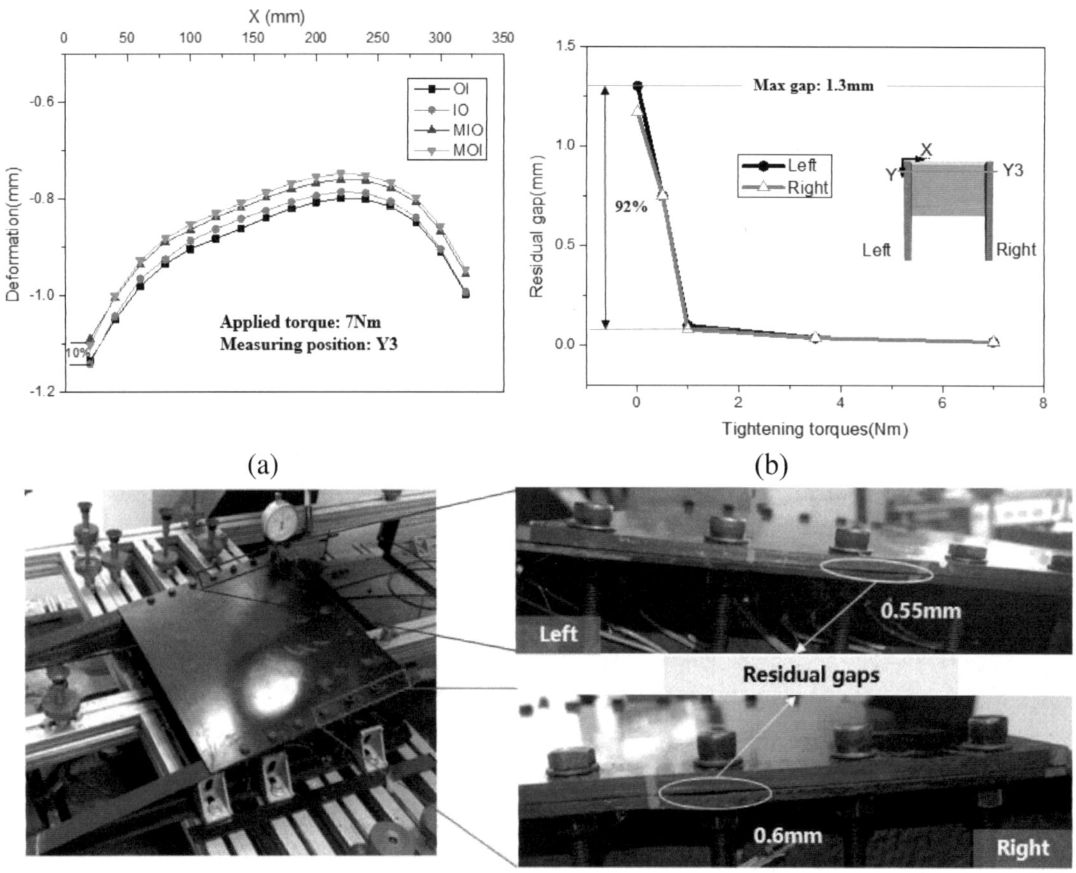

Figure 4. Residual gaps of the wing box: (a) effects of the different tightening sequence; (b) effects of the tightening torques; (c) residual gaps place of gaps generation.

4. Conclusion

In this paper, some experiments about the effects of assembly gaps and tightening operations are designed and finished. The assembly gaps have been verified in the influencing of the shape of the constructed structures. Moreover, the different tightening sequences and torques can both determine the quality of assembly. In the future, more researches will be established to improve the assembly methods

Acknowledgments
This work was supported by the National Key Basic Research Program of China (973 Project) [Grant No. 2014CB046504], the National Natural Science Foundation of China [Grant No. 51375068, Grant No. 51475073 & Grant No. 51605076] and the Fundamental Research Funds for the Central Universities [Grant No. DUT15RC(3)089]. The authors would like to acknowledge the above financial supports.

References
[1] McCarthy M A, McCarthy C T and Padhi G S 2006 A simple method for determining the effects of bolt–hole clearance on load distribution in single-column multi-bolt composite joints: *Composite Structures*, 73(1): 78-87.
[2] McCarthy M A, Lawlor V P, Stanley W F 2002 Bolt-hole clearance effects and strength criteria in single-bolt, single-lap, composite bolted joints: *Composites Science and Technology*, 1415-1431.
[3] McCarthy M A , McCarthy C T and Stanley W F 2011 Bolt-hole clearance effects in composite joints In Wood head Publishing Series in Composites Science and Engineering: *Composite Joints and Connections*, ISBN 9781845699901, Pages 112-185.
[4] Standard Test Method for Bearing Response of Polymer Matrix Composite Laminates, ASTMD5961 /D5961M-13.US: ASTM International, 2013
[5] Kumakura S and Saito K 2003 *Tightening Sequence for Bolted Flange Joint Assembly* (Cleveland: ASME Pressure Vessels and Piping Conference).
[6] Alsahan M M, Ns A M and Akeel R F 2014. The influence of tightening sequence and method on screw preload in implant superstructures: *International Journal of Prosthodontics*, 27(1): 76-79.
[7] Ioannis K. Giannopoulos, Damian Doroni-Dawes and Kyriakos I 2017 Effects of bolt torque tightening on the strength and fatigue life of airframe FRP laminate bolted joints: *Composites Part B: Engineering*, Volume 125, Pages 19-26.

IOP Conf. Series: Materials Science and Engineering **408** (2018) 012009 doi:10.1088/1757-899X/408/1/012009

A Kind of Incremental Kalman Filter in Fast Implementation

YAN Guangming and SUN Xiaojun*

Electrical Engineering Institute, Heilongjiang University, Harbin, 150080, China

*sxj@hlju.edu.cn

Abstract. Now the observation error is widespread in applications. It is usually caused by the influence of surrounding environment, the error of measuring equipment or the improper selection of model and parameters. When the observation equation of system has not been verified or corrected under certain environmental conditions, applying it will yield to unknown system errors and filtering errors. For the systems under poor observation condition, an incremental Kalman filter in fast implementation is presented in this paper, which can eliminate the unknown system errors and improve the accuracy of state estimators. Moreover, it also has the smaller computation load compared with the general incremental Kalman filter. It is in fast implementation and easy to be applied in engineering practice. Two simulation examples show its effectiveness and feasibility.

1. Introduction

Kalman filtering is an important state estimation method. This time domain method was first proposed in the 60s of last century [1-3]. Its main advantage is that its recursive form is convenient for computer realization and real-time application, and it can handle the filtering problem of time-varying system, non-stationary random signal and multidimensional signal. At present, it has been widely used in many high technology fields such as target tracking, guidance, GPS positioning, communication and signal processing, fault diagnosis, robotics and other [4].

However, the Kalman filtering algorithm requires accurate system model parameters and noise statistics [5]. In the practical application, the observation error will be caused due to the influence of surrounding environment, the error caused by the measuring equipment itself, the improper selection of model and parameters. And the traditional Kalman filtering algorithm is difficult to eliminate the error of this kind of system [6-8]. In order to solve this problem, a series of incremental filtering algorithms have been proposed. For the nonlinear systems under poor observation condition, the extended incremental Kalman filtering algorithm was proposed by [9]. The incremental particle filter was proposed by [10], and the adaptive high order unscented incremental Kalman filtering algorithm was proposed by [11]. In [12] and [13], the incremental Kalman filtering and smoothing algorithms were presented for the systems under poor observation condition, which successfully eliminate the measurement error and improve the filtering accuracy. However, the inverse operation of the matrix with the same dimension yields the effect on the speed.

In this paper, an incremental Kalman filter in fast implementation is presented. This method can effectively eliminate the unknown error of the system under poor observation condition, and greatly improve the estimation accuracy. Compared with the general incremental Kalman filter, it has a faster implementation speed. The simulation example shows its effectiveness and feasibility.

2. Problem formulation

Consider a target tracking system as follows

$$x_k = \Phi_{k-1}x_{k-1} + \Gamma_{k-1}w_{k-1} \tag{1}$$

where $x_k \in R^n$ is the state at time k, $\Phi_k \in R^{n\times n}$ is the state transformation matrix, $\Gamma_k \in R^{n\times r}$ is the system noise distribution matrix, $w_k \in R^r$ is the system noise vector.

Assumption 1 w_k is Gauss white noise with zero mean and variance Q_k. x_0 is the initial state for the target, and $\mathrm{cov}[w_k, w_j] = Q_k\delta_{kj}$, $\mathrm{cov}[x_0, w_k] = 0$.

The observation equation can be denoted by

$$z_k = H_k x_k + v_k + a_k \tag{2}$$

where $z_k \in R^m$ is the observation vector at time k, $H_k \in R^{m\times n}$ is the observation matrix at time k. a_k is the observation system error at time k. $v_k \in R^m$ is observation noise vector at time k. v_k is a Gauss white noise with zero mean and variance R_k.

Assumption 2. Suppose the observation noises are uncorrelated with each other at the same time k, as well as the observation noises at different times. i.e. $\mathrm{cov}[v_k, v_j] = R_k\delta_{kj}$, $\mathrm{cov}[w_j, v_k] = 0$ and $\mathrm{cov}[x_0, v_k] = 0$.

Based on (2), we can obtain the incremental observation equation as follows

$$\Delta z_k = H_k x_k - H_{k-1}x_{k-1} + v_k - v_{k-1} \tag{3}$$

where Δz_k is the observation vector increment at time k, $\Delta z_k = z_k - z_{k-1}$. Suppose that the stochastic vectors v_k and v_{k-1} are Gauss white noises uncorrelated with each other, then $V_k = v_k - v_{k-1}$ is also a Gauss white noise with zero mean and variance $2R_k$.

Furthermore, applying (1) and (3) yields

$$\Delta z_k = \overline{H}_k x_{k-1} + \overline{V}_k \tag{4}$$

where $\overline{H}_k = H_k\Phi_{k-1} - H_{k-1}$, $\overline{V}_k = H_k\Gamma_{k-1}w_{k-1} + V_k$. In (4), the unknown observation error is eliminated. Eq. (1) and Eq. (4) contribute to the incremental filtering fusion equation. The objectives are to find a incremental Kalman filter in fast implementation $\hat{x}_{k|k}$ for the state x_k, based on the measurement $(\Delta z_1, \Delta z_2, \text{L}, \Delta z_k)$ at time k.

Remark 1. In engineering practice, the observation system errors between two adjacent observations z_k and z_{k-1} are close to size, so that the system error of Δz_k is relatively small and can be ignored [12]. In addition, Δz_k and Δz_{k-1} are better to satisfy the independence requirement than that of z_k and z_{k-1} according to the principle of independent increment stochastic process [14].

3. A incremental Kalman filter in fast implementation

Lemmma 1 [15]. For the known non-singular matrices $P \in R^{n\times n}$, $R \in R^{m\times m}$ and $H \in R^{m\times n}$, if the inverses of $P^{-1} + H^{\mathrm{T}}R^{-1}H$ and $HPH^{\mathrm{T}} + R$ exist, we have the following matrix equation

$$[I + PH^{\mathrm{T}}R^{-1}H]^{-1}P = [P^{-1} + H^{\mathrm{T}}R^{-1}H]^{-1} = P - PH^{\mathrm{T}}(HPH^{\mathrm{T}} + R)^{-1}HP \tag{5}$$

Applying Lemma 1 and [13] yields a new incremental Kalman filter as follows.

Theorem 1. The system (1) and (4) with Assumptions 1 and 2 has the following Kalman filter in fast implementation

$$\hat{x}_{k+1|k} = \Phi_k \hat{x}_{k|k} \tag{6}$$

$$P_{k+1|k} = \Phi_k P_{k|k} \Phi_k^{\mathrm{T}} + \Gamma_k Q_k \Gamma_k^{\mathrm{T}} \tag{7}$$

$$\bar{R}_{k+1}^{-1} = \frac{1}{2} R_{k+1}^{-1} - \frac{1}{2} R_{k+1}^{-1} H_{k+1} \Gamma_k [\Gamma_k^{\mathrm{T}} H_{k+1}^{\mathrm{T}} (\frac{1}{2} R_{k+1}^{-1}) H_{k+1} \Gamma_k + Q_k^{-1}]^{-1} \Gamma_k^{\mathrm{T}} H_{k+1}^{\mathrm{T}} (\frac{1}{2} R_{k+1}^{-1}) \tag{8}$$

$$K_{k+1} = \Phi_k P_{k|k} \bar{H}_{k+1}^{\mathrm{T}} [\bar{R}_{k+1}^{-1} - \bar{R}_{k+1}^{-1} \bar{H}_{k+1} (P_{k|k}^{-1} + \bar{H}_{k+1}^{\mathrm{T}} \bar{R}_{k+1}^{-1} H_{k+1})^{-1} \bar{H}_{k+1}^{\mathrm{T}} \bar{R}_{k+1}^{-1}] \tag{9}$$

$$\hat{x}_{k+1|k+1} = \hat{x}_{k+1|k} + K_{k+1} (\bar{H}_{k+1} \tilde{x}_{k|k} + \bar{V}_{k+1}) \tag{10}$$

$$P_{k+1|k+1} = P_{k+1|k} + K_{k+1} \bar{H}_{k+1} P_{k|k} \bar{H}_{k+1}^{\mathrm{T}} K_{k+1}^{\mathrm{T}} + K_{k+1} (H_{k+1} \Gamma_k Q_k \Gamma_k^{\mathrm{T}} H_{k+1}^{\mathrm{T}} + 2R_{k+1}) K_{k+1}^{\mathrm{T}} \tag{11}$$

Proof. From [13], we have (6), (7), (10) and the filtering error variance matrix as follows

$$P_{k+1|k+1} = P_{k+1|k} + K_{k+1} \bar{H}_{k+1} P_{k|k} \bar{H}_{k+1}^{\mathrm{T}} K_{k+1}^{\mathrm{T}} + K_{k+1} (\Gamma_k Q_k \Gamma_k^{\mathrm{T}} + 2R_{k+1}) K_{k+1}^{\mathrm{T}} \tag{12}$$

where the filtering gain is given as

$$K_{k+1} = \Phi_k P_{k|k} \bar{H}_{k+1}^{\mathrm{T}} [\bar{H}_{k+1} P_{k|k} \bar{H}_{k+1}^{\mathrm{T}} + \Gamma_k Q_k \Gamma_k^{\mathrm{T}} + 2R_{k+1}]^{-1} \tag{13}$$

Setting $\bar{R}_{k+1} = H_{k+1} \Gamma_k Q_k \Gamma_k^{\mathrm{T}} H_{k+1}^{\mathrm{T}} + 2R_{k+1}$, applying Lemma 1 yields

$$K_{k+1} = \Phi_k P_{k|k} \bar{H}_{k+1}^{\mathrm{T}} [\bar{R}_{k+1}^{-1} - \bar{R}_{k+1}^{-1} \bar{H}_{k+1} (P_{k|k}^{-1} + \bar{H}_{k+1}^{\mathrm{T}} \bar{R}_{k+1}^{-1} H_{k+1})^{-1} \bar{H}_{k+1}^{\mathrm{T}} \bar{R}_{k+1}^{-1}] \tag{14}$$

Furthermore, we have

$$\bar{R}_{k+1}^{-1} = \frac{1}{2} R_{k+1}^{-1} - \frac{1}{2} R_{k+1}^{-1} H_{k+1} \Gamma_k [\Gamma_k^{\mathrm{T}} H_{k+1}^{\mathrm{T}} (\frac{1}{2} R_{k+1}^{-1}) H_{k+1} \Gamma_k + Q_k^{-1}]^{-1} \Gamma_k^{\mathrm{T}} H_{k+1}^{\mathrm{T}} (\frac{1}{2} R_{k+1}^{-1}) \tag{15}$$

In summary, we can obtain an incremental Kalman filter in fast implementation. The proof is completed.

Remark 2. In the multi-sensor information fusion Kalman filtering problem [15], it will encounter the measurement matrix H_k in $m \times n$ dimension. Here m is far greater than n such as that in simulation example 2. In [13], it is necessary to calculate the inverse matrix of the high-dimensional matrix $[\bar{H}_{k+1} P_{k|k} \bar{H}_{k+1}^{\mathrm{T}} + H_{k+1} \Gamma_k Q_k \Gamma_k^{\mathrm{T}} H_{k+1}^{\mathrm{T}} + 2R_{k+1}]$ or $(\bar{H}_{k+1} P_{k|k} \bar{H}_{k+1}^{\mathrm{T}} + \bar{R}_{k+1})$ online. This problem can be solved applying (8) and (9), which can reduce the amount of incremental Kalman computation and speed up the implementation.

4. Simulation model and result analysis

Many simulation experiments show that the system errors can not be eliminated by the traditional Kalman filtering and self-adaptive Kalman filtering when the observation data has unknown errors due to the environments. In this paper, the presented incremental Kalman filter in fast implementation can eliminate the system error and greatly improve the filtering accuracy.

Example 1. Consider the one dimension discrete system

$$x_k = 0.8 x_{k-1} + w_k \tag{16}$$

$$z_k = x_k + a_k + v_k \tag{17}$$

where w_k and v_k are independent Gauss white noises, the system noise w_k has the mean $q = 0$ and variance $Q = 0.1$, the observation noise v_k has the mean $r = 0$ and variance $R = 1$. They are all known. a_k is the unknown observation error. Suppose $a_k = 3$.

Applying the method in [13] and Theorem 1 yields two kinds of incremental Kalman filters respectively. The simulation results are shown in Figure 1, Figure 2 and Table 1. Figure 1 shows the contrast curve between the real value and filter $\hat{x}_{k|k}$ for the state, where the solid line represents the true value and the dashed line represents the estimated value.100 Monte Carlo runs are carried out and the mean square error (MSE) curves are shown by Figure 2. Table 1 gives a comparison of calculation between the method in [13] and Theorem 1 in 200 steps. It can be seen that the proposed incremental

Kalman filtering algorithm can realize the same accuracy for state estimation at a faster speed, and the method is effective and feasible.

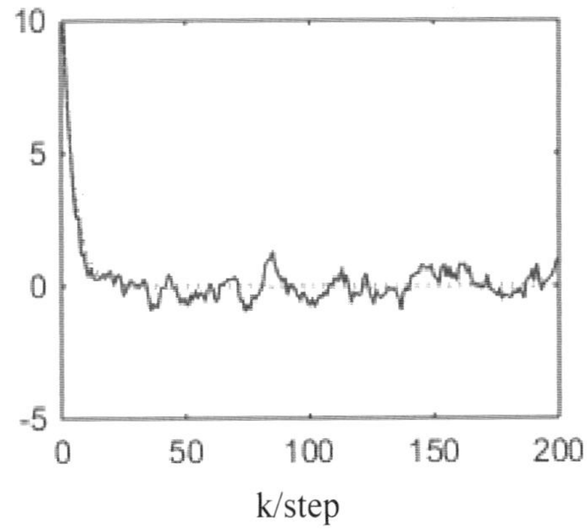

Figure 1. x_k and $\hat{x}_{k|k}$

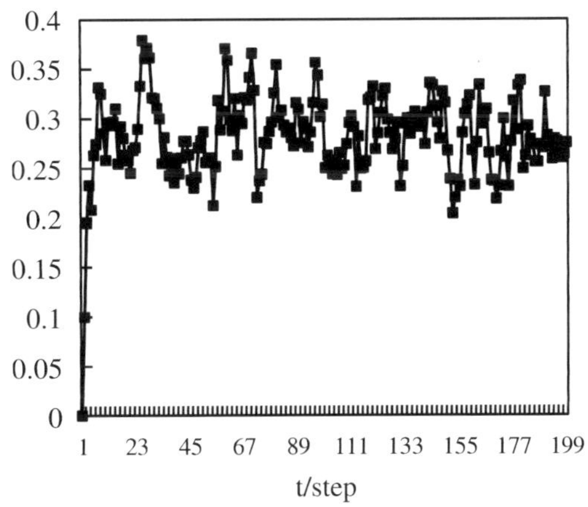

Figure 2. MSE curves for an incremental Kalman filter in fast implementation

Table 1. The comparison of calculation time between the proposed and existed incremental Kalman filters for example 1

Estimator	The calculation time used to in 200 steps (second)
Existed incremental Klaman filter	0.125
Proposed incremental Kalman filter	0.116

Example 2. Consider the one dimension discrete system

$$x_k = 0.5x_{k-1} + w_k \tag{18}$$

$$z_k = H_k x_k + a_k + v_k \tag{19}$$

where the system noise w_k is a Gauss white noise with zero mean and variance $Q=10$, the observation noise v_k is a Gauss white noise with zero mean and variance $R=1$, they are all known. The observation matrix $H_k = [1 \quad 1 \quad 0 \quad 1 \quad 0]^{\mathrm{T}}$. $a_k = [3 \quad 3 \quad 3 \quad 3 \quad 3]^{\mathrm{T}}$ is unknown observation error.

Applying the method in [13] and Theorem 1 yields the two kinds of incremental Kalman filters respectively. The simulation result is shown in Table 2. Table 2 gives a comparison of calculation between the method in [13] and Theorem 1 in 100 steps. It can be seen that the proposed incremental Kalman filtering algorithm is effective and feasible

Table 2. The comparison of calculation time between the proposed and existed incremental Kalman filters for example 2

Estimator	The calculation time used to in 100 steps (second)
Existed incremental Klaman filter	0.136
Proposed incremental Kalman filter	0.094

5. Conclusion
Considering the systems under poor observation condition, an incremental Kalman filter in fast implementation is proposed in this paper. On the one hand, the error of the system is eliminated, and the accuracy of state estimation is improved. On the other hand, compared with the existed incremental Kalman filter, it has a faster implementation speed. Two simulation examples show its effectiveness.

Acknowledgments
This work is supported by National Natural Science Foundation (NNSF) of China under Grant 61104209, Outstanding Youth Science Foundation of Heilongjiang University under Grant JCL201103, Key Laboratory of Electronics Engineering, College of Heilongjiang Province under Grant DZZD2010-5 and Youth Science Foundation of Heilongjiang University under Grant QL201212.

References
[1] KALMAN R E 1960 A new approach to linear filtering and prediction problems. *ASME Journal of Basic Engineering*, 82D: 35-45.
[2] Y. Y. Qin, H. Y. Zhang, S. H. Wang 2010 *Kalman filtering and integrated navigation theory*. Northwestern Polytechnic University,
[3] HU Zhentao, HU Yumei, LIU Xianxing 2016 Kalman Filter Based on Measurement Lifting Strategy. *ACTA ELECTRONICA SINICA*, 44(5): 1149-1155
[4] WANG Xuemei, LIU Wenqiang, DENG Zili 2015 Robust Covariance Intersection Fusion Steady-state Kalman Filter for Uncertain Systems. *Journal of Electronics & Information Technology*, 37(8): 1900-1905
[5] SUN X J, GAO Y, DENG Z L, LI C, WANG J W. 2010 Multi-model Information Fusion Kalman Filtering and White Noise Deconvolution. *Information Fusion*, 11: 163-173.
[6] L. S. Liu, B. Wu, P. Yang. Orbit precision determinination & self-calibration technique of spacecraft[M]. National Defense Industry Press, 2005.
[7] ZHANG H L. Variable dimension cubature kalman filter algorithm with incomplete measurements [J]. CHINA MEASUREMENT & TEST, 2016, 42(6): 112-116
[8] ZHANG Q H. 2017 On stability of the Kalman filter for discrete time output error systems. *System & Control Letters*, 107: 84-91
[9] FU Huimin, LOU Taishan, WU Yunzhang. 2012 Extended incremental Kalman filter method under poor observation condition. *Journal of Aerospace Power*, 27(4): 777-781.
[10] FU Huimin, LOU Taishan, WU Yunzhang. 2013 Incremental particle filter method. *Journal of Aerospace Power*, 28(6): 1201-1207
[11] ZHANG H L. 2017 Adaptive High Order Unscented Incremental Kalman Filtering. *MEASUREMENT & CONTROL TECHNOLOGY*, 36(4): 40-42

[12] FU Huimin, WU Yunzhang, LOU Taishan. 2012 Incremental Kalman Filter method under poor observation condition. *Journal of Mechanical Strength*, 34(1): 43-47.
[13] SUN X J, YAN G M, ZHANG B. 2017 A Kind of Incremental Kalman Smoother Under Poor Observation Condition. *CCC2017*, 2: 2524-2527.
[14] HAN C Z, ZHU H Y, DUAN Z S. 2010 *Multi-source Information Fusion*. Tsinghua University Press,
[15] SUN X J, YAN G M. 2013 Self-tuning weighted measurement fusion white noise deconvolution estimator and its convergence analysis. *Digital Signal Processing*, 23(1): 38–48.

IOP Conf. Series: Materials Science and Engineering **408** (2018) 012010 doi:10.1088/1757-899X/408/1/012010

Mechanical Properties of Silicone Elastomer Seals for Space Docking Vehicles

CH J Liao*, H R Lu, X H Chang, H L Sun, SH Q Fan and H L Wu

Beijing Institute of Astronautical Systems Engineering, Beijing, People's Republic of China

*liaochuanjunl@163.com

Abstract. Low impact docking mechanisms (LIDM) are developing. The LIDM seal is one of the key components of the LIDM, and play very critical roles during the docking and berthing of space vehicles. Due to the severe functional requirements, it is difficult to develop the LIDM seal that is a silicone elastomer seal. The excellent mechanical property is firstly required for the LIDM seal. Five design schemes of the LIDM seal are proposed by this paper, and their mechanical properties are studied by the proposed finite element method. The mechanical properties of the LIDM seal have important impacts on the required sealing performance, docking feature and separation feature. The findings are beneficial for the development of the LIDM seal for space docking vehicles.

1. Introduction

For expanding the docking and berthing capability of further space vehicles, many counties are developing low impact docking mechanisms (LIDM). The LIDM is the executive component of the space rendezvous and docking of two vehicles, and is used for the mechanical docking, the docking retention and the separation of the two vehicles. For the two vehicles involving in the space rendezvous and docking, one is the target vehicle, and the other is the tracing vehicle. After the docking is finished, both the target vehicle and the tracing vehicle are connected as a whole, so some interactive actions can be carried out in the two vehicles such as material exchange, astronaut delivery and fuel filling-up. In order to prevent leakage of the matters in the cabins such as atmospheres and fuels, the sealing performance of the interface of the two vehicles is very important, which is guaranteed by the LIDM seal. There are many key requirements for the LIDM seal that is the silicone elastomer seal, and it is one of the key research objects of the LIDM.

At present, the main research findings about the LIDM seal are from the research subjects funded by the National Aeronautics and Space Administration (NASA). Christopher C et al designed a candidate seal of LIDM, which was composed of a primary and a redundant elastomer seal retained by a metallic ring, and the elastomer used to manufacture the seal was Esterline ELA-SA-401 [1]. Patrick H D et al studied two types of seal designs for the LIDM seal, one is a Gask-O-seal, and the other is a multi-piece seals. A series of performance assessments and comparisons were performed between the two seals such as design feature, leak rate and adhesion load etc [2]. LIDM seals are typically made of silicone elastomers, they can suffer damage from ultraviolet radiation and atomic oxygen etc when such seals are exposed to low earth orbit conditions. Emily C I et al studied the effects of the space environment on the LIDM seal [3-5]. Patrick H D et al designed a test fixture which was used to evaluate the leakage of candidate full-scale LIDM seals under simulated thermal, vacuum, and engagement conditions [6, 7]. Ian M S et al studied a LIDM seal that is a composite design consisting of elastomer seal bulbs molded into the front and rear sides of a metal ring, and the test specimens were sub-scale seals with two different elastomer cross-sections and a 12-in outside diameter [8]. Nicholas G G et al studied experimentally the leakage rate of a LIDM seal that was manufactured from silicone elastomer S0383-70 vacuum molded in a metal retainer ring [9, 10]. Nicholas P et al designed a LIDM seal, and performed some leak tests that were completed at nominal temperatures of –30, 20, and 50 °C, and the influences of both test temperature and atomic oxygen exposure on the performance of the seal were studied [11]. Nicholas G G et al studied the compressible permeation of a LIDM seal, and purposed a physics-based model to provide a predictive methodology, by which the fundamental permeation mechanics of the LIDM seal were obtained [12].

The design of the LIDM seal is the chief topic for its development. The excellent mechanical property is required for the LIDM seal, and it decides the sealing performance, docking feature and separation feature. Using the finite element (FE) method, the mechanical properties of five design schemes of the LIDM seal will be discussed, which can be referenced by the development of the seal.

2. LIDM seal

The LIDM seal is a key component of the LIDM that is the seal at the main docking interface, which inhibits the loss of cabin air once docking is complete. The diagrammatic sketch of a LIDM and its interface seal (LIDM seal) is shown in figure 1 [1-3]. The LIDM seal is designed to maintain acceptable leak rates while the LIDM is exposed to the harsh environmental conditions of outer space. Besides the good sealing performance, the LIDM seal is required to have some other properties such as low load, easy docking, easy separation and repetitive use etc.

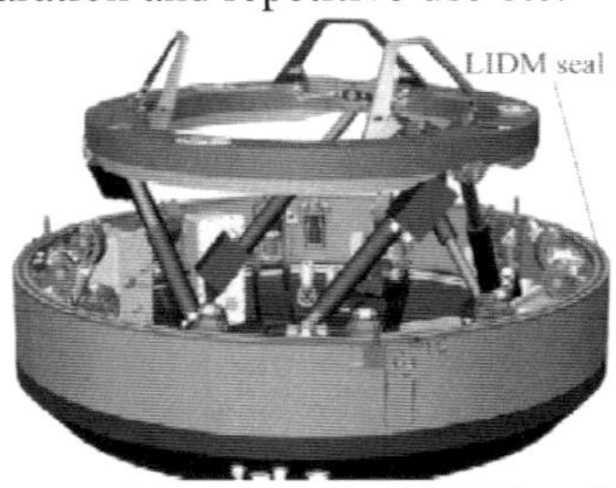

Figure 1. Diagrammatic sketch of a LIDM and its interface seal

According to the docking interface, there are generally two configurations for the LIDM seal, one is the seal-on-flange configuration, and the other is the seal-on-seal configuration. When a docking system equipped with a seal docks to a system with a flat metal flange, the seal-on-flange configuration needs to be used, for example an application when a vehicle docks to a node on the International Space Station. The seal-on-seal configuration needs to be used when two docking systems equipped with seals dock to each other, and the two types of seal are identical. Here, the seal-on-seal configuration is discussed here, and the diagrammatic sketch of a seal-on-seal configuration is shown in figure 2. A LIDM seal is constituted by two front seals, two back seals and their retainer. In the retainer, there are many tap holes, by which the LIDM seal is fixed in the LIDM when the corresponding fasteners are used.

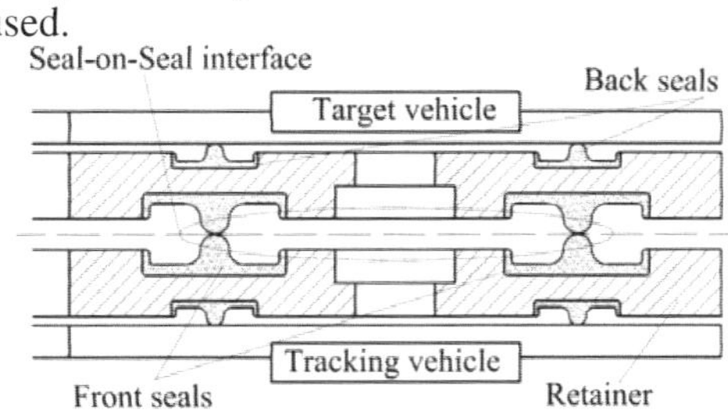

Figure 2. Diagrammatic sketch of a seal-on-seal configuration

For the designs of the LIDM seal, there are mainly two types for the seal-on-seal configuration now, one is the gask-O-seal, and the other is the multi-piece seal. Both types of seals have a pair of seal bulbs to satisfy the redundancy requirement. The materials used for both the front seals and the back seals mainly are the silicone rubber.

For the design of the LIDM seal, the design of the front seals is pivotal, which plays an important role on the combination properties. Here, five design schemes for the front seals are discussed. The diagrammatic sketches of cross-sections of the five front seals are shown in figure 3. The seal material is also a silicone rubber.

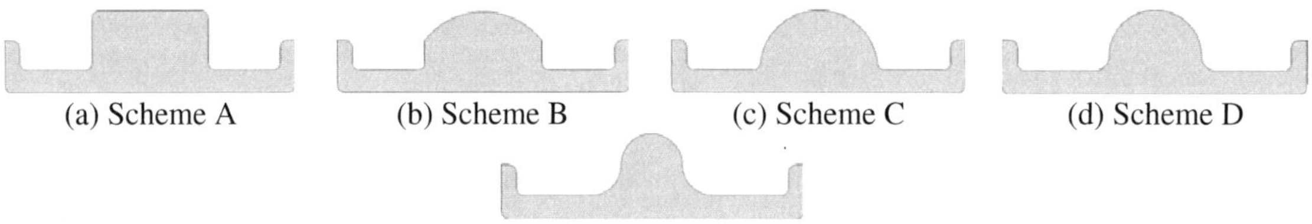

Figure 3. Diagrammatic sketches of cross-sections of five front seals

3. Mathematical methods

The FE method is adopted to study the mechanical property of the front seal based on ANSYS® software. The 2D-axisymmetric FE models are developed to investigate the five front seals, and the FE model of the scheme E is shown as figure 4. In the FE models, the PLANE183 element is used to represent the seals, and the Coulomb model is adopted for the contact and friction analysis. Both the TARGE169 element and the CONTA172 element are used for the contact elements.

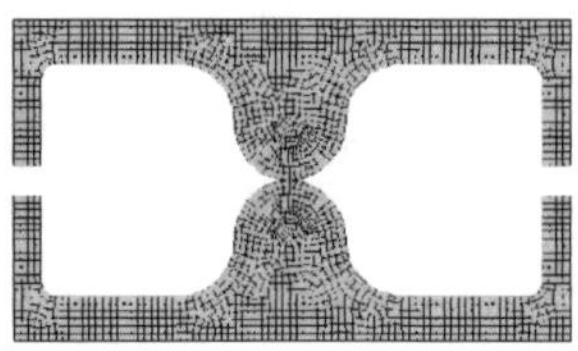

Figure 4. FE model of the scheme E

In the FE model, a Yeoh formulation is selected as the hyperelastic constitutive model where the strain energy is defined as a function of the first strain invariant and the Jacobian. The Yeoh model is taken into account as [13]

$$W=\sum_{i=1}^{N}c_i\left(\bar{I}_1-3\right)^i+\sum_{k=1}^{N}\frac{1}{d_k}(J-1)^{2k} \tag{1}$$

$$\frac{\partial W}{\partial I_1}=\sum_{i=1}^{N}ic_i\left(\bar{I}_1-3\right)^{i-1}J^{-2/3} \tag{2}$$

$$\frac{\partial W}{\partial I_2}=0 \tag{3}$$

$$\frac{\partial W}{\partial I_3}=\sum_{k=1}^{N}\frac{k}{d_k J}(J-1)^{2k-1} \tag{4}$$

Where W is the strain energy. I_1, I_2 and I_3 are the principal invariants, and J is the Jacobian. Both c_i and d_k are the material constants. N is the order of the polynomial fit, which is chosen to cubic so that the larger strain behaviour can be captured and the material response remains stable at all strains. So, the simplified Yeoh model is stated as

$$W=c_1\left(\bar{I}_1-3\right)^1+c_2\left(\bar{I}_1-3\right)^2+c_3\left(\bar{I}_1-3\right)^3+\frac{1}{d_1}(J-1)^2+\frac{1}{d_2}(J-1)^4+\frac{1}{d_3}(J-1)^6 \tag{5}$$

IOP Conf. Series: Materials Science and Engineering **408** (2018) 012010 doi:10.1088/1757-899X/408/1/012010

4. Analysis and discussions

By the FE method, the mechanical properties of the five front seals will be investigated. The key material and structural parameters are described in table 1. The sealing medium is atmospheric, and the operating pressure is the normal atmospheric pressure and the temperature is 25 ℃. The compression amounts of all the seals are all identical [13].

Table 1. Description of the seals

Parameters	Values
Inner diameter of front seals	1.24 m
Outer diameter of front seals	1.28 m
Young's modulus	3.15 MPa
Poisson's ratio	0.47
c_1	0.2693
c_2	-0.0644
c_3	0.0278
d_1	0.00112
d_2	8.38×10^{-5}
d_3	6.43×10^{-6}

The Mises stress contours of the five front seals are shown in figure 5, from which the following conclusions can be obtained.

(1) The stress of the scheme A is detrimental to both the sealing performance and the stability.

(2) The stresses of the last four schemes are similar, and the maximum Mises stress decreases with the cross-sectional area.

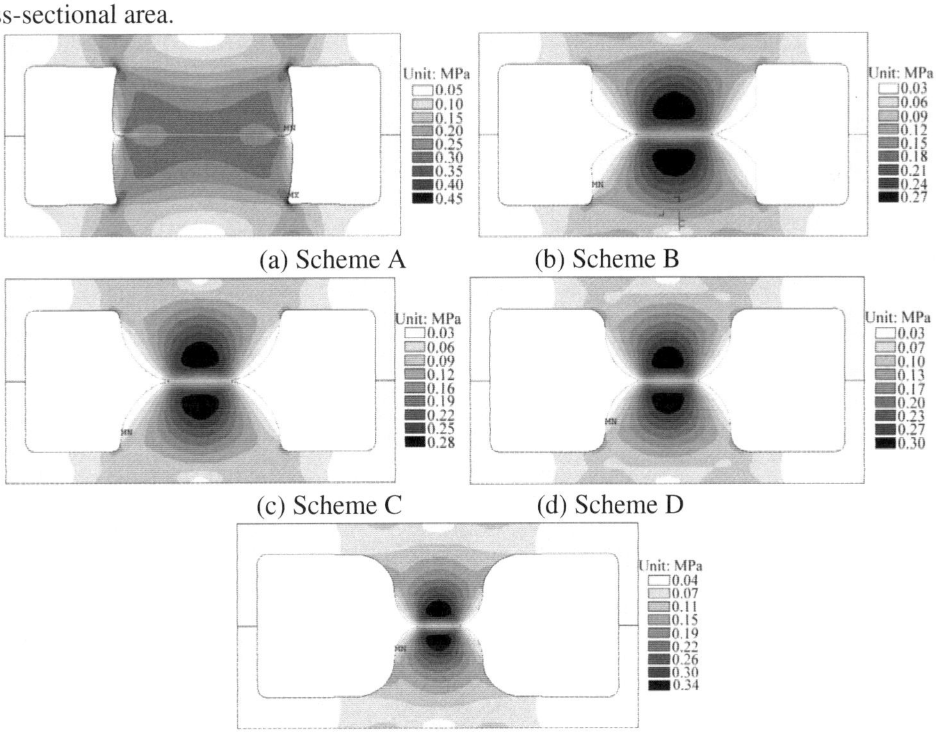

(a) Scheme A (b) Scheme B

(c) Scheme C (d) Scheme D

(e) Scheme E

Figure 5. Mises stress contours of the five front seals

The σ_{22} axial stress of the five front seal is important for the docking and separation features. The σ_{22} axial stress contours of the five front seals are shown in figure 6. Compared with the other schemes, the docking and separation features of the scheme A is poorer.

IOP Conf. Series: Materials Science and Engineering **408** (2018) 012010 doi:10.1088/1757-899X/408/1/012010

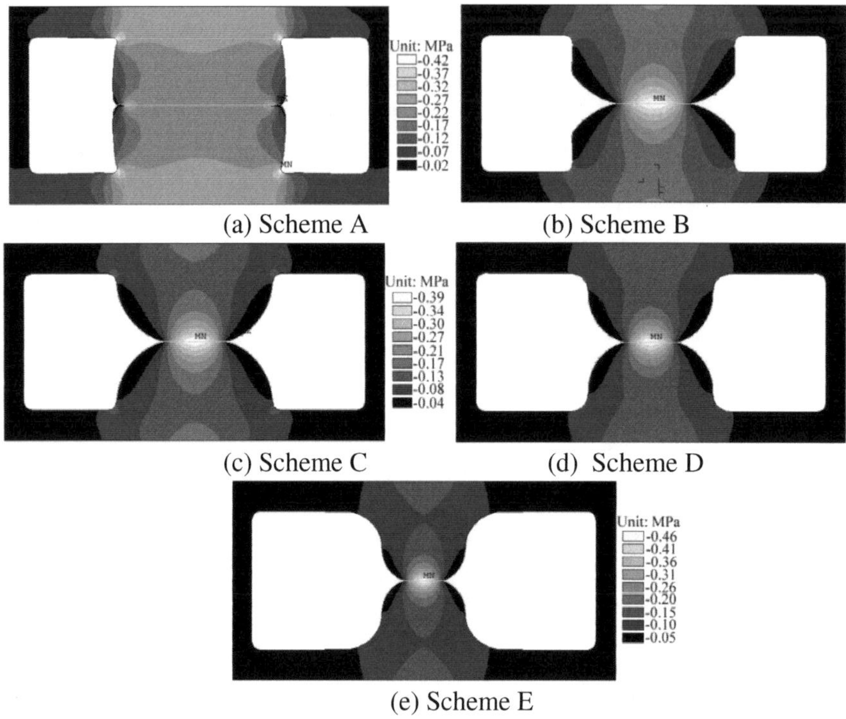

(a) Scheme A (b) Scheme B

(c) Scheme C (d) Scheme D

(e) Scheme E

Figure 6. σ_{22} axial stress contours of the five front seals

The sealing performance of the front seal is decided by the contact pressure of the contact area. The contact pressure distributions of the five front seals are as shown in figure 7. The pressure distribution of the scheme A is relatively flat, which is disadvantage for the sealing performance.

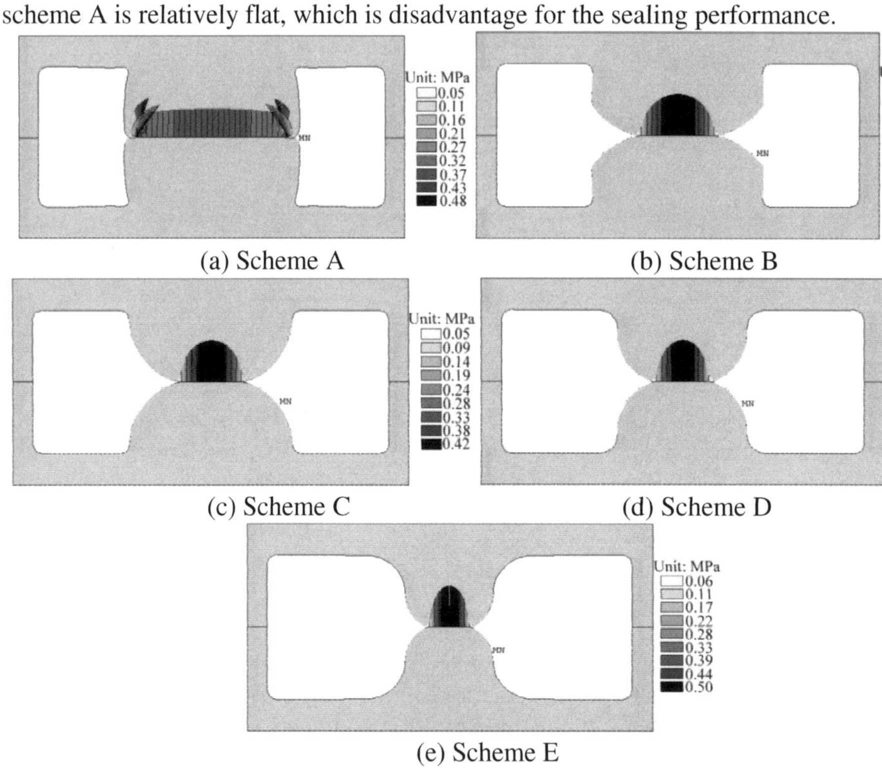

(a) Scheme A (b) Scheme B

(c) Scheme C (d) Scheme D

(e) Scheme E

Figure 7. Contact pressure distributions of the five front seals

5. Conclusions

The five design schemes of the LIDM seal are introduced, and their mechanical properties are investigated by the proposed FE method. The mechanical properties have important impacts on the required sealing performance, docking feature and separation feature. The mechanical properties of the scheme A is found to be disadvantage for the sealing performance, docking feature and separation feature of the LIDM seal. The mechanical properties of the schemes from B to E are similar, and all can be selected as the alternatives. Of course, there are some differences for the four design schemes including of the stress distribution, the sealing load, the sealing area. According to the actual development requirements of the LIDM seal as a silicone elastomer seal, the suitable design scheme will be selected and optimized further.

Acknowledgments

The authors acknowledge the financial support by the Project study on mechanism of novel space docking seal of spacecraft by National Natural Science Foundation of China (No. 51675056).

References

[1] Christopher C D, Nicholas G G and Marta B P et al 2012 Evaluation of a Novel Seal for Space Applications *Journal of Spacecraft and Rockets* **49** 83-90.

[2] Patrick H D and Bruce M S A 2012 *Comparison of Candidate Seal Designs for Future Docking Systems* NASA/TM-2012-217722.

[3] Emily C I, Olivia C A and Henry C et al 2014 *Effects of Low Earth Orbit on Docking Seal Materials* NASA/TM-2014-218435.

[4] Henry C, Christopher C D and Joyce A D et al 2010 *Space Environment Effects on Silicone Seal Materials* NASA/TM-2010-216332.

[5] Henry C, Bernadette J P and Bruce M S 2012 *Effects of Atomic Oxygen and Grease on Outgassing and Adhesion of Silicone Elastomers for Space Applications* NASA/TM-2012-217263.

[6] Patrick H D, Christopher C D and Bruce M S et al 2007 *Full-Scale System for Quantifying Leakage of Docking System Seals for Space Applications* NASA/TM-2007-215024.

[7] Patrick H D, Christopher C D and Janice W et al 2010 *Pressure Decay Testing Methodology for Quantifying Leak Rates of Full-Scale Docking System Seals* NASA/TM-2010-216244.

[8] Ian M S and Christopher C D 2008 *Performance of Sub-scale Docking Seals under Simulated Temperature Conditions* (44th AIAA/ASME/SAE/ASEE Joint Propulsion Conference & Exhibit, Hartford) AIAA 2008-4713: 1-11.

[9] Nicholas G G and Christopher C D 2011 *An Experimental Investigation of Leak Rate Performance of a Subscale Candidate Elastomer Docking Space Seal* NASA/CR-2011-216829.

[10] Nicholas G G and Christopher D 2013 Experimental Investigation of Leak-Rate Performance of a Subscale Composite Elastomer-Retainer Docking Seal *Journal of Spacecraft and Rockets* **50** 709-714.

[11] Nicholas P, Janice L W and Christopher C D 2011 *Temperature and Atomic Oxygen Effects on Helium Leak Rates of a Candidate Main Interface Seal* NASA/CR-2011-216830.

[12] Nicholas G G and Christopher C D 2012 A Compressible Permeation Approach to Elastomeric Space Seal Characterization *Journal of Fluids Engineering* **134** 051204-1-051204-9.

[13] Jay J O, Christopher C D and Bruce M S et al 2007 *Simulating Elastomer Seal Mechanics for a Low Impact Docking System* (AIAA SPACE 2007 Conference & Exposition) AIAA 2007-6206: 1-12.

IOP Conf. Series: Materials Science and Engineering **408** (2018) 012011 doi:10.1088/1757-899X/408/1/012011

Effect of Initial Crack Length of C(T) Specimen on Crack Propagation Measurement by Dc Potential Drop Method

Ni Chenqiang, Xue He* and Wang Weibing

School of Mechanical Engineering, Xi'an University of Science and Technology, Xi'an 710054, China

* xuehe1961@163.com

Abstract. The compact tensile specimens (C(T) specimens) are commonly used in crack propagation tests. The DC potential drop method is an important method for measuring crack length in crack propagation tests. To investigate the influence of the initial crack length of the C(T) specimen on the calibration curve in practice, the variation law of the potential field distribution of the compact tensile specimen with the crack propagation has been studied. The measurement features in different initial crack lengths and different current injection positions were quantitatively analyzed. The results show that the initial crack length has a significant effect on the sensitivity of the signal and the linearity of the calibration curve. The sensitivity can be improved by increasing the initial crack length, while the linearity decrease.

1. Introduction

Compact tensile specimens (C(T) specimens) are commonly used for establishing crack growth rate models, determining creep fracture properties, and determining fracture toughness of materials through experimental methods [1,2]. When it was used for the crack propagation test, the crack growth rate was obtained by measuring the amount of crack propagation in the test [3,4]. There are several methods that can be used to achieve online crack monitoring, while in the closed environment that crack cannot be directly observed the DC potential drop method is the most commonly used method to achieve long-term monitoring during the crack propagation [5]. The principle of the DC potential drop method is to pass a constant DC current on the cracked metal specimen, and measure the crack length by measuring the potential difference at a specific location of the specimen [6]. Since conductivity of metals is very excellent, the output signal of the potential drop method is very small. So the accurate calibration of the crack length and the potential difference curve is a very important. For the case of an infinitely long, limited-width center cracked plate specimen with a constant current perpendicular to the crack plane, Johnson gave the relationship between potential difference and crack length [7], which is called Johnson's formula. Afterward, the potential field of the bending specimen and the simplified compact tensile specimen (C(T) specimen) was analyzed [8]. For the three-dimensional potential field distribution of cracked specimens, the finite element method is generally used [9]. However, since the shape of the C(T) specimen is quite different from the model used in the above study, none of the above analytical solutions could be applied. According to the standard test method for linear-elastic plane-strain fracture toughness[10], the initial crack length of the specimen is counted from the pin hole, but this standard does not give its selection basis. Compared with the analytical model, a considerable part of the "crack" of a C(T) specimen is a slot, whose width is much larger than the width of the crack. So the initial crack length will significantly affect the geometry of the

IOP Conf. Series: Materials Science and Engineering **408** (2018) 012011 doi:10.1088/1757-899X/408/1/012011

specimen and thus affect its equivalent resistance. In this paper, the finite element software is used to analyze the three-dimensional potential field of CT specimens, and the influence of different initial crack lengths on the calibration curve is studied. Quantitative analysis is also provided to give a basis for practical application of crack measurement based DC potential drop method.

2. Finite Element Model

The geometry of the C(T) specimen, which is in accordance with ASTM E399-09 [10], is shown in Figure. 1. The nominal width of the specimen W=25mm, the specimen thickness B=0.5W. A 3D 1/4 finite element model of the specimen was developed using a commercial FEM software. Since the geometry of the specimen is symmetrical to the crack plane and the current injection points are symmetrical to the crack plane in practice, it is enough to develop a 1/4 finite element model(take half of the specimen on one side of the crack plane and take half the thickness of the specimen). Cracks were simulated using a rectangular slot with a width of 0.01 mm. The specimen grid model is shown in Figure. 2. The mesh near the crack was refined to ensure calculation accuracy.

To simulate a constant current flow through the specimen, a point source unit current (1A) was applied to the current injection point. According to symmetry, the ligament ahead of the crack tip is a uniform potential surface, so a ground (0 V potential) boundary condition was applied to the ligament ahead of the crack tip. For material properties, only the conductivity of the material is required, and it was set to 1388.89 S /mm [9], the conductivity of 304 stainless steel commonly used in structural materials for pressure vessels in nuclear power plants.

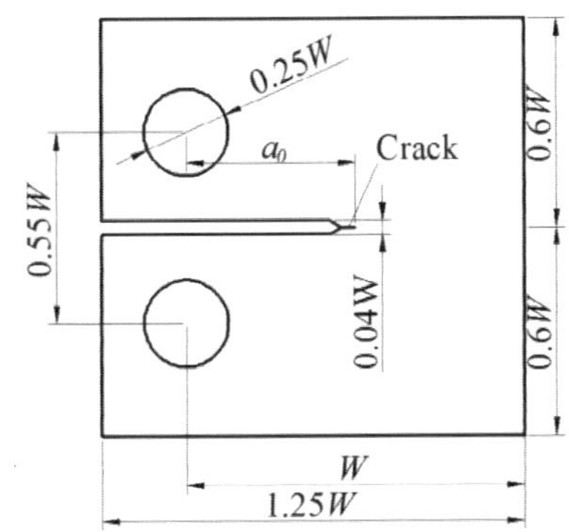

Figure 1. Geometry of C(T) specimen

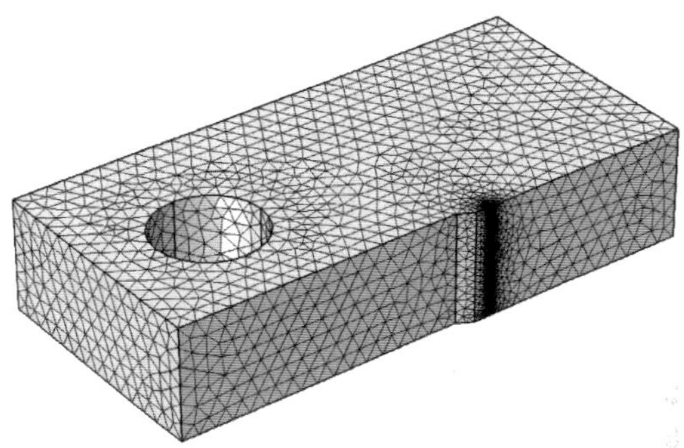

Figure 2. Finite element mesh of C(T) specimen model

3. Results and Discussion

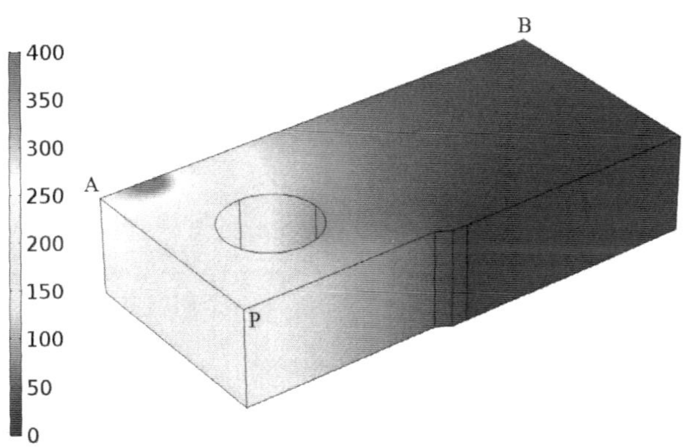

Figure 3. The potential(μV) distribution at a/W=0.35

An example of the potential distribution (current is input near the upper left corner of the specimen,at a/W=0.35) is shown in Figure. 3. With the software post-processing function, the potential of any point in the three-dimensional space of the specimen under different initial crack lengths, different potential loading positions and different crack length increments can be obtained. Because the current injection position has a significant influence on the potential field distribution, the potential fields under different current injection positions are calculated separately. All the studied current injection points is selected on the intersection line between the top surface of the specimen and the plane perpendicular to the middle plane of the thickness direction (ie, the line AB in Figure. 3), and the symbol LP ($0 \leqslant LP \leqslant 1$) represents different positions from left to right. The potential difference signal is measured from the notch mouth (point P in Figure. 3).

3.1. Sensitivity

Figure. 4 (a) and (b) show the average sensitivity of the potential difference signals corresponding to different initial crack lengths a_0 for two crack extension ranges. As can be seen from the figures, the signal sensitivity increases with the increase of the initial crack length no matter where the current is injected. The difference in current injection position has a significant effect on the signal sensitivity. The sensitivity of the potential difference signal is greatest when the current injection point is near the upper left corner of the specimen.

Table 1 shows the average sensitivity gain at different initial crack lengths based on the average sensitivity at a_0/W=0.35. It can be seen that there is a significant difference in sensitivity with different initial crack lengths. In addition, the closer the current input point is to the upper right corner of the specimen, the more obvious the influence of the initial crack length on the sensitivity.

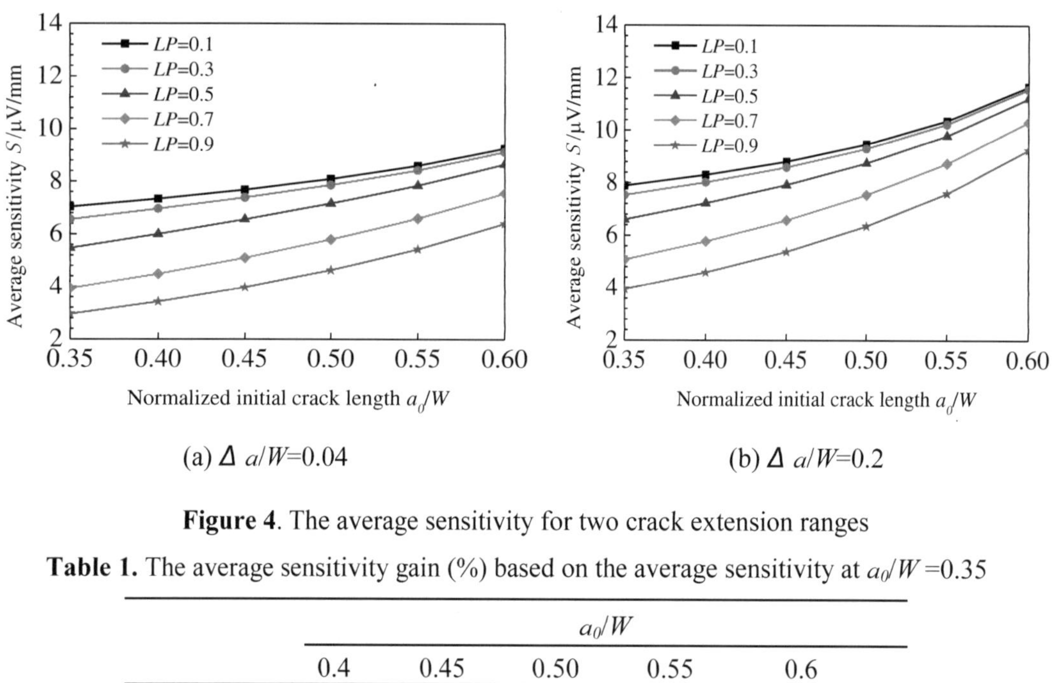

(a) $\Delta\ a/W$=0.04 (b) $\Delta\ a/W$=0.2

Figure 4. The average sensitivity for two crack extension ranges

Table 1. The average sensitivity gain (%) based on the average sensitivity at a_0/W=0.35

	a_0/W				
	0.4	0.45	0.50	0.55	0.6
LP=0.1	5.2	11.7	20.1	31.4	47.8
LP =0.3	6.5	14.1	23.5	36.0	53.6
LP =0.5	9.4	20.0	32.6	48.4	69.9
LP =0.7	13.7	29.6	48.6	72.2	103.3
LP =0.9	16.5	36.4	61.1	92.5	134.6

3.2. Linearity

Figure. 5 (a) and (b) show the ΔU-$\Delta a/W$ curves for different initial crack lengths at two current injection positions. As can be seen from the figures, the linearity of theΔU-$\Delta a/W$ curve is different for different initial crack lengths. The larger the initial crack length, the worse the linear characteristic. Figure. 6 (a) and (b) show theΔU-$\Delta a/W$ at different current injection positions for two initial crack lengths. It can be seen from the figures that the current input position also has a significant influence on the linearity of theΔU-$\Delta a/W$ curve. The closer the current input point is to the upper right corner of the specimen, the worse the linearity. For quantitative observations, all theΔU-$\Delta a/W$ curves for different initial crack lengths and different current input positions were linearly fitted using the end-base method, and the linearity was then calculated. The results are shown in Table 2.

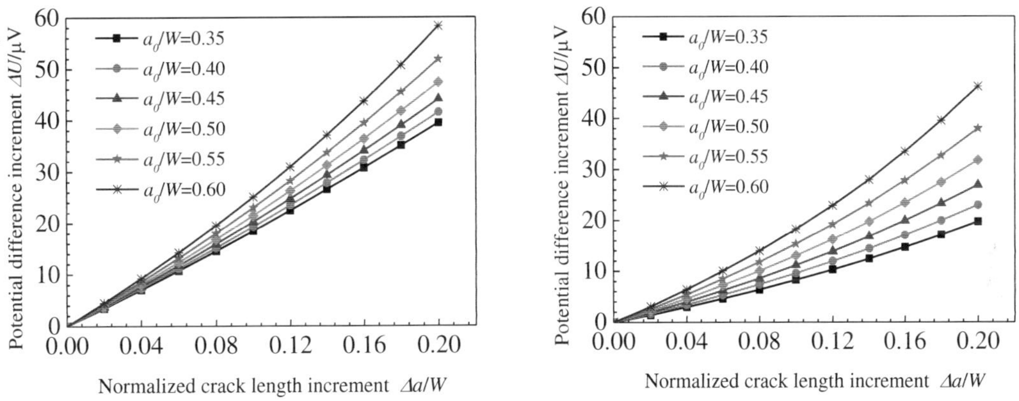

(a) LP=0.1 (b) LP=0.9

Figure 5. TheΔU-$\Delta a/W$ curves for two current injection positions

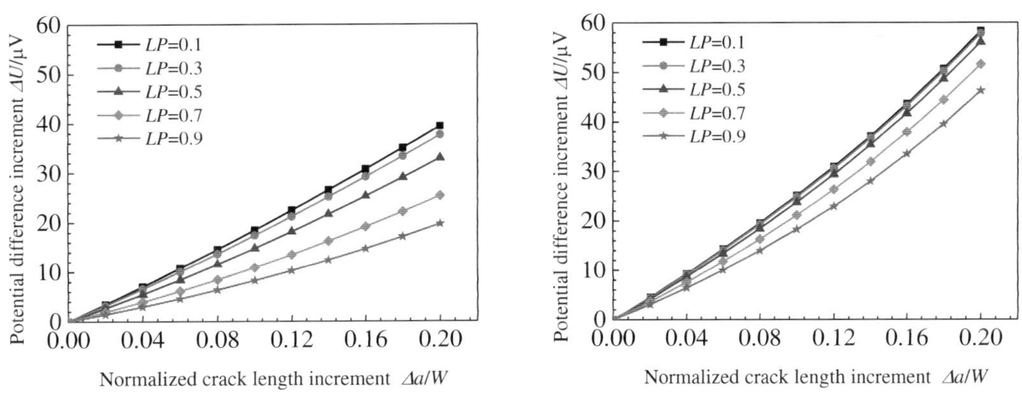

(a) a_0/W=0.35 (b) a_0/W=0.6

Figure 6. TheΔU-$\Delta a/W$ curves for two initial crack length

Table 2. The linearity(%) with different initial crack length and different current injection positions

	a_0/W					
	0.35	0.4	0.45	0.50	0.55	0.6
LP =0.1	3.2	3.5	4.0	4.6	5.6	7.0
LP =0.3	3.8	4.0	4.3	4.9	5.8	7.2
LP =0.5	5.2	5.2	5.4	5.8	6.5	7.7

LP =0.7	7.0	7.1	7.2	7.5	8.1	9.1
LP =0.9	8.1	8.3	8.5	8.9	9.6	10.5

4. Conclusions

The initial crack length of the C(T) specimen has a significant influence on the sensitivity of the potential difference signal and the linearity of the potential difference-crack length curve. Selecting a larger initial crack length can increase the signal sensitivity. However, as the initial crack length increases, the linearity of the potential difference-crack length curve becomes worse.

Acknowledgments

This work is financially supported by the National Natural Science Foundation of China (Grant nos. 51475362, 51775427, and 11502195), the Key Research and Development Project of Shaanxi Province (grant no. 2017GY-034) and the Ph.D. Research Fund Project of Xi'an University of Science and Technology(grant no. 2018QDJ009)

References

[1] Jia X, Xu-Teng HU, Song Y D. Calculation Formula for Stress Intensity Factors of CT Specimens based on Three Dimensional Finite Element Solutions[J]. Materials for Mechanical Engineering, 2015.

[2] Yao Y, Cai L, Bao C, et al. Fracture Toughness Based on Test Method for Straight-Notch CT Specimen[J]. Journal of Aeronautical Materials, 2014, 34(6):67-74.

[3] Xue H, Sato Y, Shoji T. Quantitative Estimation of the Growth of Environmentally Assisted Cracks at Flaws in Light Water Reactor Components[J]. Journal of Pressure Vessel Technology, 2009, 131(1) : 011404.

[4] Dong-Hai D U, Kai C, Lun Y U, et al. Measurement of Fatigue Crack Growth Rate of Reactor Structural Material in Air Based on DCPD Method[J]. Atomic Energy Science & Technology, 2014, 48(8):1386-1391.

[5] Wang L, Ding C F. Application of DC Current Method for Crack Length Measurement in High Temperature[J]. Journal of Aeronautical Materials, 2006, 26(3):359-360.

[6] Ritchie R O, Bathe K J. On the calibration of the electrical potential technique for monitoring crack growth using finite element methods[J]. International Journal of Fracture, 1979, 15(1):47-55.

[7] Johnson H H. Calibrating the electric potential method for studying slow crack growth[J]. Materials Research And Standards, 1965, 5: 442-445.

[8] Chen Chi. Metal Fracture Research Collection [M]. Beijing: Metallurgical Industry Press, 1978. 113-114.

[9] Z.J. Li, H. Xue. Numerical calibration of crack monitor based on direct current potential drop[J]. Journal of Xi'an University of Science and Technology, 2012, 32(1):116-116.

[10] ASTM E399-09, Standard Test Method for Linear-Elastic Plane-Strain Fracture Toughness K_{Ic} of Metallic Materials[S], 2009.

Research of Forward Design Method of Contact Dynamic Seal

Zhang Miaotian[1, *], Suo Shuangfu[2], Guofei[2], Jia xiaohong[2] and Meng guoying[1]

[1] School of Mechanical Electronic & Information Engineering, China University of Mining and Technology Beijing, China
[2] School of Mechanical Engineering, Tsinghua University, Beijing, China

[*]Email: zhangmiaotian1231@163.com

Abstract. The development of forward design of contact dynamic seals is an inevitable way to realize the independent development of core key components based on the research results of existing rubber and plastic seals. This paper proposes a multi-index comprehensive evaluation system for sealing performance. Based on the requirements of multiple performance indexes of dynamic sealing systems, a comprehensive evaluation of multiple index functions is constructed to provide a comprehensive evaluation of the performance of rubber and plastic seals. Based on the specific design examples of air contact seals, the design process of a certain seal is explored. Finally, the article elaborates on the future work of the forward design method and its development direction.

1. Introduction

Rubber and plastic seals are important basic parts of mechanical systems, and their structure and performance are crucial to the operation, performance and service life of mechanical equipment. Relevant statistics in the United States in the 1970s showed that actuators must be repaired or replaced due to the failure of actuator reciprocating seals, accounting for more than 90% of the causes of aircraft actuator failure[1]. However, the study of rubber and plastic dynamic seals in China started late and there are few researches, mainly focusing on the finite element analysis of the seal ring and how to select and combine the seal ring. There is less research on the design method of the seal ring and it is in a catch-up status[2][3]. At present, most of the design of rubber seals adopts reverse design methods, that is, selecting advanced foreign products, conducting imitation studies on materials, structures, etc., and combining experimental research, finally obtain the required sealing rings.

Nowadays, with the changes in the environment and conditions of use of rubber and plastic sealing rings, especially in the face of extreme conditions and supercritical problems, such as: high-pressure high-speed reciprocating seals, ultra-deep marine detection seals, The higher requirements are placed on the reliability and service life of the seal ring. Under these conditions, it is impossible to learn from the existing sealing ring products, and the foreign closure of key sealing technology, it is difficult to find and produce suitable sealing ring. Developed countries have formed a relatively complete industrial system for sealing technology, including design, production, testing, and application. Faced with the lack of domestic basic research, technological backwardness, and the critical status of key equipment dependent on imports, a positive design method and comprehensive performance evaluation system are urgently needed to fill the gap in domestic rubber and plastic dynamic sealing technology.

Recently, the "US blocked ZTE" incident once again sounded the alarm for us. We must

IOP Conf. Series: Materials Science and Engineering **408** (2018) 012012 doi:10.1088/1757-899X/408/1/012012

independently research and develop new types of sealing rings with intellectual property and carry out forward design method research.It lays a theoretical foundation for the development of China's high-performance rubber and plastic seals and intelligent seal design, and it not only helps to break the status quo of China's major equipment and key seals that are completely dependent on imports, but also helps to prevent major accidents and ensure the safety of major projects in China, therefore, it has important economic and social significance. In recent years, the Department of Mechanical Engineering of Tsinghua University has done relevant work on the forward design method of the seal ring in aviation and other fields, such as designing high-speed and high-pressure reciprocating seal rings[4], W-shaped metal seal ring design[5], etc.

2. Forward Design Method

2.1. Theoretical Research Content

Due to the multi-field coupling characteristics of rubber and plastic dynamic sealing system and the physical and chemical characteristics of rubber and plastic materials, such sealing products have complex working mechanisms, strong environmental dependence, and easy wear and aging. The theoretical research content of the forward design method includes: exploring the sealing mechanism, designing the sealing structure, studying the sealing performance, carrying out reliability analysis and testing, optimizing the sealing structure, and prolonging the life of the seal.

The seal lip pressure distribution is the research basis of the rubber and plastic dynamic seal system. By establishing the simulation model of the seal system, the lip pressure distribution of the reciprocating dynamic seal and the rotary dynamic seal can be obtained, and the next calculation is performed. Reverse pumping rate and friction torque can be used as an important indicator of sealing performance evaluation. Reverse pumping rate reflects the sealing effect of dynamic sealing. Friction torque reflects the degree of wear and the service life of the seal. Guo Fei of Tsinghua University[6] "Reverse pumping mechanism based on the lip seal of the rotating shaft" established a model of mixed lubrication under the steady state conditions of the simulated lip seal. The reverse pumping rate and the variation of friction torque with the rotation speed were calculated, and based on the calculation results, the design was optimized to improve the lip seal performance.

2.2. Forward design method system

The research on the dynamic seal performance of rubber and plastic involves many disciplines, including fluid mechanics, solid mechanics, contact mechanics and thermodynamics, and they interact and interact with each other (as shown in Figure 1).

First of all, the fluid mechanics analysis obtains the initial oil film pressure distribution curve of the sealing lip part through the relationship between the oil film thickness and the bearing capacity of the oil film. Considering the coupling relationship between the contact surface mechanics and the oil film pressure distribution curve, the roughness peak of the sealing lip surface is obtained, and the corrected oil film pressure distribution curve of the lip is obtained. The frictional force produced by the contact pressure influences the hydromechanical analysis results. Secondly, the oil film pressure and the rough peak contact pressure affect the results of fluid mechanics analysis through solid mechanics analysis. Thirdly, the solid-mechanical analysis using finite element analysis of the static contact pressure of the sealing lip and the change of the roughness peak will also affect the contact mechanics analysis results. Finally, as the speed of the lubricant changes, the viscosity of the lubricant changes accordingly, and changes in the viscosity affect the results of the fluid mechanics analysis.

Therefore, how to establish a multi-scale and multi-field coupled full-system sealed lubrication model is the key to the forward design method. The pumping rate and friction torque of the sealing system can be obtained through this model. Then by studying the system structure parameters, material parameters, operating parameters and micro-morphology parameters, etc. on the sealing mechanism of action and influence rules, design and optimization to get the required rubber seals.The forward design method flow is shown in Figure 2.

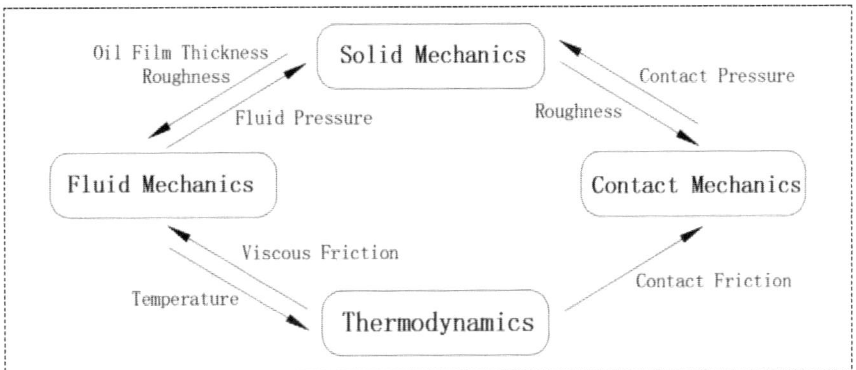

Figure 1. The contact of each disciplines.

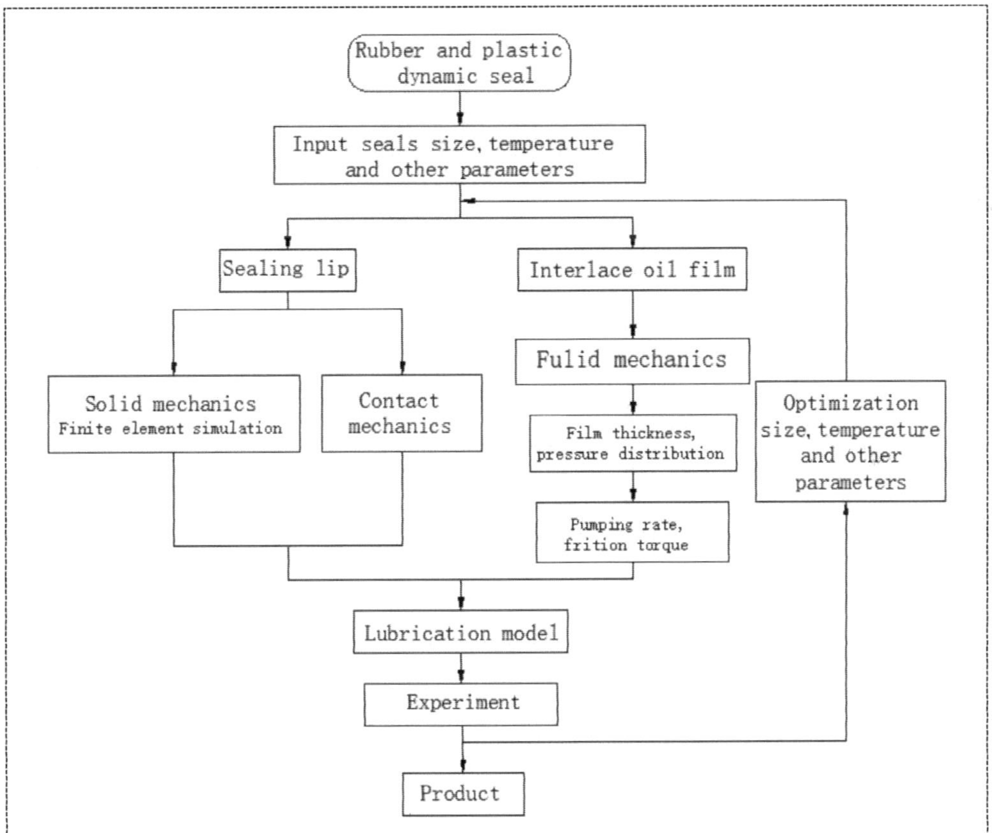

Figure 2. The forward design method flow.

3. Sealing Performance Evaluation System

Good sealing performance is the ultimate goal of the design of the sealing system. The indexes for evaluating the sealing performance include: leakage amount, sealing life, quality stability, and even leak detection system to detect the sealing performance of the sealing system. In recent years, researchers have used single performance indicators to evaluate the sealing performance of sealing systems. However, the evaluation of the sealing performance of the entire sealing system based on only one performance index is one-sided and inaccurate. Therefore, there is an urgent need for a method for comprehensively evaluating the sealing system by integrating a plurality of performance indexes, and to provide an accurate and reliable performance evaluation of the rubber-plastic dynamic sealing system.

For this purpose, we build an equilateral polygon based on multiple indicators that determine the sealing performance, where each point of the polygon represents an indicator. According to the standard and size distribution of different indicators, a polygon reflecting the measured performance index of the sealing system is constructed to construct a comprehensive evaluation index function(as shown in Figure 3). According to the index polygon and data processing of multiple indexes, the comprehensive performance of the sealing system is obtained and the performance is judged to be qualified. This method can not only accurately reflect the overall performance of the sealing system, but also compare the size of each sub-indicator, more intuitive to get the impact of a single indicator.

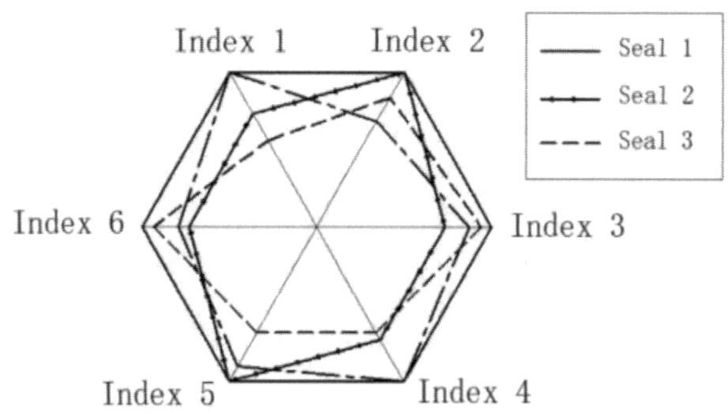

Figure 3. Example of equilateral polygon.

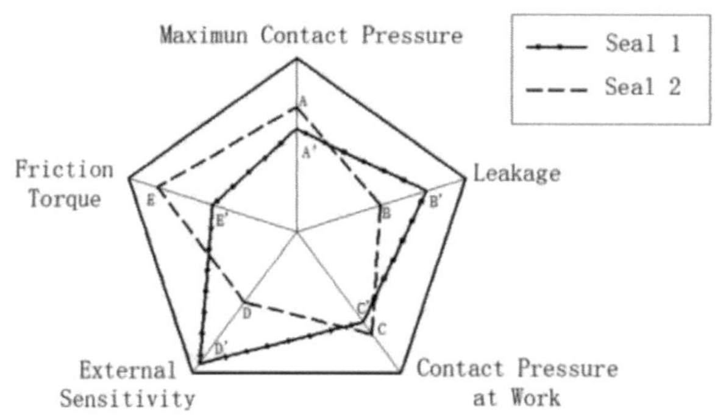

Figure 4. Instance analysis.

At present, the multi-objective optimization does not have a forming method in the study of mathematics, we have transformed the multi-objective problem into a single objective function to solve it.That is, according to the convergence areas of multiple objectives often in the feasible region, the n-multi-objective optimization problem is used to find the constraints of n-1 targets one by one, and then the layered solution is used to reduce the feasible region, the final conversion to a single-objective problem is solved.

As shown in Figure 4, set the comprehensive evaluation index to 5, namely: the maximum contact pressure, leakage, contact pressure at work, external sensitivity and friction torque, draw a regular pentagon. Set the qualified area is S. The evaluation rule is to compare the size of the pentagon. When the area is $\geq$S, the seal is not qualified. On the contrary, when the area is $\leq$S, the seal is qualified. According to the five evaluation indexes, the pentagonal patterns ABCDE and A'B'C'D'E' corresponding to the seal 1 and the seal 2 are respectively drawn, and two pentagonal areas are respectively found, Comparing with the qualified area S, the qualified seal can be obtained.

4. Conclusion

The mastery of the core component technology is one of the important contents of China's 2025 manufacturing planning. The core technology is controlled by others is the biggest hidden danger, and the core technology depends on the affinities or not, only self-reliance. Therefore, it is very urgent and necessary to carry out research on the forward design method for dynamic seal of rubber and plastic in aviation, nuclear power, military, marine and other fields.

The forward design method of rubber and plastic dynamic seal proposed in this paper can design suitable high performance rubber and plastic sealing ring according to the actual environmental conditions. This article also mentions a multi-index sealing performance comprehensive evaluation system, which can comprehensively, accurately and comprehensively evaluate the performance of rubber and plastic sealing ring based on multiple performance indexes of rubber and plastic dynamic sealing system.

Acknowledgments

The work decribed in this paper was supported by National Basic Research Program of China (973) (2014CB046404).

References

[1] OLSEN, ROBERT B 1981 *Dynamic Seals for Advanced Hydraulic Systems (Dallas:1978-198*1 *Air Force Final Report)*
[2] Wu qiong 2012 *Static Sealing and Pseudo-sealing Characteristics of Nitrile Rubber O-ring (Beijing:Lubrication Engineering)* p 5-11 21
[3] Guo zhipan 2015 *The Research of Aviation Hydraulic Sealing Designing Method (Beijing: Hydraulics Pneumatics & Seals vol 07*) p 8-21
[4] Wu changgui 2015 *Lip Contact Stress Analysis of Aircraft Actuator VL Seal (Beijing: Hydraulics Pneumatics & Seals vol 07*) p 18-21
[5] Wang chenxi 2016 *Research on Compression-resilience and Sealing Performance of Metallic W-Ring (Guangzhou: Lubrication Engineering vol 01)* p 50-54,69
[6] Guo fei 2014 *A Mixed Lubrication Theoretical Model and Experimental Verification of Rotary Lip Seals (Beijing: Journal of Mechanical Engineering vol 02)* p 137-144

IOP Conf. Series: Materials Science and Engineering **408** (2018) 012013 doi:10.1088/1757-899X/408/1/012013

Numerical Study of the Circumferential Fluctuations in Swept Compressor Cascades

Zefeng Li[1,2], Lei Wen[3], Mingzhi Tang[4], Donghai Jin[1,5,*] and Xingmin Gui[1,5]

[1]School of Energy and Power Engineering, Beihang University, Beijing, China
[2]Shen Yuan Honors College of Beihang University, Beijing, China
[3]Technical Division, Shanghai Electric Gas Turbine Co., Ltd, Shanghai, China
[4]AECC Shenyang Engine Research Institute, Shenyang, China
[5]Collaborative Innovation Center of Advanced Aero-Engine, Beijing, China

[*]jdh@buaa.edu.cn

Abstract. This paper presents comparative numerical studies to analyze the inlet flow and circumferential fluctuation(CF) source terms in compressor cascades. Forward swept cascades with different sweep angles and straight cascade were calculated numerically by a computational fluid dynamics(CFD) package to get the required three-dimensional(3D) flow parameters. A dimension reduction method was introduced and an in-house code for data reduction was conducted to obtain a circumferentially averaged flow field. The results show that different sweep angle and nominal incidence angle will change the magnitude and distribution of the CF source terms at inlet thus altering the inlet flow field. The Mach number(Ma) could also change the inlet CF source terms under the same swept angle. The trends of radial migration and the blade loading would increase with the growing of the inlet Ma.

1. Introduction

In recent years, the developing trend of aeroengine fans/compressors concentrates on high through-flow, high efficiency, high loading, and acceptable stability operation margin. Swept blades techniques are widely explored in the aerodynamic design of axial fans/compressors to meet the above requirements effectively. However, due to the complexity of 3D viscous flow field of swept blade, the mechanism of sweep has not been fully understood yet.

Lots of researches of the impact of the sweep blade on flow field have been carried out by former researchers recently. Wennerstrom and Frost[1] firstly introduce sweep design in a highly loaded transonic fan in the HTFC program in 1976. Wadia[2] applied the forward sweep rotor to multi-stage fans and found that the structure of sweep blades can improve the sensitivity of compressors to inlet distortion. McNulty[3] suggested that forward swept rotor could induce higher inlet axial velocity at blade tip and lower at blade hub. Sweep can lead to a spanwise redistribution of flow with more flow to the tip sections. The authors further concluded that forward sweep reduced tip-leakage flow blockage and tip loading.

Passrucker[4] utilized forward sweep technique in rotor design of a transonic compressor and experimental results show that forward sweep rotor possesses a higher efficiency and wider stall margin. Forward sweep could induce the flow to the blade tip, especially at the near-stall operation state, which contributes to the flow stability in the tip region. Rotor with sweep can work stably although separation happened around the hub.

Zhu Fang[5] reduced the dimension of inviscid momentum equations of compressors for analysis and suggested that the circumferential fluctuation of pressure field in the blade passage affected the flow in the form of inviscid blade force in 2013. The influence of the inviscid blade force will extend to the upstream and lead to the inlet CF of the flow field. They also suggested that the effect of forward sweep is opposite to backward sweep. Chang Hao investigated the inlet radial equilibrium from the aspect of circumferential averaging and found that the most important terms in the radial equilibrium are the GPR, and CMR. They also explained the inlet flow field alteration related to the aerodynamic behaviour of swept blades. Chang analyzed the effect of blade sweep on the inlet radial equilibrium of cascades.[6]

Tang proposed a transport model for the CF source terms[7] induced by inviscid blade force at the inlet of the blade to include more 3D effects in the circumferential averaged throughflow model in 2016. With the CF terms included, a major enhancement in the prediction capability of the throughflow analysis is obtained.

An objective of this study was to further investigate the mechanism of sweep blade. In this paper, comparative compressor cascades were studied numerically to analyze the impact of blade sweep on the CF and the flow field. The circumferential average method is generally averaged in arc direction of annular cascades and the circumferential average in this paper is particularly along the pitch direction.

2. Circumferentially averaged momentum equations

2.1. The coordinates for cascades

The Coordinates for the cascades in this paper is regulated as figure 1, where coordinate x is axial direction, y stands for circumferential direction and coordinate z is the radial direction respectively.

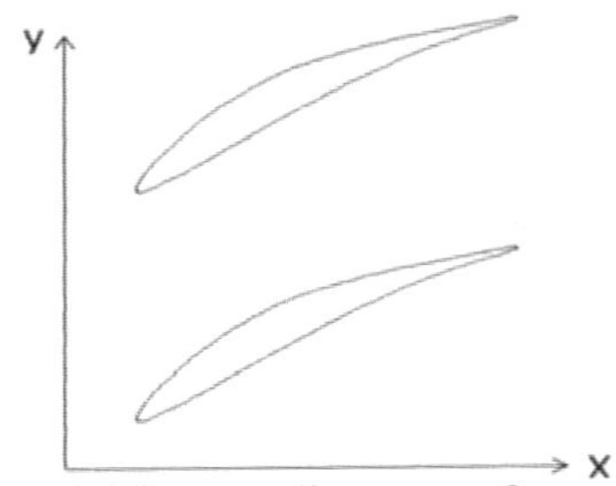

Figure 1. The coordinates set for cascades

2.2. Circumferential average method

The circumferentially averaging operation is defined as follows:

$$\overline{q}(x,z,t)=\frac{1}{y_p-y_s}\int_{y_s}^{y_p}q(x,y,z,t)dy \tag{1}$$

Where $\overline{q}$ is the circumferential averaged value of q, y_p-y_s the pitch length of the blade passage and the subscript p, s stand for the pressure side (PS)and suction side(SS) respectively. With the definition of circumferential averaging, the 3D flow parameter can be decomposed into the sum of a circumferential averaged term and a spatial fluctuation which reflects the circumferential non-uniformity of the flow field. It can be expressed as:

$$q(x,y,z,t)=\overline{q}(x,z,t)+q'(x,y,z,t) \tag{2}$$

Due to the non-linearity of the equations, the fluctuations appear in the form of the circumferentially averaged stresses, i.e. the so-called CF stresses[7], which cannot be eliminated in the governing equation after circumferential dimension reduction like Reynold Stress. The non-linear terms in incompressible flow can be expressed as equation (3) and equation (4) state.

$$\overline{q_1 q_2} = \overline{q_1}\,\overline{q_2} + \overline{q'_1 q'_2} \tag{3}$$

$$\overline{\rho q_1 q_2} = \rho(\overline{q_1}\,\overline{q_2} + \overline{q'_1 q'_2}) \tag{4}$$

2.3. The momentum equations

The Navier–Stokes (N–S) momentum equation in Cartesian coordinates can be expressed in the axial, radial and circumferential direction, with volume force and viscous term neglected and steady assumption, as follows:

$$\begin{cases} \dfrac{\partial}{\partial x}(\rho u^2+p)+\dfrac{\partial}{\partial y}(\rho uw)+\dfrac{\partial}{\partial y}(\rho uv)=0 \\ \dfrac{\partial}{\partial x}(\rho uw)+\dfrac{\partial}{\partial y}(\rho w^2+p)+\dfrac{\partial}{\partial y}(\rho wv)=0 \\ \dfrac{\partial}{\partial x}(\rho uv)+\dfrac{\partial}{\partial y}(\rho vw)+\dfrac{\partial}{\partial y}(\rho v^2+p)=0 \end{cases} \tag{5}$$

Where ρ is the density, p the pressure and u, v, w represent the axial, circumferential and radial velocity respectively. Employing the circumferential averaging operator equation (1) to equation (5) yields

$$\begin{cases} \dfrac{\partial}{\partial x}(\overline{u}\,\overline{u})+\dfrac{\partial}{\partial z}(\overline{w}\,\overline{u})=-\dfrac{1}{\rho}\dfrac{\partial \overline{p}}{\partial x}+P_x+F_{bx} \\ \dfrac{\partial}{\partial x}(\overline{u}\,\overline{v})+\dfrac{\partial}{\partial z}(\overline{w}\,\overline{v})=P_y+F_{by} \\ \dfrac{\partial}{\partial x}(\overline{w}\,\overline{u})+\dfrac{\partial}{\partial z}(\overline{w}\,\overline{w})=-\dfrac{1}{\rho}\dfrac{\partial \overline{p}}{\partial z}+P_z+F_{bz} \end{cases} \tag{6}$$

where

$$\begin{cases} P_x=-\dfrac{1}{b}\dfrac{\partial}{\partial x}\left(b\overline{u'u'}\right)-\dfrac{1}{b}\dfrac{\partial}{\partial z}\left(b\overline{u'v'}\right) \\ P_y=-\dfrac{1}{b}\dfrac{\partial}{\partial x}\left(b\overline{u'w'}\right)-\dfrac{1}{b}\dfrac{\partial}{\partial z}\left(b\overline{v'w'}\right) \\ P_z=-\dfrac{1}{b}\dfrac{\partial}{\partial x}\left(b\overline{u'v'}\right)-\dfrac{1}{b}\dfrac{\partial}{\partial z}\left(b\overline{v'v'}\right) \end{cases} \tag{7}$$

and

$$\begin{cases} F_{bx}=\dfrac{N}{\rho l b}\left(P_p\dfrac{\partial y_p}{\partial x}-P_s\dfrac{\partial y_s}{\partial x}\right) \\ F_{by}=\dfrac{N}{\rho l b}(P_s-P_p) \\ F_{bz}=\dfrac{N}{\rho l b}\left(P_p\dfrac{\partial y_p}{\partial z}-P_s\dfrac{\partial y_s}{\partial z}\right) \end{cases} \tag{8}$$

F_b is the inviscid blade force, $\overline{u'u'}$, $\overline{v'v'}$, $\overline{w'w'}$, $\overline{u'v'}$, $\overline{u'w'}$, $\overline{v'w'}$ are the CF stresses terms and they are symbolled as UU,VV,WW,UV,UW,VW respectively in this paper. CF stresses terms are related to the blade-to-blade spatial fluctuations[7], arising in the circumferential averaging of the 3D N-S equations, due to the non-linearity of the equations. P is the CF source term and b is the blockage factor that represents circumferential blade thickness distribution, which is calculated as equation (9):

$$b = \frac{y_P - y_S}{l / N} \tag{9}$$

Where l refers to the circumferential length of the blade passage and N represents the number of blades. The blockage factor b is less than 1.0 within blade area and equal to 1.0 outside.

3. Computational Methodology

3.1. Definition of sweep

Different definitions of sweep have been adopted in the literatures. In this paper, the definition of sweep defined by Ramakrishna and Govardhan[8] is used. The sweep was defined as moving airfoil sections along the chord line (see figure 2 and θ is the blade camber angle). According to Denton and Xu[9], defining sweep in this way would not introduce any spanwise blade force, whereas lean (moving the blade sections perpendicular to the chord line) would introduce the influences of spanwise blade forces.

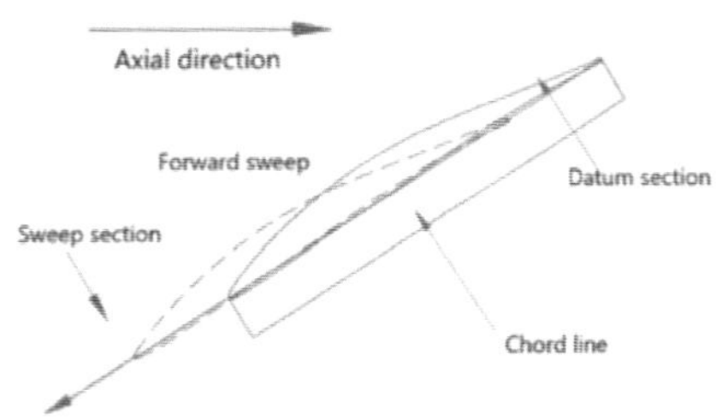

Figure 2. The definition of Sweep

The sweep angle was defined as the angle between the leading edge of the cascade and the radial direction in a radial profile section. When a higher span section was moved in the upstream direction relative to the lower section, it is defined as forward sweep. In this paper, the sweep angle keeps the same alongside all the spanwise profile sections to focus on the mechanism of sweep impact on the inlet flow.

3.2. Cascade model

The blade profile is identical at all spanwise sections in every cascade to ensure a concentrative study on the mechanism of blade sweep. A circular arc was adopted for the style of blade airfoil camber. The thickness distribution is the same as the MAN GHH 1-S1 controlled diffusion airfoil[10] and other specific parameters are listed in Table 1

Table 1. Three Scheme comparing.

Parameter	Value
Blade camber angle θ /°	30
Chord length/mm	50
Blade geometric inlet angle β_{1k} / °	57
Aspect ratio	2.4
Pitch length/mm	33

3.3. CFD method

NUMECA FINE/Turbo was used to carry out numerical simulation for 3D flow field of cascades in this paper. The mesh was drawn with around 429,135 grid points and a selected O4H topology, which is shown in Figure 3.

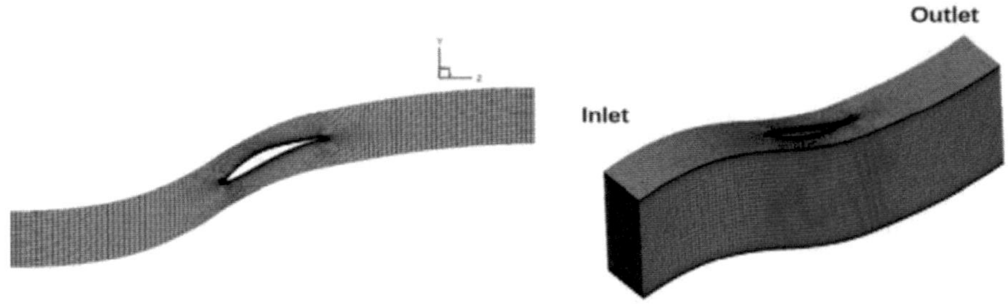

Figure 3. View of grid of computational domain

The length of the inlet and outlet domain were set to 2 times as long as the chord line. For turbulence calculation, the Spalart–Allmaras model was selected for calculation. The boundary condition was the total pressure(101325Pa), total temperature(277.15K) and direction of flow fixed at the inlet and the static pressure set at the outlet to ensure an inlet Mach number(Ma), adiabatic non-slipping condition was used for solid wall boundaries. In this study, only one cascade passage was modeled to save computational resource and periodical continuous boundary was set to ensure the repeatability of cascades.

4. Results and discussion

The circumferential fluctuation can be caused by many aspects, including the inhomogeneity of the incoming flow, the trail of upstream blade row, tip leakage, vortex, passage shock, shock-boundary layer interaction, hub corner stall and the pressure fluctuation in the blade passage[11], as well as the inviscid blade force[12]. For cascades, the main factor responsible for the inlet flow CF is the circumferential pressure fluctuation (inherent characteristic of the flow field of non-axisymmetric geometry) inside the cascade passages as the upstream flow was set to uniform. The influence of such pressure CF would extend to the upstream to a distance from the leading edge thus inducing the inlet circumferential fluctuations. A program for circumferential data reduction was developed to get the required variables to calculate the CF source terms, CF stress terms and other flow parameters.

4.1. Influence of blade sweep

Figure Figure **4** plots the axial distribution of the CF stresses terms at the mid-span section in straight cascade. The relative axial location is defined as the distance from the leading edge normalized by the chord length. 0 location stands for the leading edge(LE) and 100% represents the trailing edge(TE).

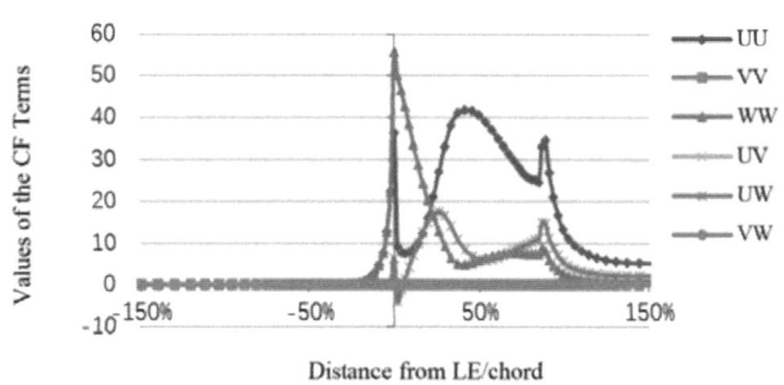

Figure 4. Axial distribution of CF stresses at mid-span in straight cascade

For straight cascades, the radial component of the inviscid blade force is 0 and the radial velocity is close to 0, which means that the CF stresses terms (VV, UV, VW) related to radial velocity V are approached to 0. However, the other CF stresses term (UU, UW, WW) are not ignorable at the location before the leading edge and the axial gradients of these terms are large, the axisymmetric assumption may fail to obtain an acceptable flow field in throughflow model[7].

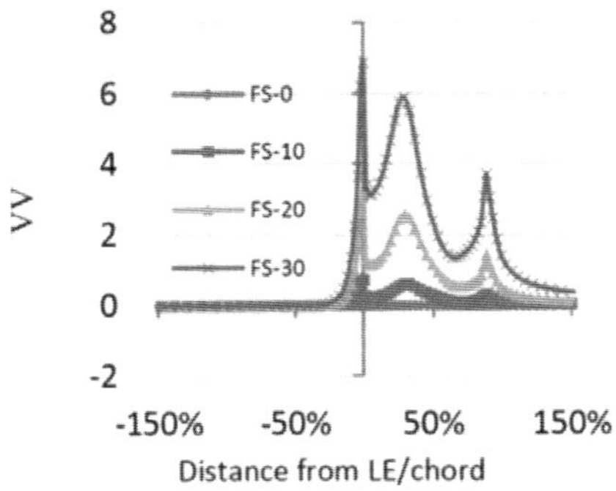

Figure 5. Axial Distribution of VV in different cascades at 50% spanwise

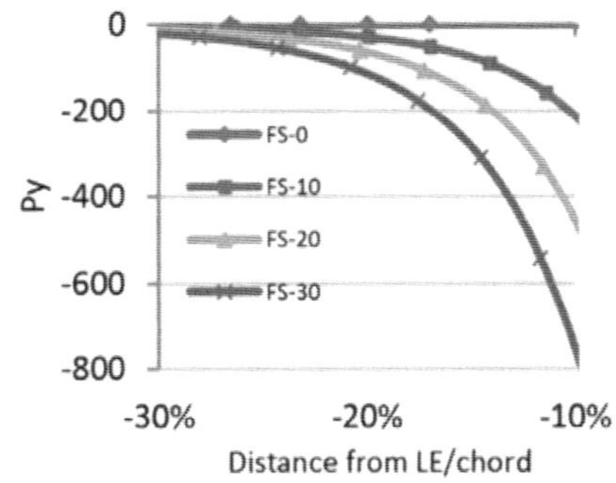

Figure 6. Axial distribution of Py in different cascades at 50% spanwise

To study the sweep effect on CF of the flow field, 3 different forward sweep angle 10 °, 20 ° and 30° were calculated respectively. FS-0 stands for the straight blade and the sweep blade are similarly named. From Figure 5 and Figure 6, the VV term and the circumferential component of CF source term P_y increase with the growing of sweep angle, which indicate that the sweep induces the CF in the flow fields. The gradient of VV increases with growing sweep angle before the LE, which also reflect the intensifying of CF.

4.2. *Influence of nominal incidence angle*

Nominal incidence angle in defined as the domain inlet flow angle minus the blade geometrical inlet angle β_{1k}, which reflects the free inlet flow direction relative to the cascade geometrical inlet angle. The spanwise distribution of the CF source term P_y at 4% chord length before LE for straight cascades with nominal incidence angle equal to 0 °, ±2 ° and ±4 ° is shown in Figure 7.

The regions below 40% of the span and above 60% of the span are omitted, due to the viscous and end-wall effect. It indicates that the magnitude of P_y increases with the growing of incidence angles, which can reflect the blade loading. Bigger magnitude of P_y means bigger pressure difference between pressure surface and suction surface, as Figure 8 illustrates.

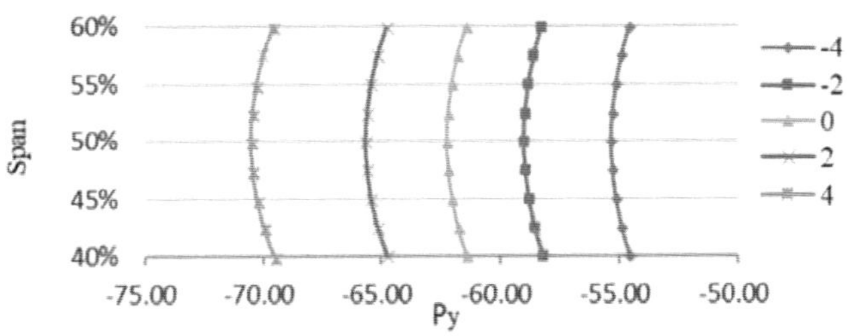

Figure 7. Spanwise Distribution of P_y at 4% chord before LE with different nominal incidence angle.

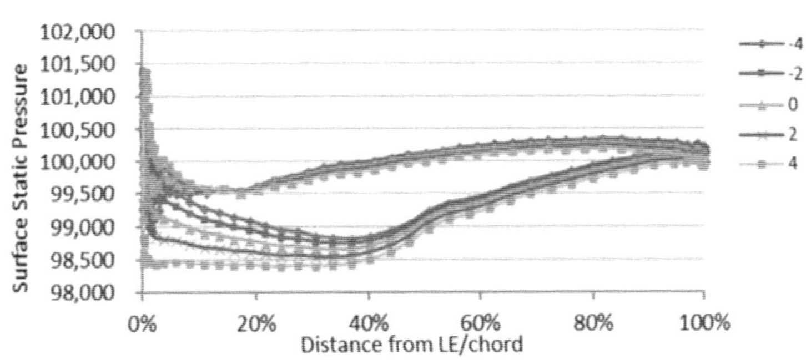

Figure 8. Pressure Distribution on blade surface at mid-span

4.3. Comparison of different Mach Number

In multi-stage compressors, the different spanwise sections of a blade row often work under different inlet velocity, thus we want to study the influence of the Mach number on the inlet CF. 30 ° forward sweep cascade was calculated under inlet Ma equal to 0.15, 0.2 and 0.25, the nominal incidence angle for all cases in this subsection are 0 °.

As the CF source terms are relevant to velocities, it is nondimensionalized as equation (10) to investigate the variation of CF under different inlet velocity.

$$P' = -\frac{P}{V_{in}^{2}}h \quad (10)$$

Where P' is the dimensionless CF source term, V_{in} the absolute velocity at the mid-span on the inlet and h is the span length.

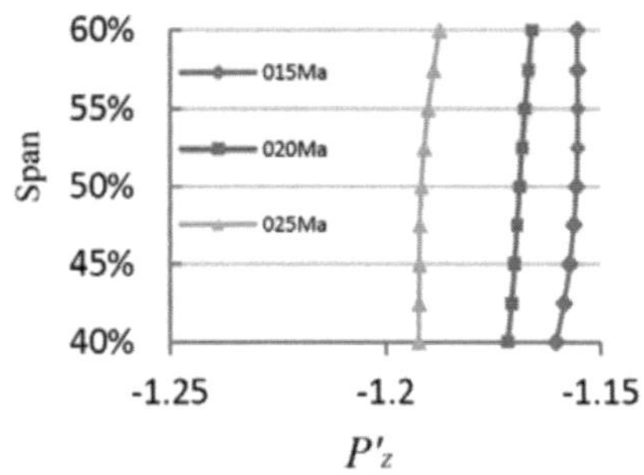

Figure 9. Spanwise Distribution of P'_z at 4% chord before LE with different Ma

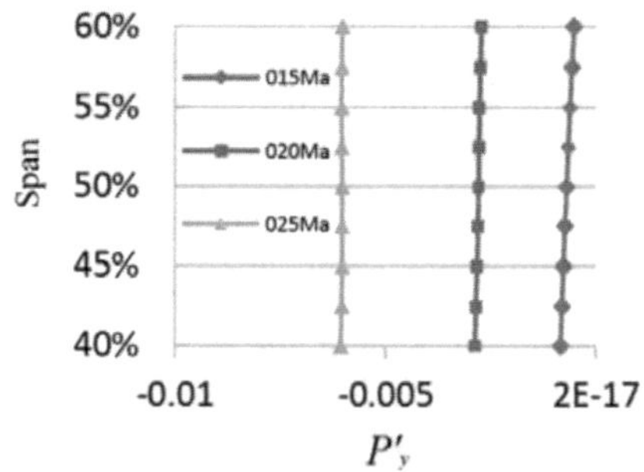

Figure 10. Spanwise Distribution of P'_y at 4% chord before LE with different Ma

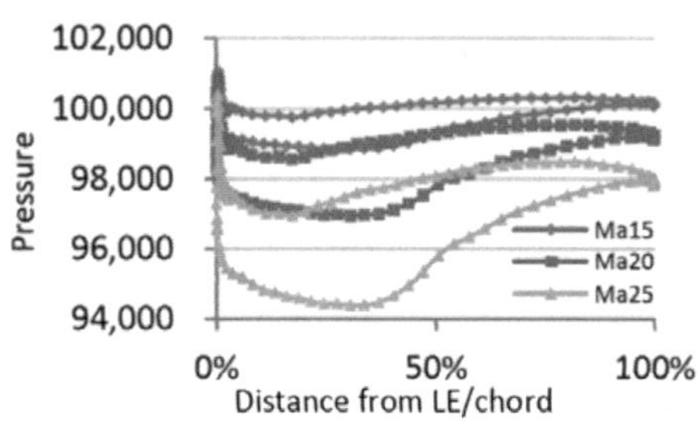

Figure 11. Axial pressure distribution with different Ma

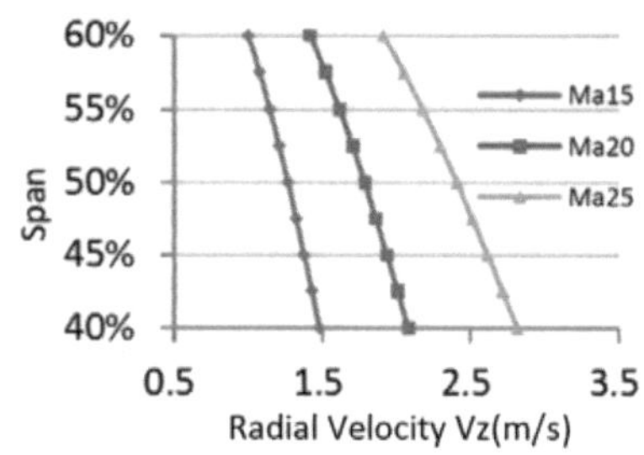

Figure 12. Spanwise Distribution of V_z at 4% chord before LE with different Ma

Figure 9 and 10 state the distribution of the radial and circumferential components of the CF source terms at 4% chord axial location before LE with different Ma. Different Ma cause different acceleration characteristic of blade surface. Higher Ma induce stronger CF, larger blade loading, thus higher pressure difference between PS and SS, which can be reflected in Figure 11. The increase of P'_z alters the inlet radial equilibrium, as is shown in Figure 12 the radial migration is enhanced.

5. Conclusion

Comparative numerical studies have been carried out on cascades to investigate the effect of blade sweep on circumferential fluctuations in axial compressor cascades. Although this work is based on an inviscid and incompressible flow, the flow mechanism is quite accurately calculated as viscous effects are secondary in free stream flows and the Ma is less than 0.3 in all cases.

The magnitude of circumferential components of CF source term P_y and the VV term increase with growing sweep angle. High sweep angle will induce radial migration of the flow from hub to tip. CF

IOP Conf. Series: Materials Science and Engineering **408** (2018) 012013 doi:10.1088/1757-899X/408/1/012013

also exists in straight cascade as discussed before, the CF terms related to radial velocity are small in straight cascades due to the low radial velocity.

The inlet incidence angles also affect the CF, bigger magnitude of P_r means bigger difference between pressure surface and suction surface and blade loading.

Under the same sweep angle, as the Mach Number increases, CF and the trend of radial migration of the flow will be intensified. The blade loading will also increase with Ma, so the inlet Ma should be carefully considered in the design process.

References

[1] Wennerstrom A and Frost G 1976 Design of a 1500 ft/sec, transonic,high-through-flow, single-stage axial-flow compressor with low hub/tip ratio Report No: AFARL-TR-76-59

[2] Wadia A, Szucs P and Crall D, et al 2002 Forward Swept Rotor Studies in Multistage Fans with Inlet Distortion *ASME Paper* GT-2002-30326

[3] McNulty G, Khalid S, Decker J, et al 2004 The Impact of Forward Swept Rotors on Tip Clearance Flows in Subsonic Axial Compressors *J. of turbomachinery* **126**(4):445-454

[4] Passrucker H 2003 Effect of Forward Sweep in a Transonic Compressor Rotor *J. of Power and Energy*, **V217**(4): 357-365

[5] ZHU F 2013 Study on design techniques of high through flow and high efficiency fan/booster of civil aeroengine Beijing: Beihang University

[6] Chang H, ZHU F, Jin D and Gui X 2015 Effect of Blade Sweep on Inlet Flow in Axial Compressor Cascades *Chinese Journal of Aeronautics*, **28**(1): 103-111

[7] Tang M, Jin D and Gui X 2017 Modeling and Numerical Investigation of the Inlet Circumferential Fluctuationsof Swept and Bowed Blades *J. of Thermal Science* **26**(1):1-10

[8] Raamakrishna P and Govardhan M 2010 Numerical Study of the Stagger Angle Effects in Forward Swept Axial Compressor Rotor Passages *ASME Turbo Expo 2010: Power for Land, Sea, and Air* pp443-453

[9] Denton J and Xu L 2002 The Effects of Lean and Sweep on Transonic Fan Performance. *ASME Turbo Expo 2002: Power for Land, Sea, and Air.* **6** p23-32

[10] Steinert W, Eisenberg B and Starken H 1991 Design and testing of a controlled diffusion airfoil cascade for industrial axial flow compressor application *J. of Turbomachinery* **4**.1991-10:583-590

[11] Baralon S, Eriksson L and Hall U 1999 Evaluation of Higher-Order Terms in the Throughflow Approximation Using 3D Navier-Stokes Computations of a Transonic Compressor Rotor *ASME 1999 International Gas Turbine and Aeroengine Congress and Exhibition* V**001** T03A013

[12] Gui X, Zhu F and Jin D 2013 Effects of Inlet Circumferential Fluctuation on the Sweep Aerodynamic Performance of Axial Fans/Compressors *Journal of Thermal Science* **22**(5):383-394

Unsteady Character and Flow Fields in an Axial Blood Pump under Pulsatile Pressure Boundary Condition

Mengyu Wang[1], Han Xu[1], Donghai Jin[1,2,*] and Xingmin Gui[1,2]

[1]Aeroengine Numerical Simulation Research Center, Beihang University, Beijing, China

[2]Collaborative Innovation Center For Advanced Aero-Engine, Beijing, China

[*]jdh@buaa.edu.cn

Abstract. Blood pumps have been widely used as a temporary method for heart failure. In recent years, the challenging factors in design have generally concentrated on how to deal with the complicated time-dependent fluid dynamics. Therefore, this research studies the impact to flow field by pulsatile boundary condition with both steady and unsteady numerical simulations. For steady computation, total pressures of 20, 40, 60, 80, 100mmHg and an average pressure of 120mmHg were given at the inlet and outlet respectively, and an unsteady simulation was then conducted with the same fixed pressure of 120mmHg at the outlet but a periodical sinusoidal pressure within 20 and 100mmHg at the inlet. Compared with a single character line drawn through the steady points, the unsteady computation resulted in a closed elliptic character curve. In the whole range of operation of the unsteady condition, flow field showed a well-attached flow in rotor passage but a large area of recirculation in the second row of stator passages. By contrast, flow field in the first row of stator passages and the pressure on the surface of rotor blades were greatly impacted by the acceleration and deceleration of the flow rate, which can be well explained by the change of incidence. Besides, flow rate acceleration or deceleration exaggerated a non-uniform distribution of velocity, which can easily form a larger area of high scalar shear stress than the steady results. A special pressure distribution between rotors and stators was also observed in unsteady flow.

1. Introduction

Heart failure is a direct death threat to human beings all the time. To deal with that, blood pump was designed to augment output pressure and flow rate of the native heart [1].

As the working fluid of blood pump, blood contains many kinds of blood cells which requires more consideration of its identified property. High velocity gradient in pump may induce supra-physiological stress to the blood. Consequently, hemolysis and activation of platelet are determined by area of stress region and blood cells exposition time [2]. Another challenge is the formation of thrombosis induced by flow separation, stagnation or recirculation, which make blood clots deposit on the surface of pump and change its geometry [3]. Therefore, evaluation to the pump should focus on separation, recirculation and high stress regions inside the pump.

In common practices, design process and experimental verification of character and flow fields of blood pumps are only evaluated under steady conditions with a fixed physical value on pump's inlet and outlet. However, with the effect by active heart contraction, the flow rate through the pump varies during the heartbeat, even while the pump is operating at a constant rotational speed. In contrast to steady operation, a closed elliptic character curve instead of a line was observed in an animal experiment [4]. Similar loops were also observed in unsteady numerical computations [5],[6],[7] and mock heart-loop experiments [8],[9],[10], [11]. Using unsteady Computational Fluid Dynamics (CFD), Song [12] evaluated the change of character when boundary pressure varied with time after designing under steady conditions. Shu [9],[13] researched the influence to flow fields by flow acceleration or deceleration and related flow fields to two non-dimensional variables, that is, flow coefficient and flow acceleration coefficient. This article studies the impact to character and flow fields by a sinusoidal pulsatile boundary condition with both steady and unsteady numerical simulations.

2. Methods

2.1. Research model

The axial pump is composed of a row of rotor and stator, as shown in **Figure 1** [14],[15]. In order to control flow separations, two splitters are employed in the latter half of the rotor, which have same profile with the main blades. In the stator row, tandem cascade is adopted with five upstream vanes and five downstream vanes to improve the pressure-rise capability. The detailed design parameters are listed in **Table 1**.

Table 1. Design parameters of the axial pump.

Design parameters	Rotor	Upstream vane	Downstream vane
Number of blades	2	5	5
Number of splitters	2	0	0
Inlet hub/tip ratio	0.32	0.83	0.80
LE tip diameter	12.7mm	12.7mm	12.7mm

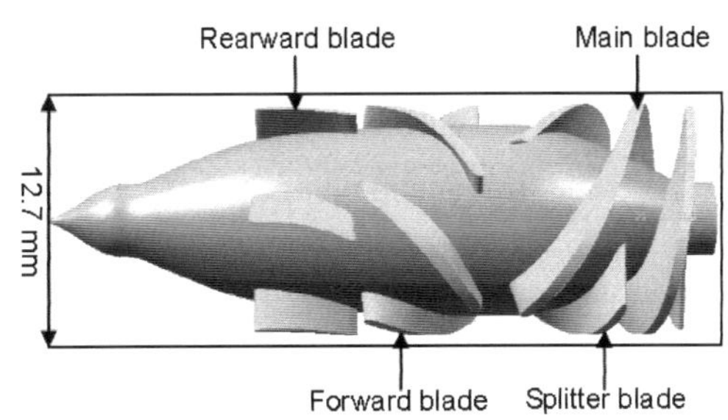

Figure 1. Structure map of the research pump.

2.2. CFD Method

Both the steady-state and time-resolved flow fields were simulated using the 3-D RANS solver ANSYS 17.1. The Reynolds number is computed using:

$$Re = \frac{V_\infty D}{\upsilon} \quad (1)$$

in which V_∞ (m/s) is the inlet free-stream velocity, D (m) the tip diameter at the rotor outlet, ν (m^2/s) the kinematic viscosity. The Reynolds number was computed to be around 1×10^4 in the simulation. The turbulent Reynolds stress is treated employing the Shear Stress Transport (SST) model to have a good prediction both inside and outside the boundary layer. The computational domain includes the inlet tube, rotor, stator and the outlet tube, as shown in **Figure 2**. The whole passage meshes are used. O-grid topology was applied around the blade surface for good orthogonality and grid refinement was

IOP Conf. Series: Materials Science and Engineering **408** (2018) 012014 doi:10.1088/1757-899X/408/1/012014

carried out near walls. Both the steady and unsteady flow fields were predicted using whole-annulus meshes, and the total number of the mesh nodes is 4924546, with 3567186 in the rotor and 1357360 in the stator. The independence of mesh and the reliability of steady simulations compared with experiments have been verified in previous articles [16],[17], [18]. The computation was first run in steady mode using the mixing-plane model to transfer flow quantities between the rotor and stator rows. A set of total pressures (relative to atmospheric pressure) of 20, 40, 60, 80, 100mmHg were given at the inlet and a fixed average static pressure of 120mmHg at the outlet. The iterations proceeded until the average residual error dropped to 10^{-6}. Considering the effects by the active heart contraction, an unsteady simulation was then conducted with the same fixed pressure of 120mmHg at the outlet but a periodical sinusoidal pressure within 20 and 100mmHg at the inlet. The steady result was used as the initial condition for the time-accurate mode with a transient interface model between the blade rows. The time-accurate simulation was run with 100 physical time steps per cycle and at least 5 inner iterations per time step. It took about 3 cycles for the simulation to converge to a periodic solution.

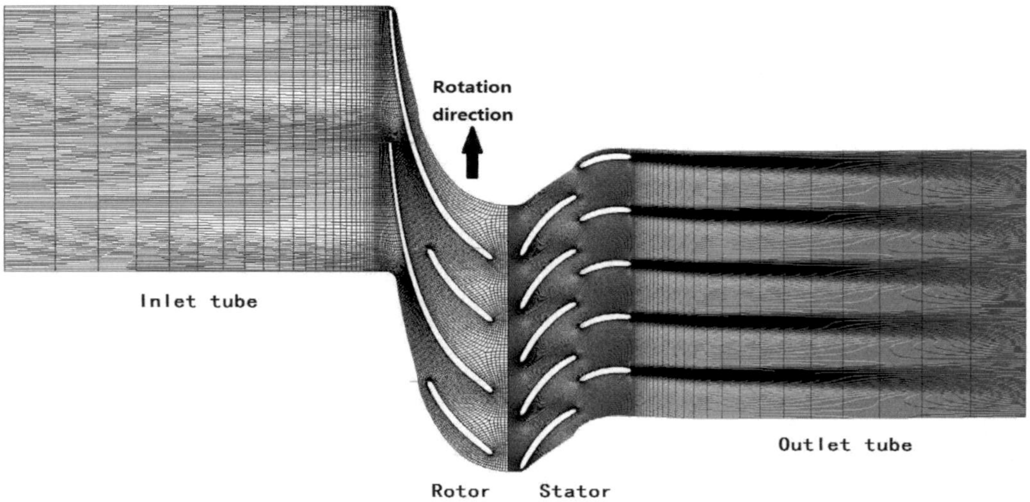

Figure 2. The blade-to-blade view of the computational domains.

3. Results

3.1. Steady and unsteady character

The pressure rises over outlet and inlet and flow rates through the pump in steady numerical computation converge to separated character points on **Figure 3**, which can be fitted to a single line. By contrast, character of the unsteady result with pulsatile boundary shows in **Figure 3** is a closed elliptic character curve. The transient character point rotates in anti-clock direction with a higher pressure rise in the deceleration of flow rate and a lower in acceleration, and the flow range of unsteady operation was totally located in that of the steady conditions with the same pressure rises. Three groups of points with same pressure rise were compared in following content. Group I is the highest boundary of unsteady character with a pressure rise of 100mmHg and flow rates of 3.278 and 3.426 L/min on steady and unsteady character respectively. Group II is the lowest boundary with a pressure rise of 20mmHg and flow rates of 5.000 and 4.865 L/min on steady and unsteady character respectively. Group III contains points near the maximum flow rate difference of unsteady character with a pressure rise of 60mmHg and flow rates of 4.292 L/min on steady line and 3.998, 4.439 L/min on unsteady loop.

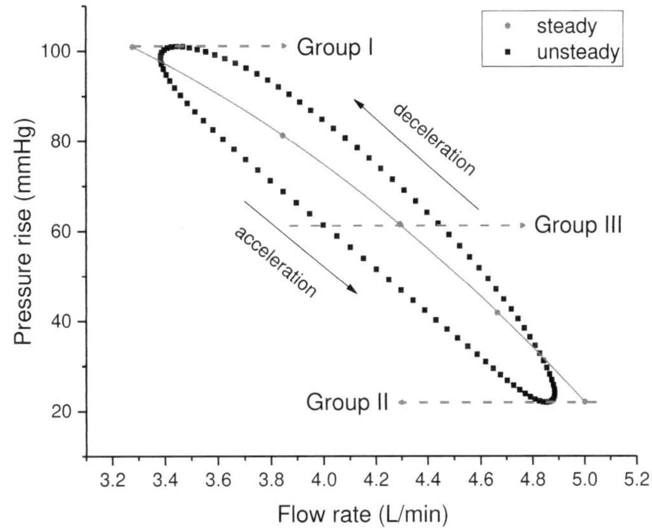

Figure 3. Steady and unsteady character.

3.2. Differences on flow fields

Flow fields on 0.5 span of the blades shown in **Figure 4** are examined as the average behavior of the three-dimensional flow. In both steady and unsteady results, there exists no area of separation or recirculation in rotor, but a large area in second row of stator. However, flow fields in first row of stator was influenced more by acceleration and deceleration of flow rates. In Group I, flow rate of unsteady is bigger than that of steady, which means a more negative incidence on stator in unsteady result. What makes it surprised, the separation points near the trailing edge (TE) differ little between both results. Due to a higher flow rate, the unsteady separation influences a larger region than steady, which even extends to the second row of stator and impacts more. Things turn more reasonable in Group II when unsteady flow rate is smaller than steady. The more negative incidence of steady stator induces a larger separation area on its pressure side. In Group III, there is no obvious separation in steady stator, but separations occur on blade's suction and pressure side in unsteady low and high flow rate point respectively, due to the change of incidence.

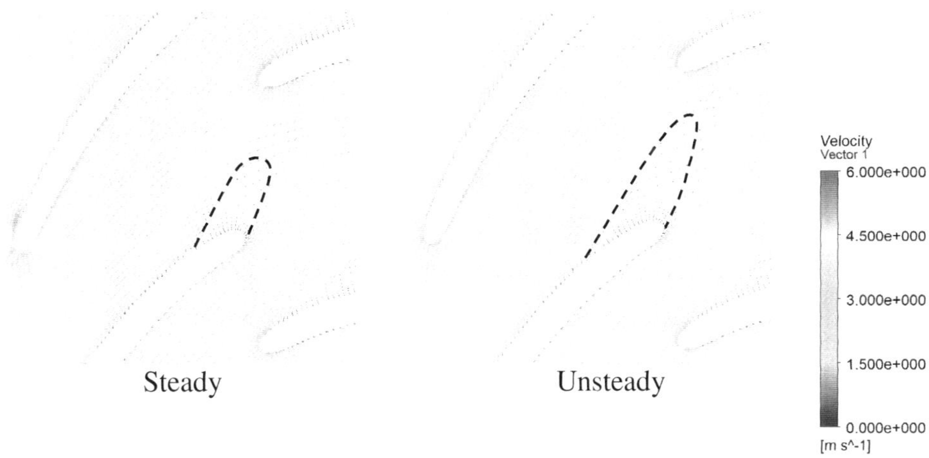

IOP Conf. Series: Materials Science and Engineering **408** (2018) 012014 doi:10.1088/1757-899X/408/1/012014

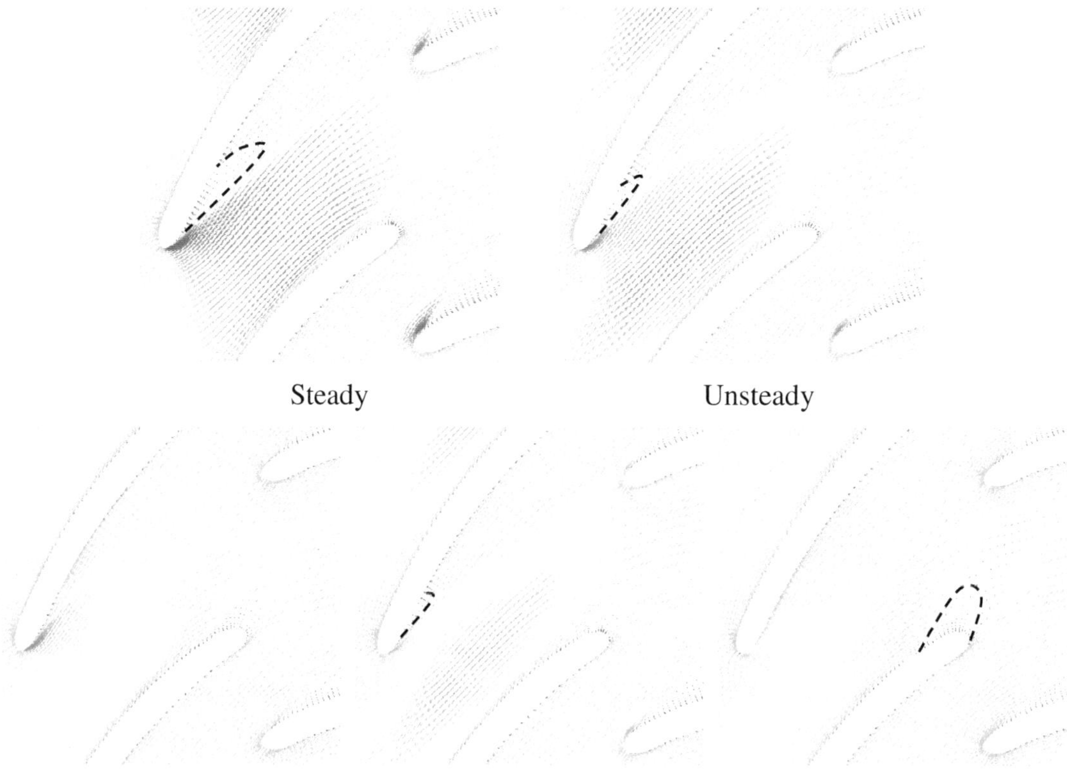

Figure 4. Flow fields in first row of stator of steady and unsteady results (top: Group I; middle: Group II; bottom: Group III. Separation region in each figure is marked with dashed line).

3.3. Scalar stress analysis

In order to describe the impact to cell rupture by shear stress, a scalar stress [19] is defined as,

$$\sigma = \left[\frac{1}{6}\sum\left(\sigma_{ii} - \sigma_{jj}\right)^2 + \sum \sigma_{ij}^2\right]^{\frac{1}{2}} \tag{2}$$

As shown in **Figure 5**, on blade-to-blade surface of 0.5 span in Group I, higher stress shows in unsteady rotor passage than steady because of less uniformity of velocity distribution circumferentially. Besides, larger velocity gradient near the TE of first stator row in unsteady result creates a high stress region which extends to second blade row. Higher stress also shows in unsteady rotor passage of in Group II, while a smaller high stress region on pressure side of unsteady stator, with consistency to previous analysis on flow field, is shown compared with steady result. In Group III, higher stress also shows in unsteady rotor passage compared with steady result. High unsteady flow rate point creates a high stress region in pressure side of stator. Due to the velocity gradient, a lower stress region is shown near the TE of unsteady stator even though the separation point in this case locates anterior to steady result, which is consistent with previous analysis.

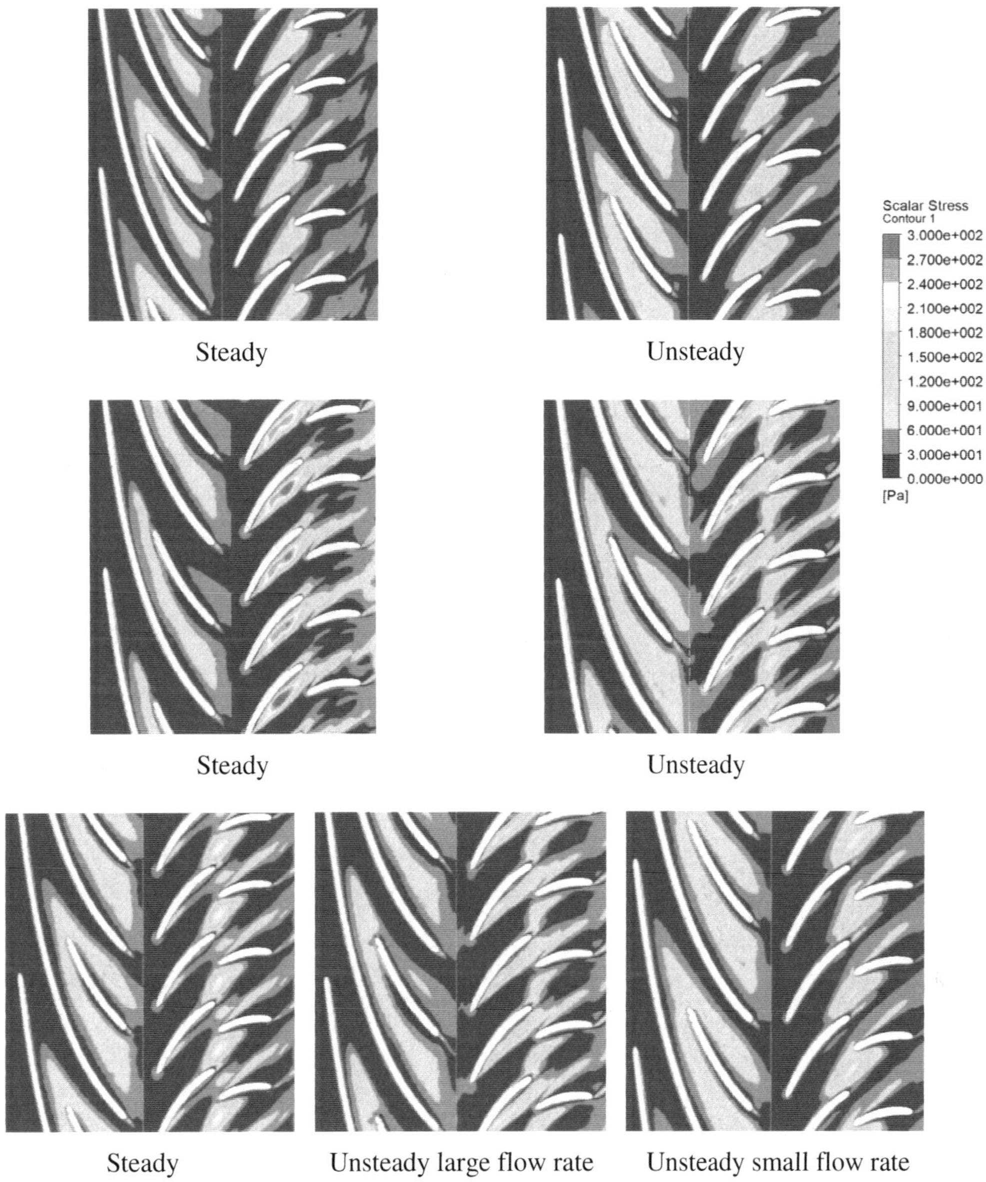

Figure 5. Scalar stress distribution in steady and unsteady results (top: Group I; middle: Group II; bottom: Group III).

3.4. Blade load and pressure rise of rotor

A more non-uniform circumferential velocity distribution in unsteady rotor passage caused by rotor blade load and pressure rise are compared with steady results. On 0.5 span of rotor blades, as the tangent velocity in rotor is much higher than axial velocity, the incidences of steady and unsteady rotor are almost the same, so the differences are mainly located in latter part of rotor. **Figure 6** compares surface pressure of rotor in each group. In Group I, pressure on unsteady rotor surface is bigger than steady, while it turns reversely in Group II. In Group III, pressure on pressure side of steady rotor is larger than that in unsteady lower flow rate and smaller than that in unsteady high flow rate, while the sequence of pressure value changes differently on suction side of rotor. However, what is in common with all three groups is that steady rotor has the largest pressure difference between its pressure and suction side, and has the lowest pressure rises.

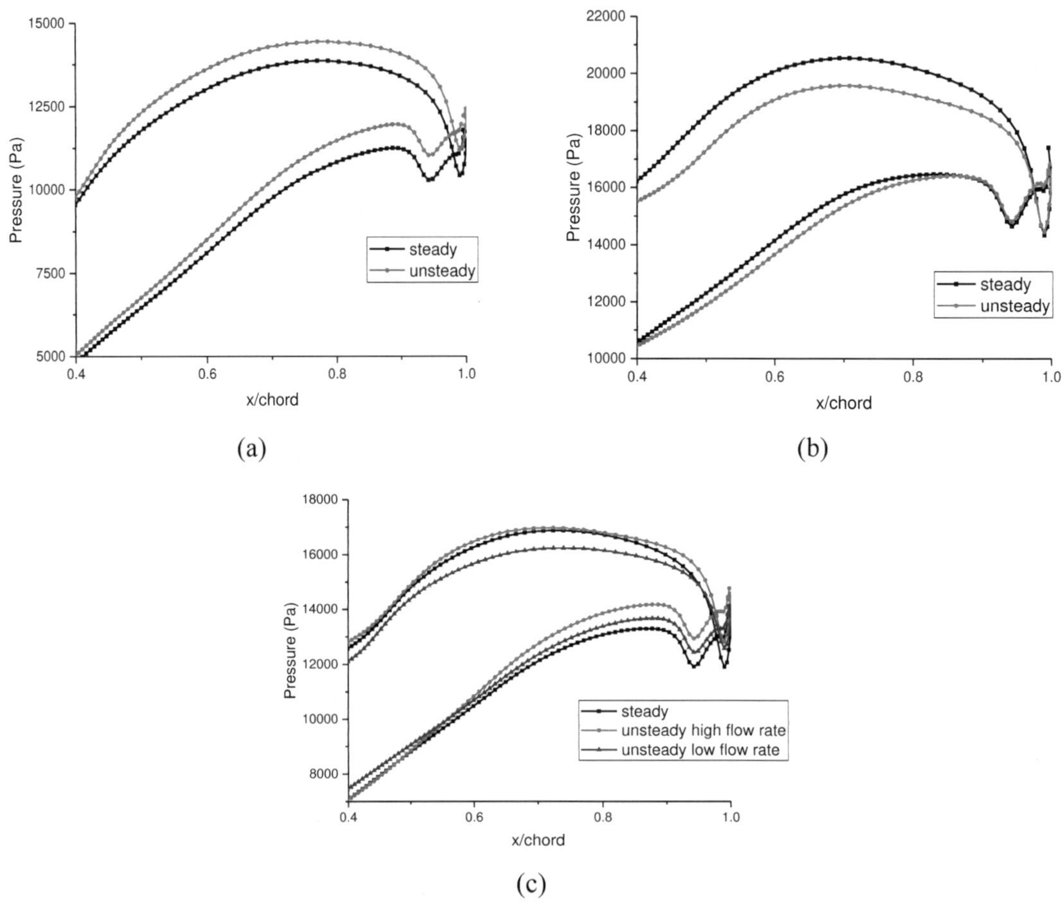

Figure 6. Distributions of the static pressure on rotor blade surface for steady and unsteady results: (a) Group I; (b) Group II; (c) Group III.

4. Conclusion

The conclusion is as following,

1) In the whole range of operation of the unsteady condition, flow field showed a well-attached flow in rotor passage but a large area of recirculation in the second row of stator passages. By contrast, flow field in the first row of stator passages were greatly impacted by the acceleration and deceleration of the flow rate, which can be well explained by the change of incidence.
2) In low flow rate, even when separation point on stator's suction side is posterior, higher velocity gradient and larger stress area may occur near the TE.
3) Generally, flow rate acceleration or deceleration exaggerated a non-uniform distribution of velocity, which can easily form a larger area of high scalar shear stress in rotor passage than the steady results.
4) A special pressure rise distribution between rotor and stator was also observed in unsteady flow. Compared with unsteady results, the pressure difference between pressure and suction side of rotor blades is larger in steady results.

References

[1] Kirklin J K, Naftel D C, Kormos R L, Stevenson L W, Pagani F D, Miller M A, Baldwin J T and Young J B 2013 *J. Heart Lung Transplant* **32** 141-56
[2] Leverett L B, Hellums J D, Alfrey C P and Lynch E C 1972 *Biophys. J.* **12** 257-73

[3] Kroll M H, Hellums J D, Guo Z, Durante W, Razdan K, Hrbolich J K and Schafer A I 1993 *J. Biol. Chem.* **268** 3520-4
[4] Konishi H et al 1996 *Artif. Organs* **20** 618-20
[5] Wu Y, Allaire P E, Tao G and Olsen D 2007 *IEEE Trans. Control Syst. Technol.* **15** 754-67.
[6] Baloa L A, Boston J R and Antaki J F 2001 *Ann. Biomed. Eng.* **29** 244-51
[7] Cox L G E, Loerakker S, Rutten M C M, de Mol B A J M and van de Vosse F N 2009 *Artif. Organs* **33** 593–603
[8] Pantalos G M, Koenig S C, Gillars K J, Giridharan G A and Ewert D L 2004 *ASAIO J.* **50** 37–46
[9] Shu F, Vandenberghe S and Antaki J F 2009 *Artif. Organs* **33(9)** 757-62
[10] Ochsner G, Amacher R, Amstutz A, Plass A, Daners M S, Tevaearai H, Vandenberghe S, Wilhelm M J and Guzzella L 2013 *IEEE Trans. Biomed. Eng.* **60** 507–16
[11] Misgeld B J, Rüschen D, Schwandtner S, Heinke S, Walter M and Leonhardt S 2015 *Biomed. Signal Process. Control* **20** 35-44
[12] Song X, Wood H G and Olsen D B 2004 *J. Biomech. Eng.* **126** 180–187
[13] Shu F, Vandenberghe S, Brackett J and Antaki J 2015 *Cardiovasc. Eng. Technol.* **6(3)** 230-41
[14] Yang X, Zhang Y, Gui X and Hu S 2011 *Artif. Organs* **35(10)** 948-55
[15] Yang X 2011 *Flow Optimization Design and Experimental Study of the Ventricular Assist Pump of the Artificial Heart* (Beijing: Beihang University Ph.D. thesis)
[16] Xu H, Wang M, Jin D, Sun D, Gui X and Sun X 2017 *ASME 2017 Fluids Engineering Division Summer Meeting (Waikoloa)* vol 1A
[17] Xu H, Wang M, Du L, Sun D, Jin D, Gui X and Sun X 2016 *6th Asian Joint Workshop on Thermophysics and Fluid Science 2016* (Guilin)
[18] Xu H, Jin D, Wang M, Du L, Sun D, Gui X and Sun X 2018 *Aerosp. Sci. Technol.*
[19] Bludszuweit C 1995 *Artif. Organs* **19** 583–89

Dynamic Performances of the Herringbone-Grooved Gas Bearing

S S Li*, H H Chen, H B Ding, G Y Zhang and M L Dong

School of Mechatronic Engineering and Automation, Shanghai Key Laboratory of Intelligent Manufacturing and Robotics, Shanghai University, Shanghai 200072, China

*03810020@shu.edu.cn

Abstract: Aiming at studying the dynamic performances of herringbone-grooved gas bearings, a model is built to analysis the relationship between herringbone-grooved gas bearings parameters and its performances such as friction torque, stiffness, stamping, etc. In this model, the hydrodynamic effect of compressible fluid and the shunting effect of herringbone groove are taken into account, and the partial derivative method which based on local integral finite difference method is used to calculate the modal at the same time. The result shows that when the average gap and the length-diameter ratio are decreasing, the rotate speed and eccentricity ratio will be increasing and the hydrodynamic effect of the gas bearing will be enhanced. Increasing groove angle, decreasing width-ridge ratio or aggrandizing length-groove ratio can improve its load capacity. The parameters such as rotate speed and eccentricity ratio are the most important factors which affect the dynamic characteristics of herringbone-groove gas bearing greatly. Then the analysis method is verified through an experiment systematically. In the experiment, the moment balance method is used. The experiment result shows that the result of model is correct s and reliable.

1. Introduction

Now, CNC machine tools are moving towards the direction of precision, high speed, intelligence, high efficiency and environmental protection. Electric spindle is the key part of CNC machine, which support the main machining power and handle the cutting tools directly. Therefore, high-speed spindle bearing is one of the key factors that affect the overall performance of CNC machine tools and other high-end equipment[1]. Gas bearing is a kind of non-contact bearing with gas as lubricant, which has many obvious features such as easy to rotate, clean, high rotary precision, long service life and so on.

It has been a long time about the study of gas bearings. In paper [2], the method for solving the transient stiffness coefficient of aerodynamic bearing was studied when the shaft neck was subjected to a constant radial velocity perturbation. Based on the small perturbation and linearizing PH methods, the equations of the static and dynamic characteristics of three-pad-tilting bearing were deduced[3]. Wang employed a hybrid numerical method combined the differential transformation method (DTM) and the finite difference method (FDM) to analyze the nonlinear dynamic behavior of a rigid rotor supported by a micro gas bearing system (MGBS)[4]. In 2017, Theisen had solved the problem to enhance the damping of gas bearing and improved its stability through robust control approaches[5].

For the study on characteristics of herringbone-groove bearing, Vohr and Chow proposed a narrow slot theory[6]. In Hris's paper, considering eccentricity, the characteristics of the herringbone groove

IOP Conf. Series: Materials Science and Engineering **408** (2018) 012015 doi:10.1088/1757-899X/408/1/012015

bearing were studied by using the narrow slot theory, and the bearing stability and load capacity under the condition of neglecting the axial deflection and incompressible fluid lubrication were calculated[7]. Different types of spiral groove thrust bearing performance were introduced in paper and the differences with traditional thrust bearing are analyzed as well[8]. Murata proposed a two-dimensional analysis of oil film lubrication of herringbone groove[9]. Kawabata put forward a new type of herringbone-groove bearing which could turn positive inversion, and analyzed its load capacity under the condition of steady state by the theory of narrow slot[10]. Bonneau used the finite element method to calculate the characteristics of herringbone-groove gas bearing which have small slot number and compare with the characteristics of smooth surface bearing based on narrow slot theory[11]. Gad obtained the maximum radial force and radial stiffness by optimizing the geometric parameters of the bearing[12]. The influence of the geometric parameters and the lubrication gas on the performance of the herringbone-groove gas bearing was analysed[13].

In this paper, based on the theory of aerodynamics and taking the hydrodynamic effect of compressible fluid and the shunting effect of herringbone groove into account, the dynamic performances of herringbone-grooved gas bearing were studied.

2. The journal governing equation

2.1 The basic form of governing equation

The dynamic characteristics of gas film reflect the change of gas film force as journal in deviating from the static equilibrium position. Therefore, the Reynolds equation in unsteady condition must be set as the basis of calculation and analysis[14]. The partial differential derivation and dimensionless form of Equation (1) is[15][16]:

$$\frac{\partial}{\partial\varphi}\left(H^3\frac{\partial P^2}{\partial\varphi}\right)+\left(\frac{R}{L}\right)^2\frac{\partial}{\partial\lambda}\left(H^3\frac{\partial P^2}{\partial\lambda}\right)=2\Lambda\frac{\partial(PH)}{\partial\varphi}+2\sigma P(\varepsilon'\cos\varphi+\varepsilon\theta'\sin\varphi) \tag{1}$$

where, φ is the circumferential angle coordinate ($\varphi=x/R$). x is *the* bearing rotating direction coordinate. R is bearing radius. λ is dimensionless axial coordinate ($\lambda=z/L$). z is axial coordinate. μ is the dynamic viscosity of the gas. U is the linear velocity of the bearing. L is bearing length. P is dimensionless pressure ($P=p/p_a$). p is pressure. p_a is ambient pressure. H is dimensionless gas film thickness ($H=h/c$). h is gas film thickness. c is bearing clearance. ε is eccentricity ratio$=e/c$. e is eccentricity of journal. Λ is compressibility number ($\Lambda=6\mu\omega^2R^2/(p_ac^2)$). σ is extrusion coefficient ($\sigma=12\mu R^2/p_ac^2$); μ is viscosity of fluid, ω is rotating speed, (e, θ) is the axis of instantaneous position, (e', θ') is the instantaneous displacement speed, $\varepsilon'=\dot{e}/c\omega=de/(c\omega dt)$; $\theta'=\dot{\theta}/\omega=d\theta/(\omega dt)$.

2.2 The expression of bearing stiffness and damping coefficients

The relationship between the changes of the force and the perturbations is nonlinear[17]. When the perturbation is small; the relationship can be linearized. The four stiffness coefficients and four damping coefficients are called the eight dynamic characteristics of aerodynamic bearing. The gas film force is consisted of pressure distribution, so the dynamic characteristics of gas film directly depend on the perturbation pressure. Therefore, the dimensionless expressions of stiffness and damping coefficients can be obtained.

$$\left.\begin{matrix}K_{ee}\\K_{\theta e}\end{matrix}\right\}=-\int_{-1}^{1}\int_{\varphi 1}^{\varphi 2}P_{\varepsilon}\begin{Bmatrix}\cos\varphi\\\sin\varphi\end{Bmatrix}d\varphi d\lambda\ ,\quad \left.\begin{matrix}K_{e\theta}\\K_{\theta\theta}\end{matrix}\right\}=-\int_{-1}^{1}\int_{\varphi 1}^{\varphi 2}P_{\theta}\begin{Bmatrix}\cos\varphi\\\sin\varphi\end{Bmatrix}d\varphi d\lambda \tag{2}$$

$$\left.\begin{matrix}C_{ee}\\C_{\theta e}\end{matrix}\right\}=-\int_{-1}^{1}\int_{\varphi 1}^{\varphi 2}P_{\varepsilon'}\begin{Bmatrix}\cos\varphi\\\sin\varphi\end{Bmatrix}d\varphi d\lambda\ ,\quad \left.\begin{matrix}C_{e\theta}\\C_{\theta\theta}\end{matrix}\right\}=-\int_{-1}^{1}\int_{\varphi 1}^{\varphi 2}P_{\theta'}\begin{Bmatrix}\cos\varphi\\\sin\varphi\end{Bmatrix}d\varphi d\lambda \tag{3}$$

Where

$$P_{\varepsilon}=\frac{\partial P}{\partial\varepsilon},P_{\theta}=\frac{\partial P}{\varepsilon\partial\theta},P_{\varepsilon'}=\frac{\partial P}{\partial\varepsilon'},P_{\theta'}=\frac{\partial P}{\varepsilon\partial\theta'} \tag{4}$$

3. Solution of pressure perturbation

3.1 Establishment of perturbation equations

The crux of calculating the eight characteristics is to solve the four perturbation pressure P_ε, P_θ, $P_{\varepsilon'}$, $P_{\theta'}$, the results can be get by solving the partial derivative of governing equations. Four perturbation equations are:

$$\mathrm{Re}\, y(P_\varepsilon^2)=\begin{cases}-\dfrac{6\Lambda\cos\varphi}{H}\dfrac{\partial(PH)}{\partial\varphi}+3H\left[\dfrac{\partial H}{\partial\varphi}\cos\varphi-H\sin\varphi\right]\dfrac{\partial P^2}{\partial\varphi}+3H(\dfrac{R}{L})^2\left(\dfrac{\partial H}{\partial\lambda}\cos\varphi\right)\dfrac{\partial P^2}{\partial\lambda}\\ +2\Lambda\dfrac{\partial}{\partial\varphi}(P\cos\varphi+HP_\varepsilon)+2\sigma\left[P_\varepsilon(\varepsilon'\cos\varphi+\varepsilon\theta'\sin\varphi)+P\theta'\sin\varphi\right]\end{cases}\tag{5}$$

$$\mathrm{Re}\, y(P_\theta^2)=\begin{cases}-\dfrac{6\Lambda\varepsilon\sin\varphi}{H}\dfrac{\partial(PH)}{\partial\varphi}+3H\varepsilon\left(\dfrac{\partial H}{\partial\varphi}\sin\varphi-H\cos\varphi\right)\dfrac{\partial P^2}{\partial\varphi}+3H(\dfrac{R}{L})^2\varepsilon\sin\varphi\dfrac{\partial H}{\partial\lambda}\dfrac{\partial P^2}{\partial\lambda}\\ +2\Lambda\dfrac{\partial}{\partial\varphi}(P\varepsilon\sin\varphi+H\varepsilon P_\theta)+2\sigma\left[(\varepsilon'\cos\varphi+\varepsilon\theta'\sin\varphi)\varepsilon P_\theta+P(\cos\varphi+\varepsilon'\theta'\sin\varphi)\right]\end{cases}\tag{6}$$

$$\mathrm{Re}\, y(P_{\varepsilon'}^{\;2})=2\Lambda\frac{\partial}{\partial\varphi}(HP_{\varepsilon'})+2\sigma\left[P_{\varepsilon'}(\varepsilon'\cos\varphi+\varepsilon\theta'\sin\varphi)+P\cos\varphi\right]\tag{7}$$

$$\mathrm{Re}\, y(P_{\theta'}^{\;2})=2\Lambda\frac{\partial}{\partial\varphi}(H\varepsilon P_{\theta'})+2\sigma\left[\varepsilon P_{\theta'}(\varepsilon'\cos\varphi+\varepsilon\theta'\sin\varphi)+P\varepsilon\cos\varphi\right]\tag{8}$$

3.2 Solution of perturbation equations by partial derivative method

The figure 1 is the expanded structure of herringbone groove surface. When use the differential calculation, it is necessary to convert the coordinate system.

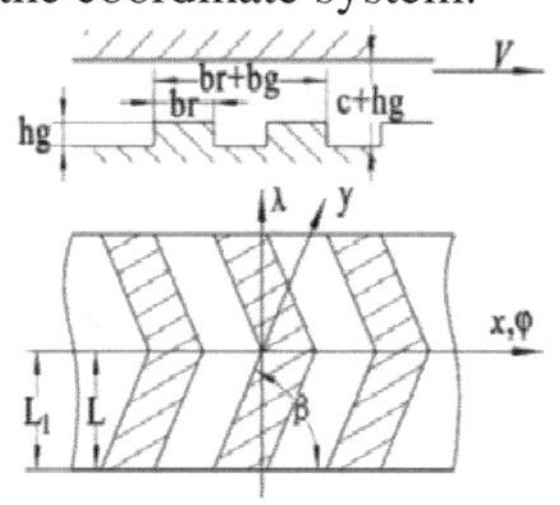

Figure 1. Expanded structure of herringbone groove surface

The corresponding partial derivative:

$$\frac{\partial p^2}{\partial\phi}=\frac{\partial p^2}{\partial x};\frac{\partial p^2}{\partial\lambda}=\frac{1}{\sin\beta}\left(\frac{\partial p^2}{\partial y}-\frac{\partial p^2}{\partial x}\cos\beta\right)\tag{9}$$

Where, φ, λ are coordinates in the rectangular coordinate system; x, y are coordinates in the skew coordinate system; β is the angle of herringbone groove.

Using partial derivative finite differential method discretize the four perturbation, and the differential forms of perturbation equations are deduced.

$$\frac{\partial}{\partial\varepsilon}\left[\frac{\partial}{\partial\lambda}\left(H^3\frac{\partial P^2}{\partial\lambda}\right)\right]=\frac{\partial}{\partial\lambda}\left[H^3\frac{\partial}{\partial\lambda}\left(\frac{\partial P^2}{\partial\varepsilon}\right)\right]+3H^2\frac{\partial H}{\partial\varepsilon}\frac{\partial^2P^2}{\partial\lambda^2}+\left(3H^2\frac{\partial^2H}{\partial\varepsilon\partial\lambda}+6H\frac{\partial H}{\partial\lambda}\frac{\partial H}{\partial\varepsilon}\right)\frac{\partial P^2}{\partial\lambda}\tag{10}$$

$$\Lambda\frac{\partial}{\partial\varepsilon}\left[\frac{\partial(PH)}{\partial\varphi}\right]=\Lambda\left(\frac{\partial H}{\partial\varepsilon}\frac{\partial P}{\partial\varphi}+H^2\frac{\partial^2P}{\partial\varepsilon\partial\varphi}+\frac{\partial P}{\partial\varepsilon}\frac{\partial H}{\partial\varphi}+P\frac{\partial^2H}{\partial\varepsilon\partial\varphi}\right)\tag{11}$$

Then the following equation can be got:

$$(P_n)_{i,j}=\frac{A_{i,j}(P_n)_{i+1,j}+B_{i,j}(P_n)_{i-1,j}}{E_{i,j}}+\frac{C_{i,j}(P_n)_{i,j+1}+D_{i,j}(P_n)_{i,j-1}+(F_1)_{i,j}}{E_{i,j}}\tag{12}$$

Where, $A_{i,j}, B_{i,j}, C_{i,j}, D_{i,j}, E_{i,j}, (F_n)_{i,j}$ are coefficient related with $H_{i,j}, P_{i-1,j}, P_{i+1,j}, P_{i,j-1}, P_{i,j+1}$; n indicates ε, ε', θ, θ', etc.

4. Calculation results and analysis

4.1 The Comparison of Dynamic Characteristics between smooth-surface gas Bearing and herringbone-groove gas bearing

Under the same condition, there is difference of the dynamic coefficients between two type gas bearings. Figure 2 to figure 5 show that under the same condition, with the change of eccentricity, the change rate of dynamic characteristic coefficient of smooth-surface gas bearing is larger, than herringbone-groove gas bearing. Therefore, we can get that the herringbone groove can make the gas film smoother, but the shunting effect of herringbone groove will decrease the load capacity.

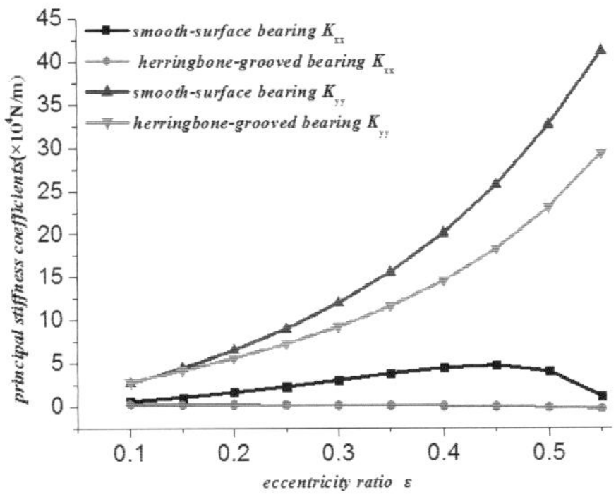

Figure 2. The relationship between principal Stiffness and eccentricity ratio

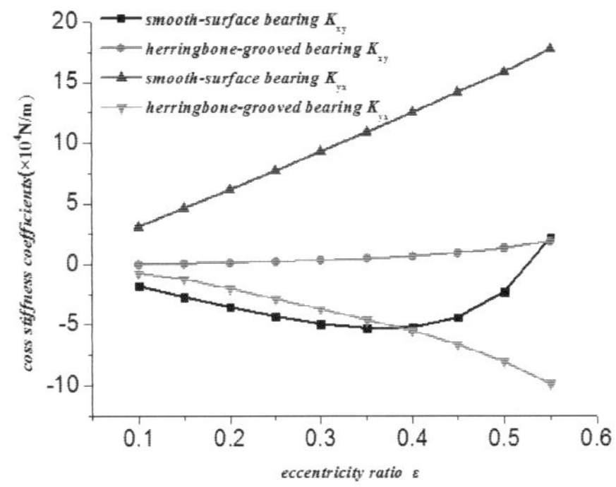

Figure 3. The relationship between cross stiffness and eccentricity ratio

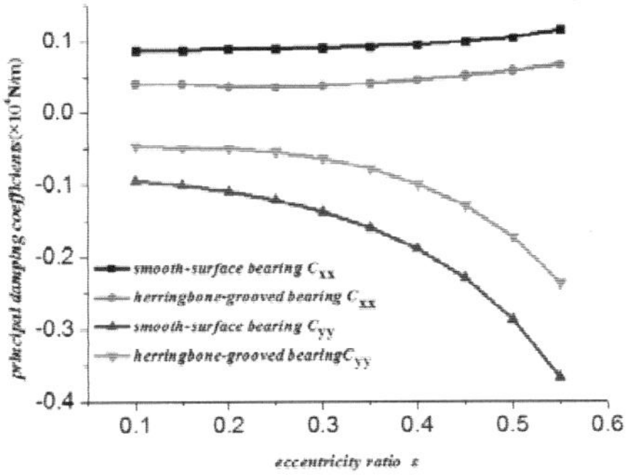

Figure 4. The relationship between principal damping and eccentricity ratio

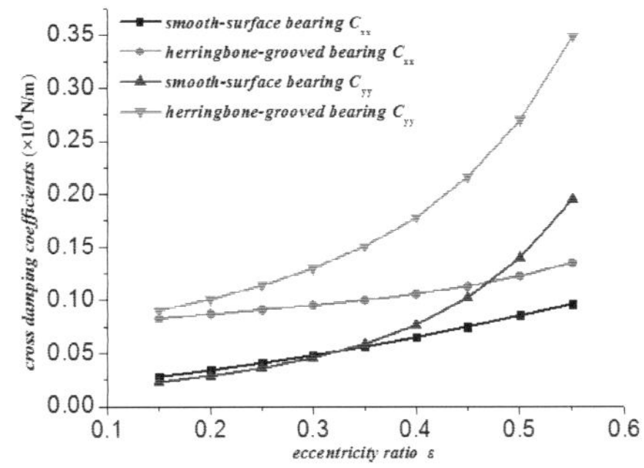

Figure 5. The relationship between cross damping and eccentricity rate

4.2 The effects of dimension parameter of bearing on dynamic characteristic coefficients

The figure 6 and figure 7 shows the variation of stiffness and damping coefficients as the bearing clearance changes, we can get that as the bearing clearance increases, the values of overall stiffness and overall damping coefficients are in the downward trend, and then keep stable. Therefore, the gas film pressure will increase as the bearing clearance decrease, and the load capacity will decrease as well.

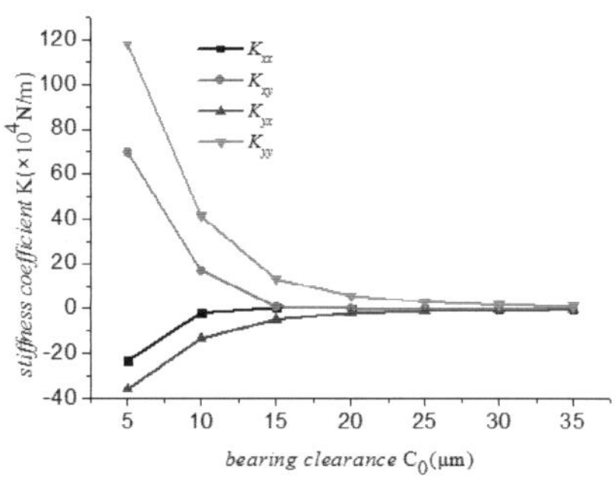

Figure 6. The relationship between stiffness coefficients and bearing clearance

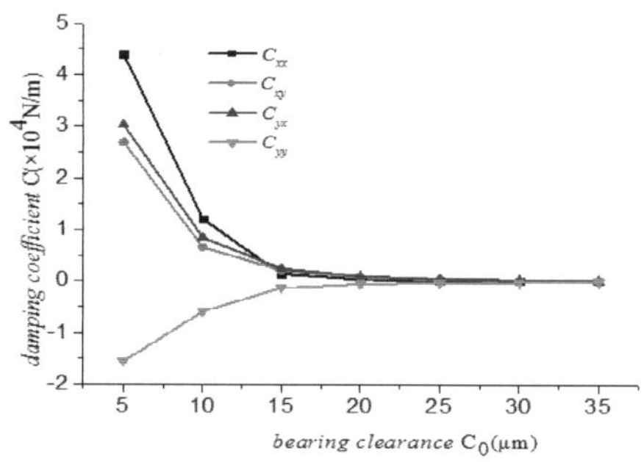

Figure 7. The relationship between damping coefficients and bearing clearance

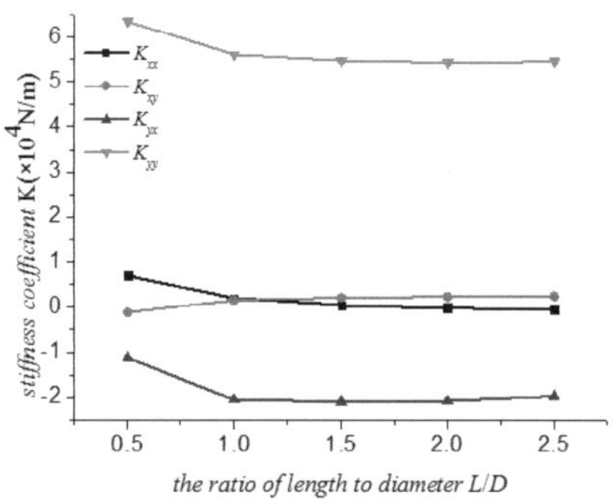

Figure 8. The relationship between stiffness coefficients and ratio of length to diameter

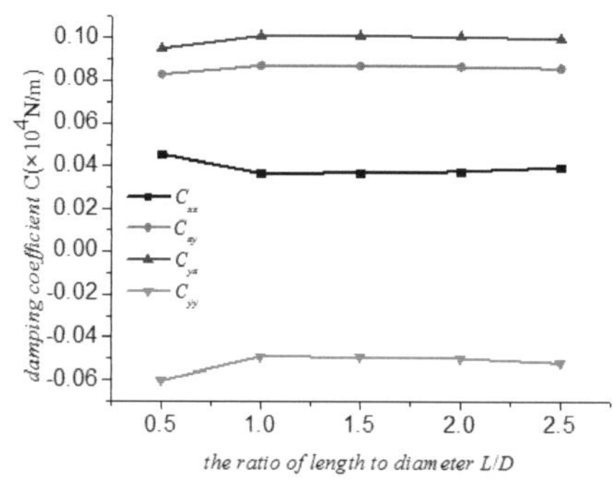

Figure 9. The relationship between damping coefficients and ratio of length to diameter

The figure 8 and figure 9 show the variation of stiffness and damping coefficients as the ratio of length to diameter changes. As the ratio changes from 0.5 to 1.0, the values of overall stiffness coefficients decreases, while the values of overall damping coefficients increase. With the upward trend continuing, the dynamic characteristic coefficients remain unchanged.

4.3 The effects of working condition of bearing on dynamic characteristic coefficients

The working condition is important for bearing, the effect of bearing working condition can be obtained by analyzing the variations of the dynamic characteristic coefficients.

The figure 10 and figure 11 shows the variation of stiffness and damping coefficients as the eccentricity ratio changes, with the eccentricity ratio increasing, the values of overall stiffness and overall damping coefficients all have the upward trend. Hence, the increase of the eccentricity ratio can improve the load capacity, and then the dynamic characteristics increase correspondingly.

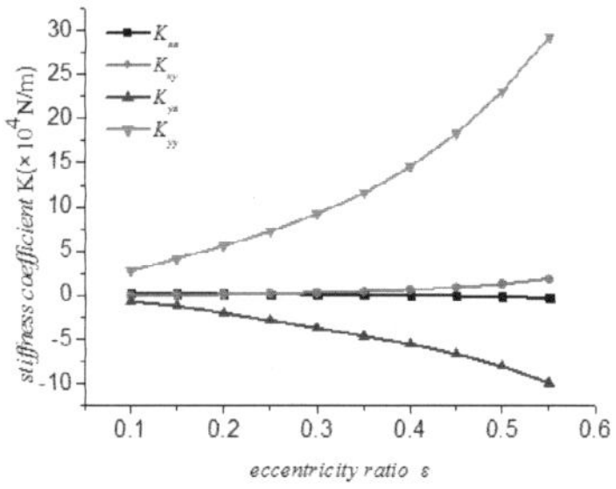

Figure 10. The relationship between stiffness coefficients and eccentricity ratio

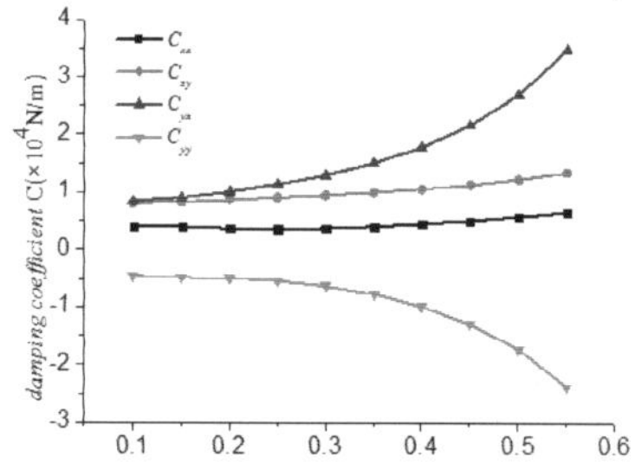

Figure 11. The relationship between damping coefficients and eccentricity ratio

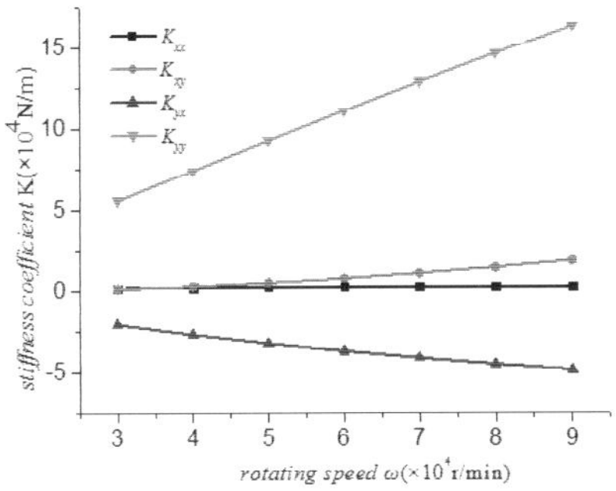

Figure 12. The relationship between stiffness coefficients and rotating speed

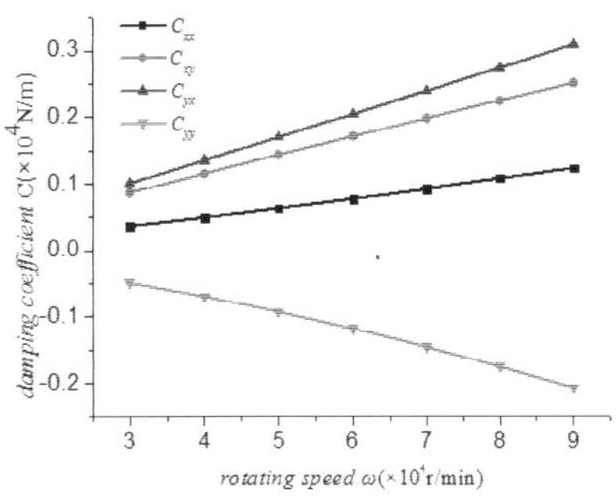

Figure 13. The relationship between damping coefficients and rotating speed

The figure 12 and figure 13 shows the variation of stiffness and damping coefficients as the rotating speed changes. As the rotating speed increase, the stiffness coefficients K_{yx}, K_{yy} and coefficient C_{yy} keep the downward trend, while the other stiffness coefficients and other damping coefficients increase. As the rotating speed increase, the hydrodynamic effect becomes more obvious. The load capacity and the dynamic characteristics will increase as well finally.

5. Experimental verification

According to the above moment balance method, the rotor of the tested bearing is directly fixed on the rotating shaft of the output end of the electric spindle, the herringbone grooves are machined on the rotor surface. The bearing is equipped on the rotor driven by electric spindle directly, so the bearing is in a rotational trend. However, due to the limitation of the string the bearing cannot rotate, the force is transmitted to the sensor though the string. Thus, the fraction torque between the bearing and rotor can be measured. As shown in figure 14, there are four kinds of bearing structures. Figure 15 is friction torque test device.

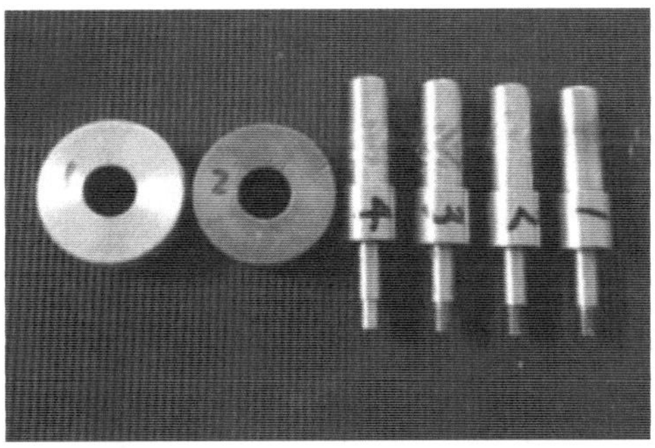

Figure 14. Herringbone-groove gas bearing

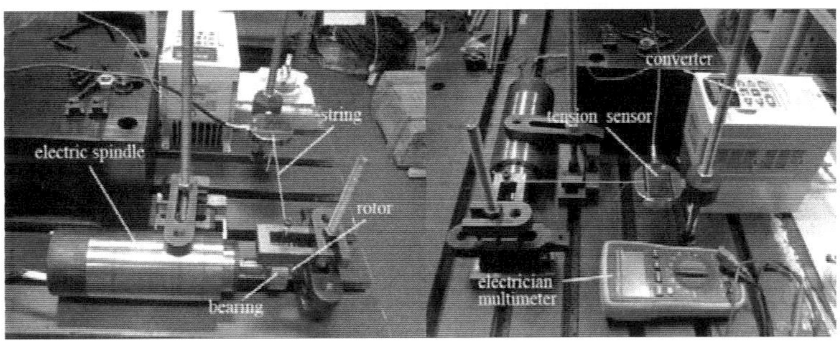

Figure 15. Friction torque test device

As presented in figure 16, when the bearing is in low speed, the friction torque decreases with the increasing of rotating speed. When the bearing is floating and stable, the friction torque has an increase trend. As to the reason of difference between the two curves, it probably is that the bearing does not float and the dry friction between the bearing and the rotor makes the measured value larger than the theoretical value when the speed is low. In addition, the differences between the experimental value and the theoretical value may be caused by the machining accuracy, the calculation error caused by the numerical method and the accuracy and sensitivity of the tension sensor.

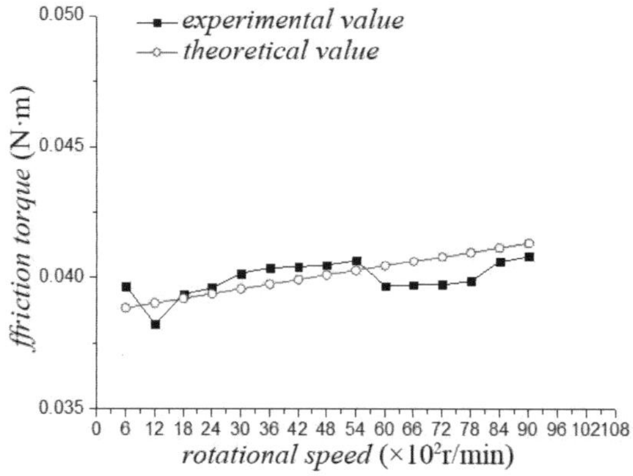

Figure 16. Comparison between experimental value and the theoretical value of the friction torque

6. Conclusions

In this paper, the dynamic characteristics of herringbone-groove gas bearings are analyzed. Through the analysis and experiments, the following conclusions can be obtained:

(1) Theoretical analysis results indicate that smooth-surface bearing will be out of stability under the condition of light load and high speed. Herringbone-groove gas bearings not only improve this defect and its pumping effect, but also increase the hydrodynamic effect.

(2) With the increase of eccentricity and rotational speed, the change of bearing dynamic characteristics is obvious. So changing these parameters properly can get better stiffness and damping coefficients.

(3) The groove depth has a significant influence on the stiffness and damping coefficients. Because the groove depth is large, the average air gap becomes large, and the air film pressure is affected. So the deeper the groove depth, the radial stiffness and damping coefficient of the bearing decreases gradually.

Acknowledgements

The project was supported by the Shanghai Science and Technology Major Project (13521103002), China and the National 'the twelfth five-year' Projects of Science and Technology of China (No. D.50-0109-15-001).

References

[1] Gao S Y, Ding H and Cheng K, "An Integrated Approach to the Design and Analysis of an Ultra-High Speed Air-Bearing Spindle." *Materials Science Forum*, Vol. 723, pp. 227-232.
[2] Nakis Y and Pinkus O, "On the Stiffness and Damping Properties of Gas Bearings", *Journal of Lubrication Technology*, Vol. 98, pp. 189-190.
[3] Wang L Z, "Static and Dynamic Study on the Three-tilting-pad Gas Bearing", *Lubrication Engineering*.
[4] Wang C C, "Application of a hybrid numerical method to the nonlinear dynamic analysis of a micro gas bearing system", *Nonlinear Dynamics*, Vol. 59, pp. 695-710.

[5] Theisen, "Enhancing damping of gas bearings using linear parameter-varying control", *Journal of Sound & Vibration*, Vol. 395, pp. 48-64.
[6] Vohr J H and Chow C Y, "Characteristics of herringbone-grooved, gas-lubricated journal bearings", *Journal of Basic Engineering*, Vol. 87, pp. 568-576.
[7] Hris G G, "The load capacity and stability characteristics of hydro-dynamic grooved journal bearing" *ASLE Trans*, Vol. 8, pp. 296-305
[8] Murata S, Miyake Y and Kawabata N, "Two-dimensional analysis of herringbone groove journal bearings", *JSME*, Vol. 23, pp.1120-1227.
[9] Kawabata N, Ozawa Y and Kamaya S, "Static characteristics of the regular and reversible rotation type herringbone grooved journal bearing", *Journal of tribology*, Vol. 111, pp. 484-490.
[10] Bonneau D, Absi J, "Analysis of aerodynamic journal bearings with small number of herringbone grooves by finite element method", *Journal of tribology*, Vol. 116, pp. 698-704.
[11] Savoulides N., Breuer N S and Jacobson S, "Low-order models for very short hybrid gas bearings", *Journal of Tribology*, Vol. 123, pp. 368-375.
[12] Schiffmann J and Favrat D, "The effect of real gas on the properties of Herringbone Grooved Journal Bearings", *Tribology International*, Vol. 43, pp. 1602-1614.
[13] Wang Y F, "Gas Lubricated Theory and Design Manual of Gas Bearing". (In Chinese: *Machinery Industry Press*, Beijing)
[14] Lu Z W, Liu B and Zhang J, "Analysis on Dynamic Performance for Conical Spiral Groove Aerodynamic Bearings", *Bearing*, Vol. 11, pp. 10-14.
[15] Zhao S X, "Computation of the performance of gas-lubricated cylindrical bearing using FDM", *Journal of Wuhan University of Science and Technology* (*Natural Science Edition*), Vol. 31, pp. 69-73.
[16] Zhang Z M, Zhang Y Y and Xie Y B, "Hydrodynamic Lubrication Theory of Sliding Bearing", (In Chinese: High Education Press, Beijing)
[17] Jiang L G and Sun H, "Study on a new type of hydrostatic bearing for high speed precision machine tools", *Jiangsu Machine Building & Automation*, Vol. 5, pp. 22-25.

IOP Conf. Series: Materials Science and Engineering **408** (2018) 012016 doi:10.1088/1757-899X/408/1/012016

Research on Parameters of the Trailing VLF Antenna Using Lumped-Mass Method

Tielin Ma[1, 2, *, a], **Zhihua Wei**[3,5, b], **Haibing Chen**[4] **and Xiangsheng Wang**[3]

[1] Institute of Unmanned System, Beihang University, Beijing, 100191, China
[2] Key Laboratory of Advanced Technology of Intelligent Unmanned Flight System of Ministry of Industry and Information Technology
[3] School of Aeronautic Science and Engineering, Beihang University, Beijing, 100191, China
[4] School of Transportation Science and Engineering, Beihang University, Beijing, 100191, China

[*, a] matielin@buaa.edu.cn; [b] wzhua@buaa.edu.cn.

Abstract. The verticality of Very Low Frequency (VLF) antenna is an important factor to decide the radiation intensity of airborne VLF communication system, and the close relation of which with the parameters of airplane, towed antenna, and drogue is an important issue. Considering from the aspect of system design, the flight speed and the drogue's mass are primary design variables whereas the bank angle constrained by airplane's maneuverability and cable's basic parameters decided by design requirement can be chosen initially. After modelling the trailing system with lumped-mass method, sensibility analyzation about relations of the verticality of antenna with the flight speed and the drogue's mass are carried on. The result is found that when a faster flight speed is chosen, it's not easy to gain enough verticality required, and leads to a conclusion that, flight speed dominates the verticality which should be decided seriously whereas the drogue's mass poses much less effect.

1. Introduction

As the sea is an electric conductor, electromagnetic wave decays quickly when spreading in it. To build communication between airplane and submarine, approaches, such as VLF signal, ELF (extremely low frequency) signal, blue-green laser, satellite delay communication, are developing. Among them, Airborne VLF submarine communication system is an important and developed communication relay approach, featured with maneuverability, anti-interference and survivability.

VLF antenna system is composed of airplane, antenna, and the drogue tied under the antenna for stabilization. For dynamic research, it's regarded as a towing system which is highly nonlinear and complex. Modeling of antenna is the primary job of the simulation for this system. It can be divided into two ways, continuous model [1-3] presented by partial differential equation, and lumped-mass method including mass-spring model [4, 5] solved by ordinary differential function, mass-rigid-rod model [6-7] solved by multi-body formulation. They are mature methods to simulate the towing system, among which the lumped-mass method is proved to be an efficient way to simulate complex cable system [4].

As for the trailing antenna, there are some researches have been done. Zheng Xiaohong et al. [8-10] solved the steady dynamic model by shooting technique and Newton-Raphson iteration respectively to discuss some effect factors influencing the steady state of the cable and had provided some guidance for system designing. But the effect of the factors is discussed separately. As for the practical experience, some factors would dominate whereas some would be secondary. Because of this, the comparison of the effect from different factors is needed. Furthermore, they researched on the parameters too extensively. However, for a system towed by an airplane, the parameters are constrained by practical condition. For example, it's hard to meet the multivalued solution, since the rotational velocity is small [7]. For a specific case, the range of the parameters should be smaller.

Therefore, this paper discussed about the working condition from the aspect of practical design, to get the appropriate range of parameters, and find out the factors which dominate the system and deserve more attention. The simulation of the system is done by using lumped-mass method which discretizes the towed antenna with mass-spring model. And the Runge-Kutta method is used to solve the nonlinear equations. The research on the sensibility of the verticaliy is carried on by building response surface.

2. The Lumped-Mass Method

2.1. Coordinate Frame

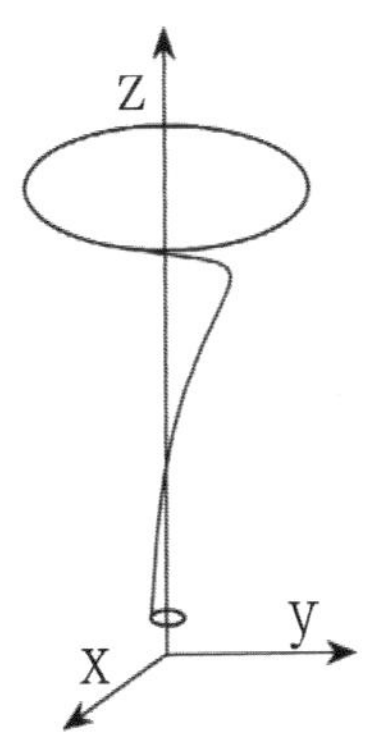

Figure 1. Sketch of the coordinate frame

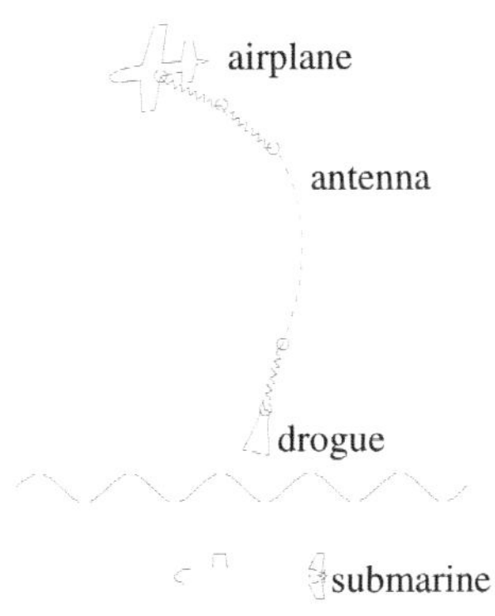

Figure 2. Sketch of the segment model

The calculation is based on an inertial frame (figure 1), the Local Tagent Plane Coordinate System. The ground is assumed as a horizontal plane. The origin is set at the sealevel point of the rotation axis. The z axis points up. The x direction is aligned with an arbitrary direction in the horizontal plane, and the y axis can be decided by the right-hand principle.

2.2. Discretization

The antenna is discretized into a number of segments which is enough to gain satisfactory accuracy. By mass-spring model, each segment is modeled as damping spring connected with mass points (figure 2). Each mass point has 3 degrees of freedom and suffers from half of the gravity and aerodynamic force act on its adjacent segments.

2.3. Force on the Segments

2.3.1. Gravity. The gravity acceleration is assumed as constant with altitude.

$$\vec{G} = m_0 \cdot \vec{g} \cdot l \tag{1}$$

where m_0 is the linear density of the antenna, l is the initial length of the segments.

2.3.2. Aerodynamic Force. The flight altitude is in the troposphere. So the variation of air density with the altitude is

$$\rho = \left(\frac{288.15 - 0.0065 \cdot h}{288.15}\right)^{4.25588} \cdot \rho_0 \tag{2}$$

Where ρ_0 is the air density at sealevel.

The aerodynamic force act on the segments can be calculated by

$$\vec{D}_t = 0.5 \cdot \rho \cdot |\vec{v}| \cdot \vec{v}_t \cdot S_t \cdot C_{dt} \tag{3}$$

$$\vec{D}_n = 0.5 \cdot \rho \cdot |\vec{v}| \cdot \vec{v}_n \cdot S_n \cdot C_{dn} \tag{4}$$

$$S_t = \pi \cdot d \cdot (l + dl) \tag{5}$$

$$S_n = d \cdot (l + dl) \tag{6}$$

Where $\vec{D}_t$ is the tangential friction, $\vec{D}_n$ is the normal drag, C_{dt} is the coefficient of tangential friction,. C_{dn} is the coefficient of normal drag, $\vec{v}$ is the airspeed of the segments, $\vec{v}_t$ and $\vec{v}_n$ is the tangential and normal component of $\vec{v}$ along the segments respectively, S_t is the superficial area of the segments, while S_n is the area of longitudinal section.

2.3.3. Tension and Damping Force. Each segment is considered as a linear damping spring. Thus the tension $\vec{T}$ and the damping force $\vec{T}_{zn}$ can be calculated by

$$\vec{T} = k \cdot A \cdot d\vec{l} \,/\, l \tag{7}$$

$$\vec{T}_{zn} = k_{zn} \cdot A \cdot \frac{d\vec{l}}{dt} \,/\, l \tag{8}$$

$$\vec{T} = \vec{T}_{zn} = \vec{0} \qquad (dl < 0) \tag{9}$$

Where k is the Young's modulus, $d\vec{l}$ is the deformation quantity of the segments, l is the original length of the segments, k_{zn} is the damping coefficient, A is the area of cross section. The change of the diameter is neglected.

2.3.4. Dynamics Equations for the Mass Points.

$$\vec{F} = m \cdot \vec{a} \tag{10}$$

$$\begin{aligned}&(\vec{T}_{zn1} + \vec{T}_{zn2}) + (\vec{T}_1 + \vec{T}_2) + (\vec{G}_1 + \vec{G}_2)/2 + \\ &(\vec{D}_{t1} + \vec{D}_{t2})/2 + (\vec{D}_{n1} + \vec{D}_{n2})/2 \\ &= (m_{01} + m_{02}) \cdot l/2 \cdot \vec{a}\end{aligned} \tag{11}$$

For each mass point, forces with the subscript 1 and 2 represent the forces act on the two segments connected with it.

2.4. Modelling of the airplane

In keeping with the assumption of the steady-state model, a constant bank angle, circular orbit was assumed, with the effect from the antenna neglected. The orbiting radius and angular velocity is calculated by bank angle

$$r = \frac{V^2}{g\sqrt{(1/\cos\phi)^2 - 1}} \tag{12}$$

$$\omega = V / r \tag{13}$$

Where V is the flight speed of the plane, ϕ is the bank angle.

2.5. Conclusion of the method

Up to now, the simplest towing system, composed of airplane and cable, has been modeled. As for the trailing antenna system, the boundary condition, drogue in this case, should also be concerned. Its aerodynamic force, gravity and mass are inherited by the first mass point.

As shown by figure 3, the result of simulating the TACAMO system is aligned with that given by the AD report [11] using second order accurate central differencing techniques. Therefore, the model's accuracy is ensured.

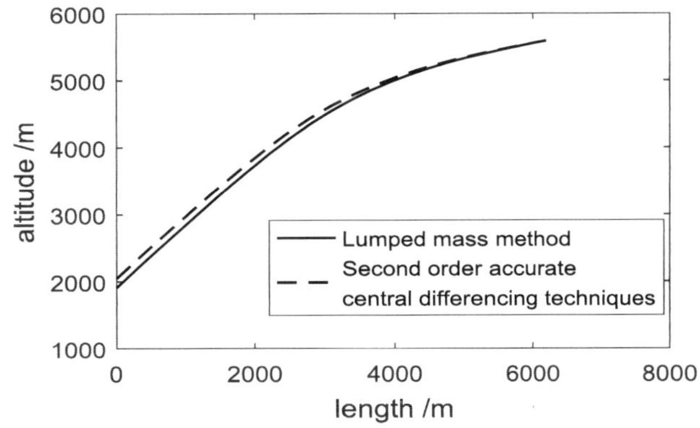

Figure 3. Altitude along the antenna

3. Research on the parameters

3.1. Discussion of the primary parameters

The primary parameters of airborne VLF communication system include the length, diameter, aerodynamic characteristics of the antenna, the flight speed, altitude, bank angle of the airplane, the mass, aerodynamic characteristics of the drogue. But it's not necessary to discuss all the parameters listed here.

Some parameters can be decided by working condition. The length of antenna related to the wave length of the signal ranges from 5km to 10km. The diameter of antenna is normally 4mm to 6mm. Airplane used is usually medium transport or communication relay aircraft, such as E-6A used in TACAMO system. The flight altitude is set to match the length of antenna. Additionally, there are trustable results about the aerodynamic characteristics of the antenna and the drogue [11]. As for normal cases, with bigger bank angle, lower flight speed, heavier drogue, the verticality increases. In fact, the bank angle, which should be almost 40°, is constrained by the airplane's maneuverability. Therefore, the design space of the flight speed and drogue mass is greater, which deserves more researches.

The working condition decided initially for a practical case is specified that, the antenna's length is 6km, its diameter is 5mm, the flight altitude is 6km, and the bank angle is 40°. The requirement is that the drogue's mass and the airplane's speed should be designed to gain a verticality of 75%. To compensate for some simplification, such as steady atmosphere, the verticality required in simulation result is set to 80%.

The cruise speed is usually about 550km/h, the minimum speed can be 300km/h. So the range of the flight speed discussed is set from 300km/h to 550km/s. As reference to the AD report [11], the small drogue's mass can work, of which the range is set from 30kg to 90kg.

3.2. Analyzation

First, to analyze the effect from the two parameters on the verticality, the response surface of the flight speed, the drogue's mass and the verticality is built. As shown in the 3D response surface (figure 4), under the practical situation that needs to be discussed, the effect from the airplane's speed and the drogue's mass on the verticality is aligned with the normal cases, in which the lower speed and the heavier drogue lead to a larger verticality. It's easy to get that the flight speed plays a dominant role. Furthermore, when a high flight speed is chosen, such as 150m/s, it's hard to meet the requirement even with a heavier drogue. So, the airplane's airspeed should be decided first.

Then, to distinctly observe the effect shown above, the steady states under different conditions are calculated and visualized. As shown by figure 5, the steady state changes little with different mass of drogue under both low and high flight speeds. And it's obvious that, a lower speed can lead to a

IOP Conf. Series: Materials Science and Engineering **408** (2018) 012016 doi:10.1088/1757-899X/408/1/012016

visibly higher verticality. More steady states under different speeds (with a 40-kg-drogue) are shown by figure 6 and figure 7. The verticality decreases dramatically from 79% to 22% and the drogue's orbiting radius increases from 190m to 1832m when the flight speed changing from 400km/h to 500km/h,

At last, based on the requirement specified above, a flight speed of 400km/h, and a 40-kg-drogue is chosen, under which the verticality can reach 79%. The steady state of this case can be seen in figure 6.

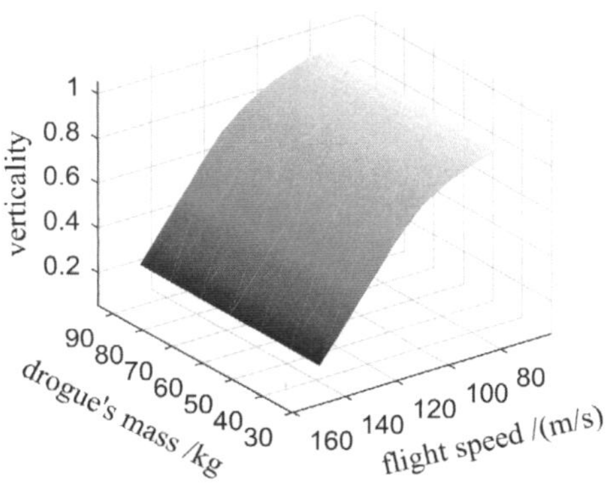

Figure 4. The response surface of verticality, the airplane's speed and the drogue's mass

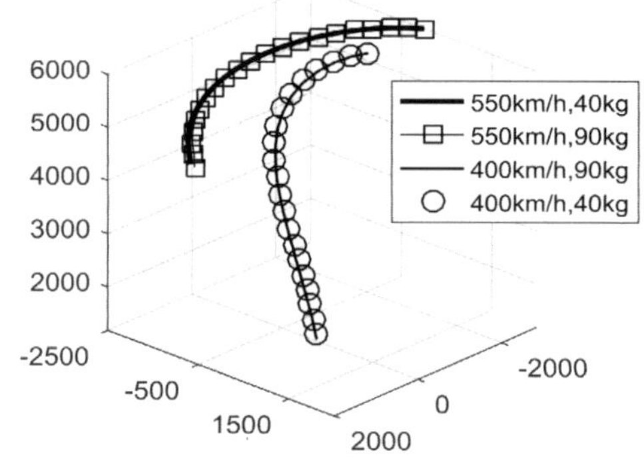

Figure 5. Steady states under different mass of drogue

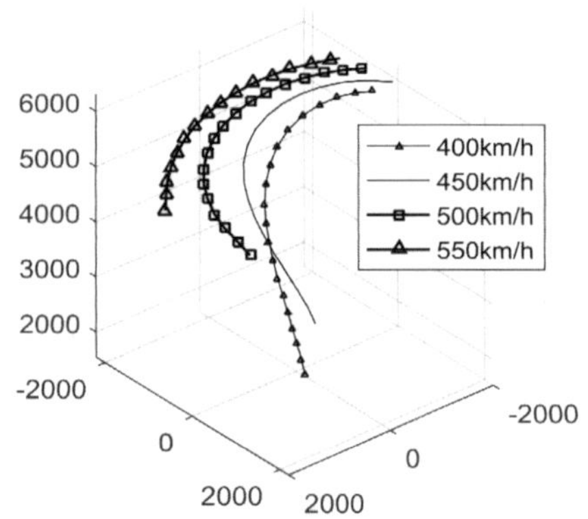

Figure 6. Steady states under several flight speeds with 40-kg-drogue

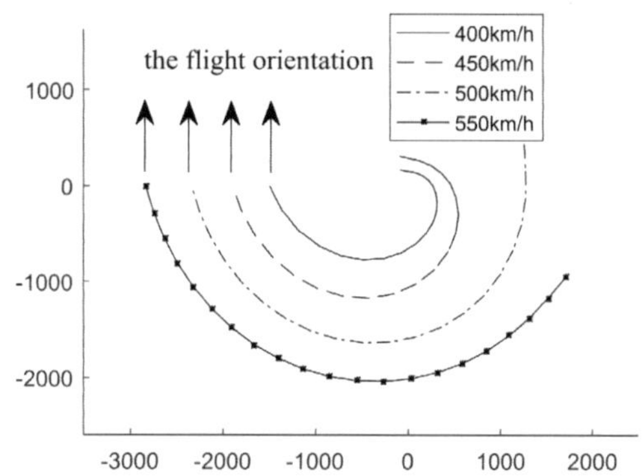

Figure 7. X-Y plot of the steady states under several flight speeds with 40-kg-drogue

3.3. Estimation of the flight speed

Taking into consideration that the antenna's length and the bank angle constraint may differ to a big extent in different cases, the flight speed chosen to reach 75% verticality under the antenna mentioned above with different length and bank angle constraint is calculated and shown by figure 8. Apparently, the flight speed for different cases differs considerably. For bigger bank angle and longer antenna, the verticality required is easier to reach with higher speed. It's obviously that the points meeting requirement can be fitted linearly. A law can be drawn roughly is that, when the antenna's length is 1km longer, the flight speed can be about 7m/s higher with the same bank angle. What is more, the bank angle should be 1 degree bigger when the flight speed increase by 1.5 m/s with the same antenna's length. By this law, when another working condition required, the approximately correct flight speed can be estimated rapidly based on the known case, so that it's more efficient to decide the exact parameters.

The factors which set constant, such as antenna's diameter, drogue's mass, and the aerodynamic characteristic of them, exert much less effect on the verticality [12]. Because of this, although this law is gained by simulating specific antenna with solely different length, it could work for an extensive range of trailing antenna system.

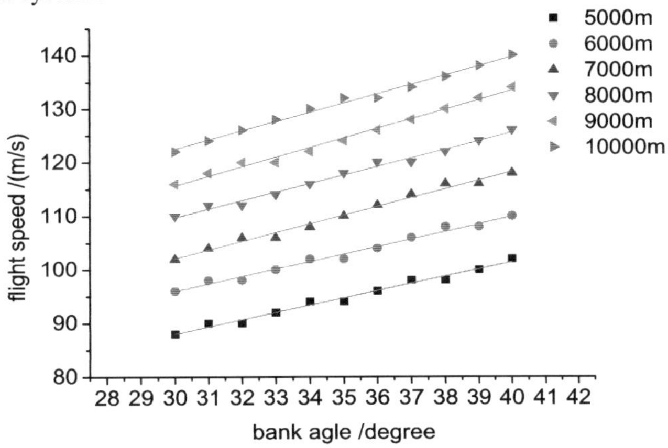

Figure 8. Flight speed chosen for different working conditions

4. Conclusion

(1) The lumped-mass method is used to simulate the trailing antenna system under constant circular motion, of which the accuracy is ensured.

(2) Considering the primary system design procedure, the flight speed and the drogue's mass is distinguished as the primary design variables, of which the former is found to dominate the verticality of antenna whereas the latter influences little. A higher flight speed would make the verticality hard to meet the requirement.

(3) Taking into account that the appropriate flight speed differs for variable bank angle constraint and the antenna's length, a law to estimate flight speed is given, and is suitable for an extensive range of trailing antenna system.

References

[1] S A Crist, J G Eisley 1970 *J.Spacecraft* **7** p 1352
[2] F Zhu, C D Rahn 1998 *J. Sound Vib.* **217** p 435-452
[3] Ma Dongli, Wang Shaoqi, Yang Muqing, Dong Yongpan 2016 *Chin. J. Aeronaut.* **29** pp 1484-1495
[4] S. A. Crist, J. G. Eisley 1969 *J.Spacecraft* **6** p 819-824
[5] Williams P, Lapthorne P, and Trivailo P 2006 *AIAA Modeling and Simulation Technologies Conference and Exhibit*
[6] González F, Prada A D L, Luaces A. and González M 2017 *Ocean Engineering* 131 pp 295-307
[7] Paul Williams 2010 *Journal of Guidance Control & Dynamics* **33** pp 1251-63.
[8] Zheng Xiaohong, Hou Zhiqiang and Li Jixin 2011 *Journal of Naval Aeronautical & Astronautical University* **26** pp 628-632
[9] Zheng Xiaohong, Han Wei, JIA Zhonghu and Guo Weigang 2013 *Flight Dynamics* **31** pp 234-238.
[10] Zheng Xiaohong, Han Wei, JIA Zhonghu 2013 *Acta Aeronautica ET Astronautica Sinica* **31** pp 234-238.
[11] James M C. 1992 *ADA256450*
[12] Zhu Yanjuan and Luo Dan 2018 *Journal of Tongji University (natural science)* **46** pp 81-86

IOP Conf. Series: Materials Science and Engineering **408** (2018) 012017 doi:10.1088/1757-899X/408/1/012017

Design and Calculation Analysis on a New Type of Aircraft Hydraulic Chocks

Meng Kesheng[1,2], Li Fengyu[2], Xia Bin[1], Chen Pandeng[1], Lin Qizhao[2,*] and Chen Chen[3]

[1]AnHui Civil Aviation Airport Group, Department of Maintainance Engineering, AnHui, 230086
[2]Department of Thermal Science and Energy Engineering, University of Science and Technology of China, Hefei, China, AnHui, 230026
[3]ShenZhen Branch of China Southern Airlines, Aircraft Maintenance Plant, Shenzhen, China, 518000

*Corresponding author's E-mail: 18255107780@163.com

Abstract: In order to facilitate the rapid drainage of airport apron in the special weather (such as heavy rain and snow), the apron is usually designed to have a certain slope. However, this slope leads to a difficult problem that chocks are not easy to take out when the airplane leave airport. In order to solve this problem, a hydraulic chocks design scheme is firstly proposed. The aircraft hydraulic chock involves two systems, which are named the slider system and the rod system Taking Boeing 737-800 as an example, this paper firstly makes reasonable assumptions. And then uses a hydraulic chocks to analyze the aircraft's stress situation. Through the calculation and analysis of the scheme, the feasibility of the scheme is verified, it also completely solve problem that chocks are not easy to take out. At the same time, the use of the hydraulic wheel block also avoids injury to personnel and waste of time.

1. Introduction

At present, the design of the domestic airport apron needs to meet a certain slope, which is to strengthen the rapid drainage in the rain and snow weather. This slope can avoid gathering water apron, which is good for the airplane to slide. The design of this slope usually needs to consider the local rain and snow amount and relevant design specifications. Taking Hefei as an example, the design slope of the airport apron is design slope is α=arctan0.05. However, the aircraft needs to release the pressure of brake system after aircraft stop slide. The purpose is to release the huge amount of heat which generate by using the brakes. At the same time, it result the displacement of the aircraft after setting brake. The displacement can lead to collisions between aircraft and boarding bridge and other injuries to personnel. In order to avoid collisions and the injuries, it is required to adopt a measure that enables the aircraft to stop slide after the aircraft arrival the apron stop line. At present, metal or rubber chocks are used as shown in figure 1 and figure 2.

A new type of aircraft hydraulic chock is designed, which aim is to solve the problem when the flight need takeoff. It makes use of the characteristics of the knob-type hydraulic pump which can rapidly relief pressure. Meanwhile, it also utilize lever principle which can reduce the pressure of the hydraulic pump. The force of chock was calculated and analyzed [1-4] in detail when the airplane stop slide.

IOP Conf. Series: Materials Science and Engineering **408** (2018) 012017 doi:10.1088/1757-899X/408/1/012017

Figure 1. metal chock

Figure 2. rubber chock

Take Boeing 737-800 aircraft as an example, through the calculation and analysis, it is found that the design of the hydraulic chock achieves the purpose of quickly and easily removing the chock when the aircraft is ready to take off.

2. Design method of hydraulic chock

The design of the aircraft hydraulic chock involves two systems, which are named the slider system and the rod system. The slider system is composed with slider 1, slider 2, slider 3 and a stop block, which supports aircraft in quiescent condition when the aircraft need to stop movement. At the same time, when the hydraulic pump releases pressure, the relative position of the slider changes, thus relieving the slider pressure. The function of the stop block is to avoid the problem that the wheel block is out of control when the hydraulic system fails. After the pressure of the hydraulic pump is released, a gap is created between the tire and the slider 1, so as to facilitate removal of the chock. The rod system includes a metal rod 1, a metal rod 2, a metal rod 3, a metal rod 4, a knob type hydraulic pump and a plurality of metal steel pipes. The main function of the rod system is to support the slider system and the second function is to reduce the pressure of hydraulic pump through lever principle. At the same time, it also reduce the weight of the hydraulic pump. Structure of airplane hydraulic chock is shown in figure 3 and amplification figure is shown in figure 4.

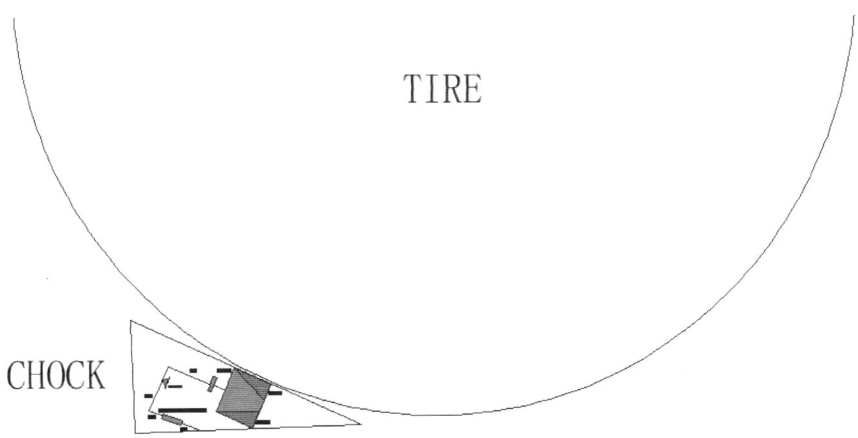

Figure 3. Structure of airplane hydraulic chock

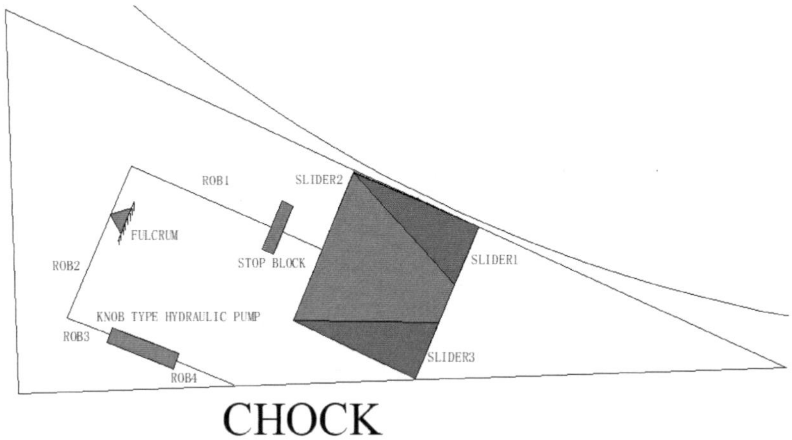

Figure 4. Amplification figure of hydraulic chock

3. Force analysis of hydraulic chock

3.1. Force analysis of slider 2

When the aircraft stop on the airport apron and the brake is released, the force of the tire acting on the slider 1 forces the slider 1 to generate a perpendicular force on slider 2 which name N1. At the same time, the slider 3 is pressed by the slider 2 and it also generate a perpendicular force on slider 2 which call N_2. Besides, the slider 2 also receive own gravity G_2 , the friction forces f_1 and f_2 , the rod 1 thrust F. Force analysis of the slider 2 as shown in figure 5, the slider 2 remains stationary under these forces.

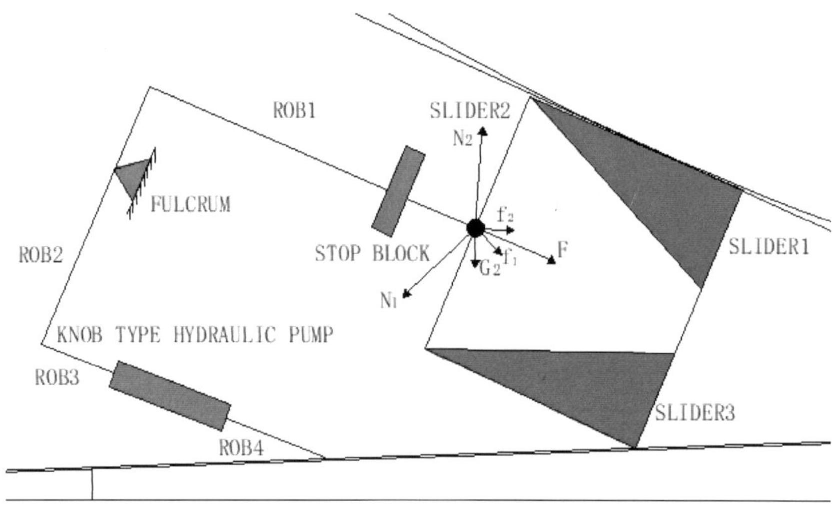

Figure 5. Slider 2 force analysis diagram

3.2. Force analysis of slider 1

After the commander guide the aircraft on stop line, the service personnel start to inform the cockpit to release the brakes. The aircraft will move forward a little distance because the cockpit release the brakes. And then the slider 1 receive a force N_3 from tire. Besides, the slider 1 also receive own gravity G_1, friction forces f_1' and f_3, the force N_3 from slider 2. The slider 1 remains stationary under these forces. Force analysis of the slider 1 as shown in figure 6.

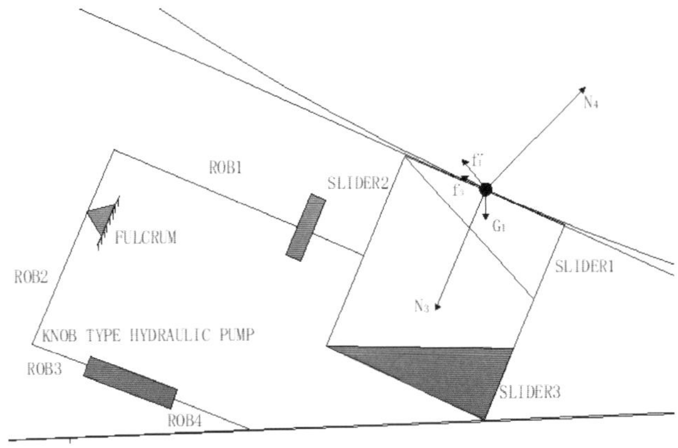

Figure 6. Slider 1 force analysis diagram

4. B737-800 case calculation and analysis

At present, the apron design slope is about 3°. Take Hefei airport for example, design slope is α=arctan0.05 . Slider size and angle are shown in figure 7.

Considering the computational problems, simplifying assumptions are made as follow.

①As the maximum takeoff weight of the B737-800 aircraft is 78245kg[5],the maximum takeoff weight of the B737-800 aircraft can be set at 80000kg;

②As the aircraft tire pressure and the pressure between the slider is far greater than the chock gravity and friction, so the calculation ignores the chock gravity and friction.

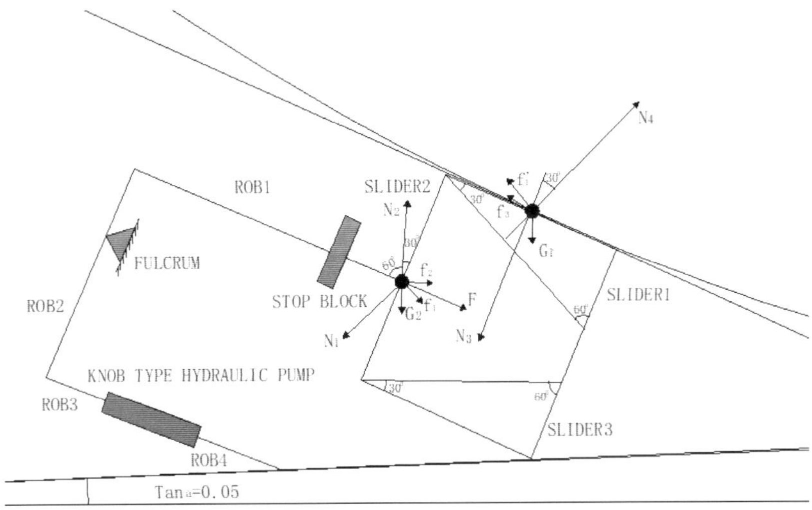

Figure 7. Slider size and angle

4.1. Tire force calculation

When the chock is squeezed by the tire, B737-800 aircraft remain static under the action of gravity G_0 and supportiveness.

The gravity G_0 include the weight of B737-800 aircraft and the supportiveness include ground support force N_8 and the chock resistance N_3'. Force analysis of the tire as shown in figure 8.

The tire force is decomposed along the N_8:

$$G\cos\alpha = N_3^{'}\sin 30^0 \tag{1}$$

Put $G = 8000(\text{kg})$ and α=arctan0.05 into (1)

$N_3^{'} = 4000(\text{kg})$

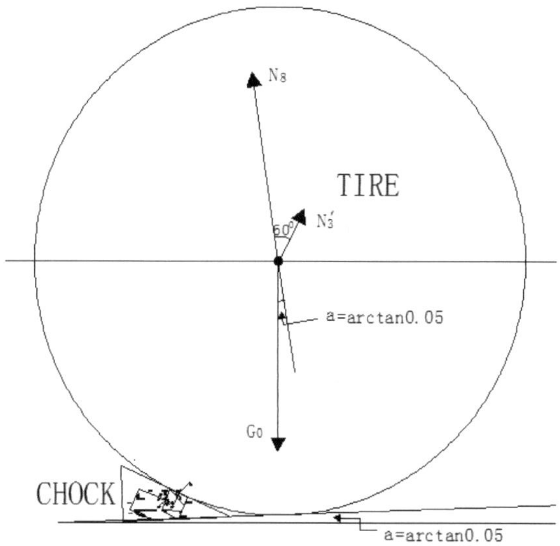

Figure 8. Force analysis and calculation of the tire

4.2. Slider 1 force calculation

The slider 1 force is decomposed along N4. The following formula can be obtained according to figure 7 and simplified assumptions;

$$N_4 = N_3 \cos 30^0 \tag{2}$$

Put $N_3^{'}=N_3=4000(\text{kg})$ into (2);

$N_4 = 2000(\text{kg})$

4.3. Slider 2 force calculation

The slider 2 force is decomposed along F. The following formula can be obtained according to figure 7 and simplified assumptions;

$$F = N_1 \cos 60^0 + N_2 \cos 60^0 \tag{3}$$

Put $N_1=N_2=N_4=2000(\text{kg})$ into (3);

$F = 2000(\text{kg})$

It can be concluded from the calculation that the B737-800 aircraft can remain static on the airport apron as long as there is a 2,000 kg force on the rob 1. Moreover, the power of the knob type hydraulic pump can be further reduced by adjusting the position of the fulcrum.

5. Conclusion

The design starting point is the problem that the chock is very difficult to remove when the airplane is ready to takeoff. Through the analysis of the reasons for the problem, a device was designed, which can quickly remove the chock when the aircraft need depart airport. This paper also takes 737-800 as an example, the function of chock confirmed by the calculation and analysis.

References

IOP Conf. Series: Materials Science and Engineering **408** (2018) 012017 doi:10.1088/1757-899X/408/1/012017

[1] Zhang Guang-guo, Tang Yuan-zhi. The design for the assembly press of side ring for trucks [C].2011 International conference on Intelligent Computation and Industrial Application(ICIA2011).
[2] Cheng Mei-hua, Yuan Suo-xian. Analysis and investigation on influence of nonprofessional factors in Parameter Selection of mechanical design manufacturing[J]. Machinery Design & Manufacture,2012, 11: 152-154.
[3] Wang Guang-lin,Pan Xu-dong,LI Yue-feng. Disposable Mechanical Element Design Theory and Applications[J]. Journal of mechanical engineering, 2014, 50 (1): 152-156.
[4] 2017AnHui civil aviation airport group department of maintainance Engineering work program manual [G].
[5] Boeing aircraft maintenance manual [G].

A FEM Based on the Parametric Variational Principle for the Cable-network Antenna's Form-finding Design

Tao Liu[1,2], Hao Li[2], Guoan Tang[1], Xin Zhou[2] and Liang Zhang[3]

[1]Department of Aeronautics and Astronautics, Fudan University, Shanghai 200433, China
[2] Aerospace System Engineering Shanghai, Shanghai 201109, China;
[3]College of Aerospace Engineering, Chongqing University, Chongqing 400044, China

*Corresponding author's Email: liutaobj@163.com

Abstract: According to the form-finding of the flexible cable-network antenna, a new form-finding analysis approach based on FEM is presented. A parametric variable and Co-rotational formulation are used to model antenna's constitutive nonlinearity and geometrical nonlinearity, and then the FEM governing equation is established. In order to improve the convergence ability, the Lemke algorithm combined with the improved Newton method is employed to solve the governing equation. The method is applied for nonlinear form-finding analysis of cable-network antenna and an ideal configuration is obtained. The research shows that the method can also be applied for studying other similar antennas.

1. Introduction

Cable-network antenna can be used as UHF, microwave and millimeter wave communications satellite antenna. It has many advantages such as high gain, light structural mass, high stowing ability and large aperture application rang. According to the nonlinear characteristics of flexible cable-network antenna, the form-finding of cable-network antenna is a great challenge to the design of deployable cable-network antenna [1]. Many researchers have done a lot of work on the form-finding of cable-network antenna. Their research methods mainly can be divided into two categories: linear form finding method and nonlinear form- finding method.

The force density method firstly given by Linkwitz [2] is the most widely used linear form-finding method. The basic principle of this method is to establish a force balance equation on each node of the cable-network and introduce the force density to transform the nonlinear equilibrium equation into a linear system of equations. On this basis, Morterolle [3] developed the force density method with iterative algorithm for the isotension design. Yang [4] proposed an optimized iterative design method combined with form-finding and force-finding based on the force density method. Lin [5] proposed a form-finding method of combing the improved fish swarm algorithm with the force density method. However, when solving large or irregular tension structures, the force density method is not efficient.

The finite element method is one of the general application nonlinear form- finding methods. Xia [6] used the structural reset method to optimize the pretension of the cable structure and proposed an iterative method for the optimal design of the pre tension. However, the distribution of the pretension in the iterative method depends on the initial value of the pretension, and the distribution of the pretension is difficult to control during the iterative solution, which makes it difficult to guarantee the uniformity of the pretension of the cable. Li [7] proposed a method for the cable pretension optimization and surface adjustment of combing genetic algorithm and finite element method combined. However, for large-scale cable antenna, the method suffers from several problems such as too many variables and convergence difficulty. The geometrical nonlinearity of cable antenna and the constitutive nonlinearity of cable are the difficulties of finite element calculation. Traditionally, The NR iteration technique is employed to solve the nonlinear finite element equations of the structure. In each iteration step, the tensile or compressive modulus is re-selected according to the current stress state of the cable. In the case of more cables, it is easy to bring instability and convergence difficulties. In order to improve the convergence of the finite element method to calculate the cable structure, many scholars have also carried on the corresponding research. Zhang [8] developed a new stable algorithm based on the parametric variational principle (PVP) for the geometric nonlinear analysis of bi modular truss and tenegrity structures. The method can improve the stability of the numerical calculation. Tan [9] applied PVP to the finite element analysis of cable-network antenna. He also used Total-Lagrangian column to describe the geometric nonlinearity of the cable net and NR iteration method to solve the equation. But the TL method is proved to be of poor accuracy and slow convergence speed in large deflection calculation [10]. NR method has local convergence property, if the initial configuration of the cable antenna and the initial stress are not ideal; the convergence of the iterative process is difficult to guarantee [11].

In this paper, according to the constitutive nonlinearity and geometrical nonlinearity of the flexible cable-network antenna, a finite element method based on parametric variational principle and Co-rotational formulation is used for the internal force calculation. Then the method combined with iterative algorithm is used for the modular antenna and the parabolic cylinder antenna' form-finding. The research shows that the method has good form-finding accuracy, and can be widely applied to the design of the cable-network antenna.

2. Description of the parametric variational principle

The antenna's cables can only sustain tension. In the presence of the initial pretension, the constitutive relationship can be expressed as:

$$\sigma = \begin{cases} E^+(\varepsilon + \dfrac{\sigma_0}{E^+}) & \varepsilon \geq -\dfrac{\sigma_0}{E^+} \\ 0 & \varepsilon < -\dfrac{\sigma_0}{E^+} \end{cases} \tag{1}$$

Where E^+ is the tensile elastic modulus; ε and σ are the axial strain and axial stress.

For the Eq. (1) to introduce a parametric variable λ , The formula can be expressed as follows:

$$\sigma = E^+(\varepsilon + \frac{\sigma_0}{E} + \lambda) \tag{2}$$

And λ should satisfy the following condition:

$$\lambda = \begin{cases} 0 & \varepsilon \geq -\dfrac{\sigma_0}{E^+} \\ -(\varepsilon + \dfrac{\sigma_0}{E^+}) & \varepsilon < -\dfrac{\sigma_0}{E^+} \end{cases} \tag{3}$$

By introducing a non-negative slack variable, the cable's tension and compression state control equation can be obtained:

$$\begin{cases} E^+(\varepsilon + \dfrac{\sigma_0}{E^+} + \lambda) - \nu = 0 \\ \lambda \geq 0, \nu \geq 0, \lambda\nu = 0 \end{cases} \tag{4}$$

where $\lambda v = 0$ is a complementary condition, when $\lambda = 0, v = 0$,The cable is in a non-deformed state, when $\lambda > 0, v = 0$,The cable is in a compressed state, when $\lambda = 0, v > 0$,The cable is in a stretched state.

3. A uniform description of the parameter variable and CR formulation

In this paper, the CR formulation is used to describe the geometric nonlinearity of the cable element. The CR formulation has a more concise column and higher computational efficiency in large deflection, large angle and small strain analysis. The tangent stiffness matrix of the cable element combined with the parametric variable and CR formula can be expressed as:

$$\boldsymbol{K}_T = \boldsymbol{K}_L + \alpha \boldsymbol{K}_N \tag{5}$$

$$\alpha = 1 + \frac{\lambda}{\varepsilon + \varepsilon_0} \tag{6}$$

Where $\boldsymbol{K}_L$ and $\boldsymbol{K}_N$ are linear and geometric stiffness matrix, ε is the element strain of current configuration, ε_0 is the initial strain caused by initial pretension.

By assembling the unit stiffness matrix $\boldsymbol{K}_T$, parametric variable λ and slack variable v. The governing equations of the nonlinear finite element of the cable-net structure can be obtained

$$[\Delta P] = [K_T]\{\Delta\delta\} \tag{7}$$

$$\begin{cases} A(\lambda + \varepsilon + \varepsilon_0) - v = 0 \\ v^T \lambda = 0, \lambda \geq 0, v \geq 0 \end{cases} \tag{8}$$

where $\boldsymbol{A}$ is the $diag(E_i^+)$, $\boldsymbol{\lambda}$ is the $diag(\lambda_i)$, $\boldsymbol{\varepsilon}$ is the $diag(\varepsilon_i)$, $\boldsymbol{\varepsilon}_0$ is the $diag(\varepsilon_{0i})$, $\boldsymbol{v}$ is the $diag(v_i)$, i is the corresponding ith cable element. Eq. (8) is a linear complementarity problem and can be solved by Lemke algorithm. The NR method is the most commonly used iterative method for solving Eq. (7). However, the NR method is locally convergent, and it would be difficult to converge when the given initial pretension is not near the equilibrium state of the cable. In order to relax the selection of the initial value, the improved NR method is used in conjunction with the linear search method [11].

4. Form-finding of cable-network antenna

According to the above theory, the nonlinear finite element program was compiled on the Matlab. Then the program is used to the form-finding of the modular cable-network antenna shown in Fig.1. The model has a total of 181 nodes and 408 elements. The tensile modulus of the cable is 20GPa and the compressive modulus is zero. The nodes on the main radial direction of the back cable net are constrained.

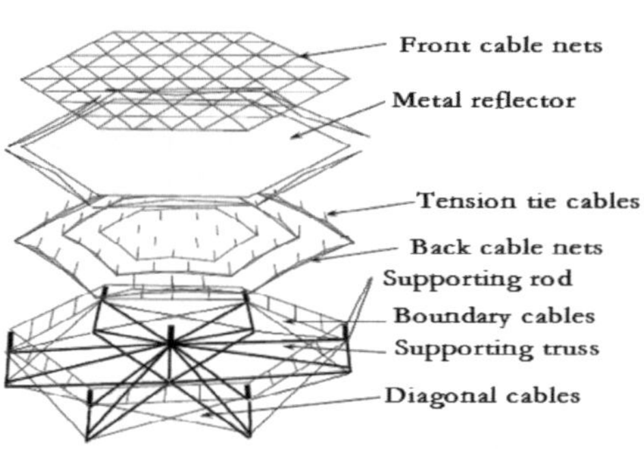

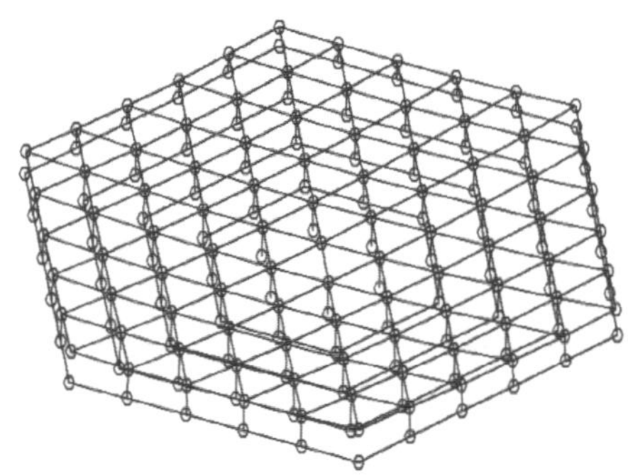

Figure.1 Structural configuration of the modular cable-network antenna

The form-finding process is as follows：

IOP Conf. Series: Materials Science and Engineering **408** (2018) 012018 doi:10.1088/1757-899X/408/1/012018

Step 1: Give the initial pretension of cable network T_0 .

Step 2: Calculate cable network node coordinates (x,y,z) and cable tension T

Step 3: Calculate the error of the front cable network node coordinates between before and after $tol = \sum_{i=1}^{n} norm((x_i, y_i, z_i)_{i+1} - (x_i, y_i, z_i)_k)$, and calculate the rise-span ratio $\rho = \frac{\delta}{l}$.

Step 4: Determine whether T , ρ and tol sastify the constraint condition, if the condition is satisfied, turn to the follow. Otherwise, update the initial tension T_0 and coordinates of cables to and (x, y, z) , then turn to step 1.

Step 5: Output cable node coordinates and tension T, end.

Under the above-mentioned index, the form-finding result shown in Fig. 2(a) is obtained. The results of the finite element method are compared with the traditional force density method. The internal force distribution of the front cables are shown in Fig. 2(b), it can be seen that the internal force of the two methods is consistent. It suggests that the calculation method adopted in this paper is feasible.

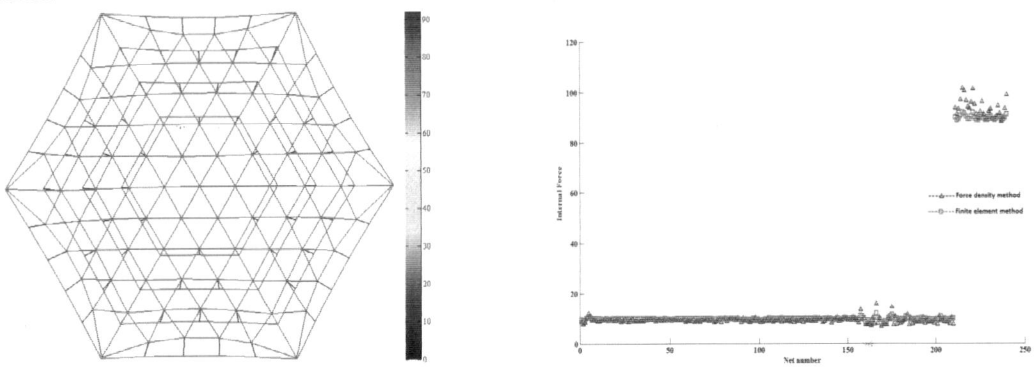

(a) Form-finding results of cable nets (b) Internal force of front cables

Figure.2 Internal force distribution of form-finding results of cable-network antenna

In order to prove the above method is also applicable to a more complex antenna. The form-finding design of the parabolic cylinder antenna has been carried out. Fig.3 (a) is the initial state of antenna with total of 4802 nodes and 11809 elements and Fig.3 (b) is the form-finding result. The result shows that the method can also be applied for studying other antennas.

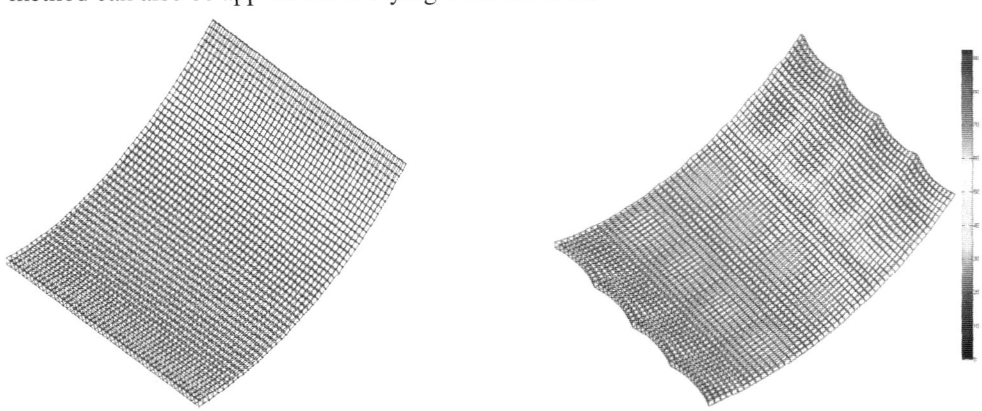

(a)Initial state of cable nets (b)Form-finding result of cable nets

Figure.3 Form-finding design of the parabolic cylinder antenna

5. Conclusion

According to the constitutive nonlinearity and geometric nonlinearity characteristics of flexible cable-network antenna, the nonlinear FEM governing equation combine with the parametric variational principle and CR formulation is established. The Lemke algorithm combined with the improved Newton method is employed to solve the governing equation. The method is applied to the form-finding of the cable-network Antenna. Through analysis, The ideal configuration is obtained. In comparison with the force density method, it is proved that the method is feasible and has better uniformity of internal force distribution.

Further, the method in this paper can also be applied to form-finding under the flexible boundary and pretension optimization of cable-network antenna.

Acknowledgement

This work is supported by the National Natural Science Foundation of China (Grant No. 11202052).

References

[1] Ma XF, Song YP, et al. Mesh reflector antennas: form-finding analysis review. 54th AIAA/ASME/ASCE/AHS/ASC Structures, Structural Dynamics, and Materials Conference, AIAA 2013-1576,2013.
[2] Linkwitz K. Form finding by the "direct approach" and pertinent strategies for the conceptual design of pre-stressed and hanging Structures. International Journal of Space Structures, 1999, 14(2):73-87.
[3] Morterolle S, Maurin B, Quirant J, et al. Numerical form-finding of geotensoid tension truss for mesh reflector. Acta Astronautica, 2012, 76(4):154-163.
[4] Yang GG, Yang DW, Du JL, et al. Method for deployable mesh antenna cable network structures' form-finding design based on force density. Journal of Mechanical Engineering, 2016,52(11):34-41.
[5] Lin M, Li TJ, Ji ZF. Form-finding of tensegrity structures based on IAFSA. Journal of Xidian University, 2014,41(5):112-117.
[6] Xia JS. The conformation theories of tensility structures and the structure analysis of flexible antennae. ZheJiang University, 2005.
[7] Li TJ, Zhou MH, Duan BY. A method of form-finding analysis for flexible cable net structures of deployable antennas. Journal of Astronautics, 2008, 29(3): 794-798.
[8] Zhang L, Gao Q, Zhang HW. An errcient algorithm for mechanical analysis of bimodular truss and tensegrity structures. International Journal of Mechanical Sciences, 2013, 70:57-68.
[9] Tan SJ, Hou J, Wu ZG, et al. The parametric varational principle and non-linear finite element method for analysis of astromesh antenna structures. Chinese Journal of Theoretical and Applied Mechanics, 2014,45(5):770-775.
[10] Crisfield M A, Moita G F. A consistent co-rotational formulation for non-linear, three-dimensional, beam-elements. Computer methods in applied mechanics and engineering, 1980,81:133-150.
[11] Jin HS, Zhu AB, Chen W. Skidding analysis of high-speed cylindrical roller bearings using improved newton-raphson method. Journal of Xi'an Jiaotong University, 2015,49(1):133-138.

Integrated System Conception of Low-altitude Air Defense and Air Traffic Control Based on Cellular Mobile Communication Network

Zhao Guhao[1,2], Wu Yarong[2], Yang Yuan[2], Zhu Jie[2], Yao Dengkai[2], Yang Nan[2] and Hao Chenlu[2]

[1] State Key Laboratory of Air Traffic Management System and Technology
[2] Air Force Engineering University, College of Air Traffic Control & Navigation

Email:zghlupin@163.com

Abstract: Low-altitude airspace as a national valuable strategic resource, because of the lack of effective monitoring method, and can not be fully utilized. Integration of air defense and air traffic control is an important problem in air traffic control establishment and this Paper has discussed a kind of integrated system of low-altitude air defense and air traffic control based on cellular mobile communication network, which can effectively meet demand of early warning detection in low altitude air defense and at the same time can meet demand of air traffic control to air surveillance, communication and navigation. The proposed method can effectively solve the low-altitude airspace detection, monitoring and communication problems, and give full play to the economic value of low-altitude airspace.

1. Introduction

Low-altitude airspace is airspace scope [1] with vertical true height lower than 1000m and is main area of general aviation activities. In recent years, general aviation develops rapidly in China, annual increase of freight gross reaches more than 10%, industry scope increases increasingly, application fields expand increasingly and flight kinds increase increasingly, and flight demand is vigorous increasingly. With continuous and quick development of economy and continuous improvement of people's living level, it is expected that annual increase of general aviation in China will reach more than 15% in future 10 years [1]. Low altitude airspace management reform and utilization of low-altitude airspace to the maximum is important system support to promote quick development of relevant industry fields, such as general aviation industry, aviation manufacture industry and traffic transportation industry, etc., to complete emergent rescue ability to accident and disaster and to promote emergency reaction for explosion protection and counterterrorism.

Although general aviation in China develops rapidly, there is still relatively large distance compared with America. As of the end of 2016, according to statistics, there were 230 thousand planes in American general aviation fleet and annual flight hours were 28 million hours, which are respectively 200 times and 40 times of China [2]. It is caused by seriously insufficient general airports and seriously shortage of aviators, but the most important reason is low airspace utilization rate caused by air-defense pressure. It is related to difference of national defense environment between China and America and America is bordered by ocean to the east and west and by ally Canada in the north and by "backyard" Mexico in the south. America nearly does not have air-defense pressure of the homeland

and can set almost all airspace lower than 1200 feet as uncontrolled airspace. However, safety environment near out country is complex and it is difficult for ground-based radar early-warning detection system to find out target on ground object surface and background because of earth curvature function and uneven ground surface, etc.. Low-altitude airspace cannot be "opened" thoroughly for national defense pressure and only part of low-altitude airspace can be opened conditionally on the premise of complete supervision and management and control to low-altitude airspace aircrafts. First and foremost, we shall solve supervision and management and control problems to low-altitude airspace aircrafts to give full play to economic value of low-altitude airspace.

2. Difficulty and problem of low-altitude airspace defense and air traffic control

2.1 Difficulty of Ground supervision

Low-altitude airspace aircrafts are characterized by small model, slow speed and low height, etc. and is hard to be found out by ground-based radar for limit of earth curvature and terrain. Therefore, for "invisibility, not hearing and loss of communication" phenomenon in these airspaces, it is difficult for aviation management and control departments to master movement status of aircrafts. To ensure homeland air-defense safety and flight safety of aircraft, management method of prohibition use and strict review and approval shall be adopted for these airspaces[3].

2.2 Difficulty of low-altitude communication guarantee

Firstly, limit of coverage, at present, very high frequency and superhigh frequency are mainly adopted for military-civil-aviation ground-air communication and there are a lot of dead zones of communication in low altitude scope of ground-air communication base station for the influence of earth curvature and ground barrier. Secondly, complex electromagnetic environment, compared with mid-high altitude, there are a lot of military and civilian wireless communication systems in low-altitude area, so that ground-air communication will be influenced easily, which will seriously influence communication ability in addition to serious multipath effect. Thirdly, relatively low data chain ability, at present, general aircraft does not have data chain communication ability and there is very large difference of military and civil aviation on data chain system, which is a difficult problem for integrated system of air defense and air traffic control in low altitude airspace.

3. Problems of new air traffic control technology in low-altitude airspace air defense and air traffic control

With development of navigation, communication and network technology, various new technologies are continuously applied to air defense and air traffic control field and relatively favorable effect has been reached. However, there are still some problems of these new technologies in low-altitude airspace air traffic control and air defense application.

3.1 Application difficulty of automatic dependent surveillance-broadcast (ADS-B) in low-altitude airspace

Compared with traditional radar supervision, ADS-B technology is characterized by low cost and high precision and is thought as development direction of future air traffic control field. However, there are still certain application difficulties and defects of ADS-B in integrated system of low-altitude air traffic control.

Firstly, transitive dependency of supervision and communication information depends on density of low-altitude in-air aircrafts. ADS-B mainly transmits supervision and communication information by air-to-air data chain and completes summary and download of data by space data chain in part of area and supervision and communication covering to large area can be completed by several limited ground stations when in-air aircraft density reaches certain degree. However, general aviation scope of China cannot reach networking density required by ADS-B in short term and integrated system of air-defense and air traffic control cannot obtain ADS-B supervision information when civil aviation suspends flying comprehensively in war zone under special condition, such as war. Secondly, characteristics of ADS-B technology do not have supervision ability to non-cooperative target and contribution to low-altitude airspace air defense is limited. Thirdly, dependence on satellite navigation technology is serious and it is easily to be influenced by deception jamming and present research shows that there is a kind of deception jamming bug in satellite navigation system, which reach the purpose of deception and interference by making receiver work out false position information by emitting false navigation code sequence. Low-altitude airspace aircraft adopting ADS-B is easier to be influenced by malignity deception jamming than plane in air route; in this sense, low-altitude air defense and air traffic control system that depends too much on ADS-B is a great threat to no matter flight safety or air-defense of the homeland.

3.2 Application difficulty of photoelectric detection technology in low-altitude airspace
Photoelectric detection technology is system to realize detection and identification to target by utilizing optical signal difference reflected or radiated by target and background. Advantages of the system are no active radiation, well concealing performance and high tracking measurement precision. However, when it is applied to integrated system for low-altitude air defense and air traffic control, there is certain limit: firstly, low information update frequency, when low-altitude target is scanned and detected by completely utilizing photoelectric detection system, image is required to be disposed by complex algorithm, which seriously influences detection rate. Secondly, influenced by self characteristics of target and environment background, when target coating is consistent with background or self infrared radiation of target is relatively weak, it is difficult for photoelectric detection system to find out target. Thirdly, constrained by meteorological condition, visible light wave band detection system is greatly influenced by meteorological condition, exploration distance and detection precision under weather of rain, snow, fog and haze, etc. will decrease urgently.

3.3 Application difficulty of floating platform detection technology in low-altitude airspace
Floating platform detection technology rises detection equipment to air by means of captive balloon, so that detection scope of system to low altitude target will be larger, which has improved early-warning ability. However, in practical application, floating platform detection performance is limited by several factors: firstly, it is seriously influenced by weather clutter, which is mainly cloud and rain clutter and to ensure detection scope, floating platform shall be generally risen to 2000~3000m air [4], which is higher than general cumulonimbus and it is easy to generate clutter signal that is hard to be distinguished from target echo for movement of clouds and self movement of platform. Secondly, influence of platform noumenon and detection equipment by weather is relatively large; for example, balloon-board radar is easily influenced by wind; in ground air inflation, ground mooring, air berthing and releasing recycle phases, it has different degree requirements to wind speed, that is to say that it will lead to unstability of in-air posture and violent swing of sphere, so that it will influence detection precision and even lead to tearing and crashing of platform when wind speed is larger than certain value. Thirdly, existence ability of battlefield is poor and floating platform is characterized by relatively large general volume, poor concealing performance, insufficient maneuvering ability and weak anti-strike ability, so that it is easy to be destroyed by antipersonnel weapons.

4 Low-altitude airspace management and control network based on cellular mobile communication

Development of cellular mobile communication in China started from 1987 and the first simulated cellular mobile communication system was opened in Guangzhou in November of 1987. Later, mobile communication network has been established in Shenzhen, Zhuhai, Shanghai, Beijing and Shenyang, etc. After development of 27 years, China has established the largest cellular mobile communication network in the world. According to data issued by Main Index Completion Condition of Communication Industry in April 2014 (II), until April 2014, the total number of mobile cellphone users in China is 1.251867 billion, of which total number of 3G users is 454.195 million. In the end of 2013, total number of cellular mobile communication network base stations exceed 1.4 million and the number of 4G network base stations that increase in this year will exceed 500 thousand. Therefore, in the end of 2014, total number of cellular mobile communication base stations of China in operation will reach more than 1.9 million, of which the number of 4G network base stations reaches 1 million. At present, more than 98% area with people in China has been covered by cellular mobile communication network.

4.1 Low-altitude covering characteristics of cellular mobile communication network

At present, cellular mobile communication network mainly aims at ground users; therefore, mobile communication terminal on ground and in building shall be taken into prior consideration for design and position placing of its communication base station antenna, etc. To utilize electromagnetism resources reasonably and avoid interference of service area, declination setting is usually adopted for communication base station antenna and common declination modes include mechanical declination and electric declination, as shown in Fig. 1.

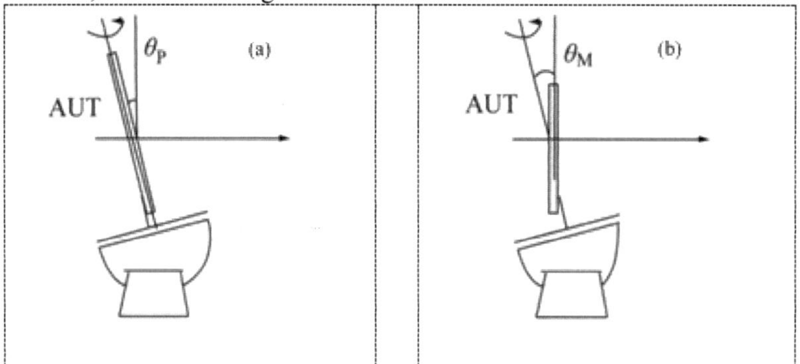

Figure 1. Declination Modes of Cellular Mobile Communication Base Station Antenna (a) Mechanical Declination; (b) Electric Declination

In addition to improving covering of base station to ground by adopting declination setting of antenna, antenna radiation of mobile base station is wide in horizontal direction and narrow in vertical direction and antenna direction figures of typical communication base station is shown as Fig. 2 and Fig. 3.

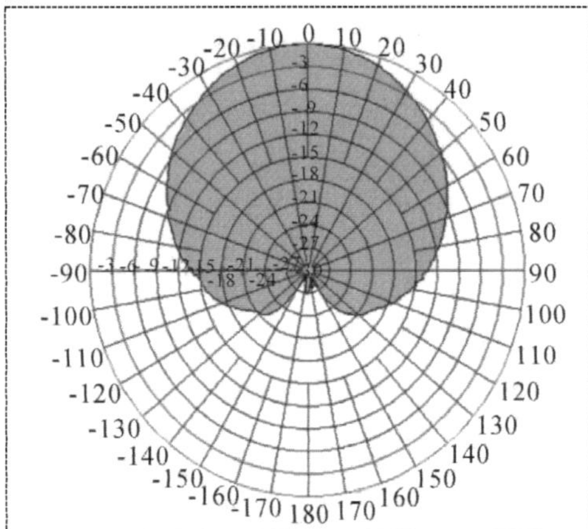

Figure 2. Section of Electromagnetic Wave Beam in Horizontal Direction

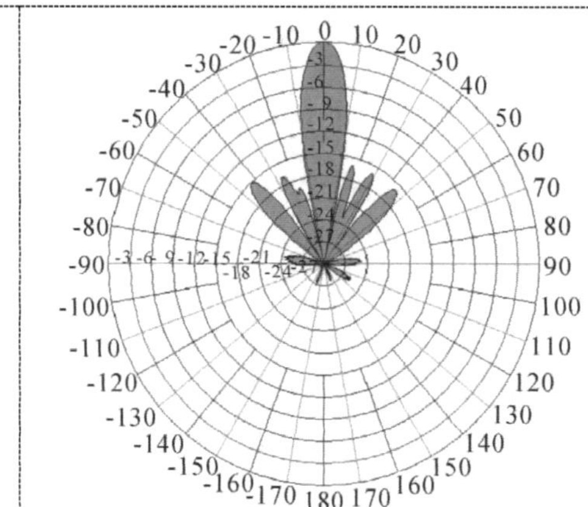

Figure 3. Section of Electromagnetic Wave Beam in Vertical Direction

It is not hard to see that directional pattern of mobile communication base station antenna in horizontal direction presents peach shape and covers relatively wide area to make antenna cover more areas. However, in vertical direction, main lobe of directional pattern of antenna presents spindle shape and there are several minor lobes with relatively strong power on the top and bottom. In the perspective of directionality, covering of mobile communication base station to air area that is lower than 1000m is insufficient. However, seen from a lot of practical cases, mobile communication signal in low-altitude airspace is still relatively strong [5]. It may be caused by reasons in two aspects; firstly, several minor lobes of electromagnetic wave beam in vertical direction can provide signal with certain strength; secondly, dipping electromagnetic wave beam covers low-altitude airspace after reflection of ground and building. In addition, there is no barrier in air, so it has provided favorable condition for spreading of mobile signal. It is obvious that antenna still can provide certain degree signal covering to low-altitude airspace even if no any supplementary establishment is made to cellular mobile communication network. However, signal covering performance of the whole cellular mobile communication network in low-altitude airspace refers to problems of wireless multipath effect and multi-network fusion, etc. and at present, there is no targeted research and verification report.

4.2 Location ability of cellular mobile communication network

Research on wireless positioning technology starts from automatic vehicle location system in 1960s and later the technology is widely applied in public transportation, taxi dispatching and public security tracking, etc. [6]. Wireless location can be divided into wireless location of satellite and wireless location of ground and wireless location based on cellular mobile communication base station belongs to wireless location of ground. This paper only discusses application of cellular mobile communication network to aircrafts location. Location technologies of cellular mobile communication network mainly include:

(1) Location technology based on cell identification code (Cell-ID)

Cell-ID location technology is the first-generation network location technology and has been largely applied to most mobile operators. The technology is the easiest technology with lowest cost in network location technology and location to mobile terminal can be realized in area as long as it is covered by base station. Location precision in the mode is low and when base stations are relatively dense, its location precision is nearly 250m~400m and it can only be provided to application occasion with relatively low location precision when base stations are few and location precision reaches more than 500m.

(2) Location technology of time advance

Time Advance (TA) is a parameter in GSM network and can be used to ensure time slot synchronization of information frame between Mobile Station (MS) and Base Transceiver Station (BTS) and distance between mobile terminal and base station can be obtained according to TA value. TA location technology can be used together with Cell-ID and it can be confirmed that terminal is located in an arc in base station center. TA location precision is not high and is related to dimension of cell [7].

(3) Location technology to angel of arrival

Angel of Arrival (AOA) is to measure direction of electromagnetic wave by utilizing antenna array and to calculate position of mobile terminal of target by utilizing relative angel of several base stations that is provided by two or more than two antenna arrays.

(4) Location technology for time of arrival

Time of Arrival (TOA) is to measure time of arrival of signal transmitted by mobile terminal by several base stations and spreading distance of signal is calculated by accurate transmission moment included in signal. The technology requires strict time synchronization between terminal and base station and precise location and at least three base stations are required for measurement and calculation. Advantage of location technology of time of arrival is relatively high measurement precision, but difficulty and cost are relatively high if strict time synchronization between terminal and base station is required.

(5) Location technology of time difference of arrival

Location Technology of Time Difference of Arrival (TDOA) calculates time difference that signal transmitted by mobile terminal reaches different mobile base stations by calculating difference value based on TOA and works out position of mobile terminal by mathematical method and advantage of the technology is not to require strict time synchronization between terminal and base station and time difference of two base stations confirms a hyperbola and at least two rubber hyperbolas are required to confirm a point of intersection; therefore, at least three base stations are required for location to confirm position of mobile terminal.

In foresaid 4 kinds of mobile communication base station location technology, location precision of TA location technology is unfavorable; TIA location technology requires strict time synchronization and realization difficulty and cost are relatively high; AOA location technology requires two or more than two precise intelligent antenna arrays in communication terminal. Realization difficulty of TDOA location technology is relatively low and relatively high location precision can be realized. It is not difficult to found out according to its location principle that location precision is the highest when location signal of mobile terminal is received directly by base station rather than by reflection. Position of low-altitude aircraft is relatively high and at the same time it can observe several ground base stations directly; therefore, location precision is relatively high and TDOA location technology is more suitable to location to low-altitude airspace aircraft comprehensively. Low-altitude airspace aircraft transmits TDOA location information externally during air flight and after base station receives location information and ground air traffic control network can work out position information of target aircraft at location information transmission moment by time difference that location information reaches base station and can transmit position information and relevant aviation information to target aircraft by data transmission function of cellular mobile communication network.

4.3 Communication capability of cellular mobile communication network

Our current cellular mobile communication network is in coexistence of 2G, 3G and 4G. Te concept and performance of every mobile network are as follows:

(1) 2G network

At present, our main mobile access way still is 2G network. 2G network normally refers to GSM network (CDMA network for China Telecom), works at 900MHz frequency band and supports the voice communication and short message communication, with its data transmission rate at 9.6kB/s.

(2) 3G network

Refer to the cellular mobile communication network using and supporting high speed data transmission. Apart from supporting the voice communication short message communication, in the data transmission, it can support a transmission speed at least 2Mbps, 384kbps and 144kbps respectively in indoor, outdoor and driving environment. What's more, it also supports the simultaneous transmission of sound and data information.

(3) 4G network

Integrated with 3G and WLAN, 4G technology can transmit the data, high-quality audio frequency, video and image, and support the downstream and upstream network bandwidth respectively at 100Mbps~150Mbps and 50Mbps.

The comparison of data communication rates of every generation of communication network under different systems is as shown in Table 1.

IOP Conf. Series: Materials Science and Engineering **408** (2018) 012019 doi:10.1088/1757-899X/408/1/012019

Table 1. Comparison of 2G, 3G and 4G Data Communication Rates

Communication standard	2G	3G			4G		5G
Cellular system	GSM	CDMA2000 (EVDO RA)	TD-SCDMA (HSPA)	WCDMA (HSPA)	TD-LTE	FDD-LTE	CCFD
Downstream rate	9.6 kbps	3.1 Mbps	2.8 Mbps	14.4 Mbps	100 Mbps	150 Mbps	20Gbit/s
Upstream rate	2.7 kbps	1.8 Mbps	384 kbps	5.76 Mbps	50 Mbps	40 Mbps	10Gbit/s

Viewing from the data communication rate, 2G network can achieve the mobile terminal location, but cannot satisfy the need of high-speed aeronautical data transmission; Although 3G has a data chain rate (about 1Mbps) slightly higher than that of ADS-B system, considered about the movement features of aircraft flying at low-altitude airspace, these two basically hold the line; At present, the data rate of 4G network is far higher than all existing aeronautical data chain system; However, the existing 5G communication experiment system shows that, the communication rate of 5G communication network even reaches 10Gbit/s, far higher than any one of existing mobile wireless communication systems, which completely can support the real-time transmission of high-definition video data. Therefore, the cellular mobile communication network completely can meet the data transmission demand of aviation activities at low-altitude airspace.

5. Advantages and disadvantages of air traffic control system based on cellular mobile communication network

According to above data, analysis and description, the advantages and disadvantages of air traffic control system based on cellular mobile communication network are listed as follows:

Advantages:

(1) Matured network: Three commercial networks have been built and cover most territory and all populated areas of our country, which can backup for each other, in a high reliability.

(2) High communication rate: At present, the commercial cellular mobile communication network has entered into 4G era, at a theoretical downstream data communication rate up to 100Mb/s. Following the input to use of new cellular mobile communication network technology, the data rate will further improve.

(3) Support air-air/air-ground communication mode: By utilizing the communication capacity of cellular mobile communication network, it can support the ground-air, air-air voice and digital communication business.

(4) High compatibility: Three data chains of 1090ES, UAT and VDL-4 are mutually incompatible. While, the air traffic control network based on mobile communication base station does not rely on the communication chains among machines, and the airborne terminal of different system can achieve the interconnection through the ground network. Meanwhile, different network data can be uniformly controlled.

(5) Strong extendibility: Support all extensive applications based on cellular mobile communication network and mobile communication terminal.

(6) Easy to be popularized and generalized: Airborne equipment is equivalent to one cell phone, so that the airborne location and communication equipment is relatively cheap, which is easy to be widely popularized and generalized.

(7) Independent of GPS: Both of the aircraft location and communication in the air traffic control system based on cellular mobile communication network are only rely on the ground network system, and the aircraft is at moving state in the air, requiring switching the accessed base station constantly. If a malicious attacker wants to control one aircraft by cheating, he/she must control the whole communication network at first.

Disadvantages:

(1) Unknown air coverage performance: Although the original intention to build the cellular communication network is to meet the ground mobile communication demand, but a large number of aviation flight practices show that, the air lower than 1000m still is covered by the signal. However, at present, there is no specific research and laboratory report on the coverage performance of low-altitude mobile communication signal.

(2) Limited network coverage at no man's area: In the cellular mobile communication network, the network coverage is often subject to the coverage rate of the population. However, the general flight activities are not limited to the populated area, and the flights at no man's area also require the control and information service. Researching the cellular mobile communication base station network topology at no man's area, with low input and high coverage rate, is an important link to achieve complete control coverage of low-altitude airspace within our territory range.

6. Opinion and suggestion to the construction of low-altitude airspace network based on cellular mobile communication network

(1) Renew the existing mobile communication base state and increase the air antenna

The current mobile communication users mainly are from the ground, so the base station antennas all have a downward inclination. Although, some time, the wireless communication signals can be received in the air at a true height about 1000m, these signals mostly come from the reflection of surface feature or the side lobe of antenna, which are at low stability. If the professional air antenna can be set on the cellular mobile communication base station, the signals sent by the antenna will not be reflected by the buildings or other tall objects, and may have no unnecessary multipath effect at the same time, so as to obtain good communication effect. The single mobile base station antennas are less than RMB 200, at a low cost to the upgrading and reconstruction of existing mobile communication system.

(2) Improve the coverage rate of mobile communication base station at no man's area

Comprehensively consider the demand of mobile communication at no man's area and low-altitude airspace aviation control, research the cellular mobile communication network topology at no man's area with reasonable structure, high coverage rate and acceptable preliminary input, coordinate with mobile communication operator to build the mobile communication base station, and achieve the air traffic control network coverage of low-altitude airspace at no man's area, which are the key to finally achieve the low-altitude airspace control network coverage all over the country on the basis of cellular mobile communication network.

(3) Develop the new generation of air traffic control technology closely based on new generation of mobile communication network

Our country has already become the international standard dominant power of 5G mobile communication technology. In the short-term foreseeable future, it is inevitable that 5G communication will be the backbone network of our mobile communication. The air traffic control information network built relying on 5G mobile communication networks will certainly be far superior to the existing air traffic control information network in the communication rate, networking performance, maximum containing capacity and other aspects. Besides, some key technologies in 5G, like space division multiplex technology, can provide the brand-new mature technical support for the cooperative target location of low-altitude airspace.

7. Conclusions

After 30-year construction, our cellular mobile communication network has relatively completely cover the most part of our territory and population, which has 3 great basic operators, with 2G, 3G and 4G networks in simultaneous operation, mobile communication network rich in resources and 5G technology predominant in the world. If the control and information service can be provided to the general aircrafts by taking full advantage of the location function of cellular mobile communication network, it can efficiently make up the insufficiency of existing air traffic control radar and communication system, and perfect the control on low-altitude airspace, to provide the important basic network facility support for the low-altitude airspace opening.

The conception of building the aircraft control network in low-altitude airspace based on cellular mobile communication network is of great realistic meaning to the development of both homeland air defense and general aviation. However, for the moment, this conception is at a distance from the realization and remains many specific problems to be researched, verified and solved. For example, the low-altitude airspace coverage effect problem of mobile communication signal, fusion and utilization problem of network signals of different systems, and high-speed accurate location in flight, etc.

Acknowledgment
Supported by the National Natural Science Foundation of China (Grant No. 61601497)

References

[1] Central Military Commission of State Council. Opinion about Deepening Our low-altitude Airspace Management Reform
[2] China Air Transportation Association. China General Aviation Development Report in 2012 [R]. 2013(3): 29~31.
[3] Li Xingchao, Application and Analysis of New Air Traffic Control Technology in Low Altitude Airspace Management [J]. Electronic Test, 2015(12), 126~127
[4] Zhan Lixiao, Tang Ziyue, Zhu Zhenbo and Fu Ying. Modeling and Simulation of Rain Clutter for Tethered Aerostat Borne Radar [J]. Radar Science and Technology, 8(1), 2010, 15~19.
[5] http://news.sina.cn/?sa=t124d6753233v71&from=mbaidu&stun=20007&vt=4
[6] Gu Jing and Li Mingyuan. Technology of Cellular Phone Position [J]. Modern Electronics Technique, 168, 2004, 68~70
[7] Li Jing. Research and Implementation of Marine Rescue System Based on Mobile Location [G]. Dalian. Dalian Maritine University. 2011

IOP Conf. Series: Materials Science and Engineering **408** (2018) 012020 doi:10.1088/1757-899X/408/1/012020

A Multiple Projectiles Weapon System Scheme and Hardware In-the-loop Simulation for Air-defense Rolling Missiles

Sha De[a], Zhang Jinsheng, Liu Shuo and Hua Chao

Xi'an Research Institute of High Technology, Xi'an, China

[a] Email: nwpu_2012sha@sina.com

Abstract. In order to deal with future systematized, networking and intelligent warfare, the overall technology of a multiple projectiles weapon system for air-defense rolling missiles is studied. The system guidance model, the missile model and the stability control model are introduced. The three point guidance law and dual channel control equation of state in quasi-body coordinate system for rolling missile are emphatically analyzed, and the reliability and stability of the system are verified by constructing a Hardware In-the-Loop Simulation experiment joined with rolling gyro and damping gyro. The simulation result shows that the system has good stability and consistency with different conditions of interference and radar modes. The pitch-yaw channels are decoupled. There are no lost frames and false frames in the simulation cycle.

1 Introducion

Rolling missiles are a typical air-defense missiles. They are characterized by the stable rotation of missiles around their longitudinal axes. They have symmetrical aerodynamic layouts, which makes the pitch channel control model and yaw channel's have similarities. The control model of rolling missiles is simple and facilitate engineering implementation[1]. But with the development of modern missile defense technology and radar jamming technology, it is more difficult for a rolling missile to hit a target independently. In order to adapt to systematized, networked and intelligent future war, multiple projectile of rolling missiles can be used to break through missile defense system of enemy and increase target hit rate.

To effectively deal with the radar jamming of the enemy, the radars of the guidance station can use the photoelectric tracking radar or the phased array radar in this multiple projectiles weapon system scheme. The transmitter can choose the appropriate radar mode according to the actual situation of the battlefield, or choose double-radar modes. The radar modes are combined and the missiles coordinate with each other in flight to improve anti radar jamming capability. At present, this weapon system scheme has begun to carry out engineering practice. This paper mainly analyzes the guidance control system of this multi-strikes rolling missiles, where rolling gyros and damping gyros are added to build a Hardware In-the-Loop simulation(HILS).The simulation results are stable and reliable, and the pitch-yaw channels are decoupled, which verify the correctness and stability of the scheme.

2 Guidance control system

The control principle of rolling missile can be simplified into three parts of Fig. 1: the guidance system, the missile model and the stability control system. The guidance system is used to search, discover, track and guide the missile to hit the target in accordance with the predetermined guidance law. The missile model includes the dynamic model and the kinematics model of the missile which adopt the 'instantaneous balance' hypothesis[2].The missile model studies the ballistic problem. It is the controlled object of the stability control system. The stability control is used to stabilize the flight attitude of the missile, and control the missile to maintain the predetermined flight state under various interference actions and change the missile flight status in time response to the order of the steering mechanism. The following is a brief analysis of the three subsystems of rolling missile.

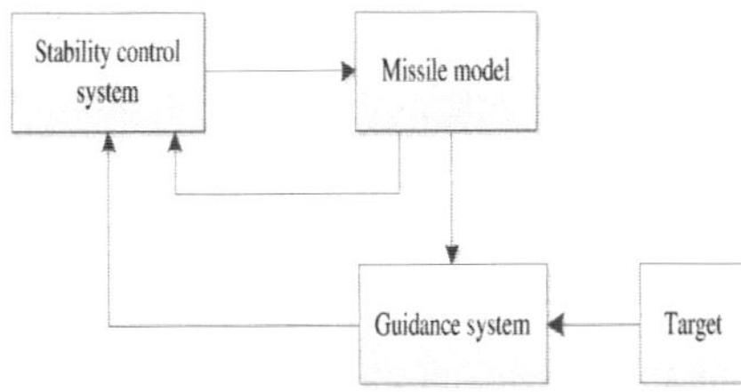

Figure 1. Simplified control principle

2.1 The guidance model

This system adopts the three-point method among classical guidance laws which belongs to the remote control guidance. The law of three-point guidance is to guide the missile centroid always on the connecting line between the radar guidance station and the target position in the process of attacking the target. The three-point method is simple and easy to implement in engineering, however, it takes disadvantages that when approaching the target the trajectory is more curved and the required normal overload increases rapidly, and when missile meets the low-altitude target the launch angle is small and the trajectory is low. According to the definition of the three-point method, the altitude and azimuth angles of the missile during flight are equal to the altitude and azimuth of the target. Therefore, the guidance equation for the three-point method is:

$$\varepsilon_M = \varepsilon_T, \beta_M = \beta_T \tag{1}$$

Where ε_M , ε_T are missile altitude and target altitude, β_M , β_T are missile azimuth and target azimuth.

Fig. 2 is plumb plane coordinate. We can get three point method of relative motion equations:

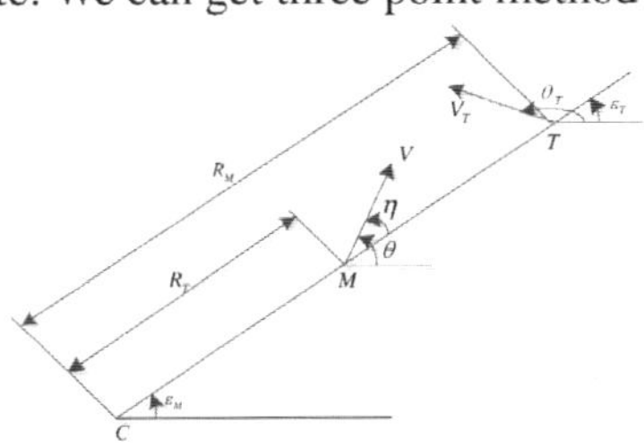

Figure. 2 Plumb plane coordinate

$$\left.\begin{aligned}
&\frac{dR_M}{dt} = V\cos\eta \\
&R_M\frac{d\varepsilon_M}{dt} = -V\sin\eta \\
&\frac{dR_T}{dt} = V_T\cos\eta_T \\
&R_T\frac{d\varepsilon_T}{dt} = -V_T\sin\eta_T \\
&\varepsilon_M = \theta + \eta \\
&\varepsilon_T = \theta_T + \eta_T \\
&\varepsilon_M = \varepsilon_T
\end{aligned}\right\} \tag{2}$$

Where R_M , R_T are missile range and target range, V , V_T are missile velocity and target velocity, θ , θ_T are missile trajectory tilt angle and target trajectory tilt angle, η , η_T are missile lead angle and target lead angle. The line deviation of missile deviating from ideal trajectory is:

$$h_\varepsilon = R_m(\varepsilon_m - \varepsilon_t) = R_m\Delta\varepsilon \tag{3}$$

In order to improve the altitude of the missile and prevent the overshoot of the missile too large from causing the danger of falling, the three-point method can be improved. We can lead into a initial preposition deviation $\Delta\varepsilon_0$.

Here, $\Delta\varepsilon$ can be defined as:

$$\Delta\varepsilon = \Delta\varepsilon_0 e^{-K(1-\frac{R_m}{R_t})} \tag{4}$$

Where K is a constant. As the prefrontal deviation attenuated with time, the ballistic trajectory of the improved method is higher than that of the traditional three point method. Because the prefrontal deviation is positive, but its derivative is negative, and the required normal overload of missile reduces along with the decrease of the deviation. So, the trajectory of the improved method is more straight than the traditional three-point when approaching the target.

2.2 The missile model

For a general missile, the missile movement can be decomposed into longitudinal motion and lateral motion, and the decomposed longitudinal motion and lateral motion are independent of each other. To rolling missiles, due to the spin of the missile body, the symmetry plane of the missile body also spins. The attack angle and sideslip angle defined in body coordinate system will generate periodical alternations, thus the missile posture cannot be directly reflected. In body coordinate system, state variables, such as force and moment, attitude angle, also need to perform rolling projection. The kinematics equation is nonlinear and cannot be directly solved. The spin of the missile results in the cross coupling of the longitudinal and lateral movements of the missile, which mainly include the inertia coupling caused by the gyro effect, the aerodynamic coupling due to the Magnus effect, the control coupling caused by the steering engine delay, and the coupling caused by accelerometer not on the centroid. And the aerodynamic coefficients obtained from the wind tunnel tests in the equation of motion are difficult to describe in body coordinate system. Therefore, in order to describe the missile model, the quasi-body coordinate system and the quasi-velocity coordinate system need to be introduced. The definition of them can be found in [2]. The quasi-body coordinate system and the quasi-velocity coordinate system do not spin with the missile body. And can be linked with body coordinate system and velocity coordinate system by the spin transformation matrix $\boldsymbol{H}$. The expression of $\boldsymbol{H}$ is:

$$\boldsymbol{H} = \begin{vmatrix} \cos w_x t & \sin w_x t \\ -\sin w_x t & \cos w_x t \end{vmatrix} \tag{5}$$

The guidance problem of missile is described in inertial coordinate system. For a missile with an axisymmetric configuration, it can be considered that the body coordinate system is its inertial spindle system[3]. The equation of motion of a rolling missile in a projectile coordinate system can be described as:

$$\overset{g}{\boldsymbol{X}} = \boldsymbol{AX} + \boldsymbol{B}\delta \tag{6}$$

Where

$$\boldsymbol{X} = \begin{vmatrix} w_{y_1} \\ w_{z_1} \\ \beta \\ \alpha \end{vmatrix}; \boldsymbol{A} = \begin{vmatrix} -a_1 & a_6 w_x & -a_2 & a_7 w_x \\ -a_6 w_x & -a_1 & -a_7 wx & -a_2 \\ 1 & 0 & -a_4 & w_x \\ 0 & 1 & -w_x & -a_4 \end{vmatrix} \boldsymbol{B} = \begin{vmatrix} -a_3 & 0 \\ 0 & -a_3 \\ -a_5 & 0 \\ 0 & -a_5 \end{vmatrix} \tag{7}$$

Where w_x is the body angular velocity of rotation, w_{y_1}, w_{z_1} are the projection component of the body angular velocity of rotation in body coordinate system, α, β are the attack angle and sideslip angle defined in body coordinate system, δ is rudder angle, $a_1 \sim a_7$ are dynamic coefficients whose definition are in [4].

Derivation of state equation in quasi-body coordinate system from the state equation in body coordinate system, we can define

$$X^*=\begin{bmatrix} w_{y4} & w_{z4} & \beta^* & \alpha^* \end{bmatrix}^T, H=\begin{bmatrix} H & 0 \\ 0 & H \end{bmatrix} \tag{8}$$

Where X^* is state variable in quasi-body coordinate system.

Thus,

$$X = HX^* \tag{9}$$

Substituting equation (9) into equation (6), we can get

$$\overset{g}{X}{}^* = \begin{bmatrix} H^{-1}AH - H^{-1}H \end{bmatrix} X^* + H^{-1}B\delta \tag{10}$$

It can be proved that[3]

$$H^{-1}AH = A, H^{-1}B=BH \tag{11}$$

Substituting equation (11) into equation (10),we can get

$$\overset{g}{X}{}^* = (A - H^{-1}H) + BH\delta \tag{12}$$

In addition

$$\delta = H\delta^* \tag{13}$$

So

$$\overset{g}{X}{}^* = (A - H^{-1}H) + B\delta^* \tag{14}$$

The equation (14) is the state equation of pitching-yaw channels in quasi-body coordinate system.

2.3The stability control model

From Section 2.2, it can be known that there are cross couplings in the pitch-yaw channels' state equation of rolling missile. In addition, the steering engine segment and the accelerometer pre- segment will also be coupled from body coordinate system spin to quasi-body coordinate system. So, it is necessary for the control system to perform decoupling control. For the rolling channel, it only need to be stable but not controlled.

To achieve decoupling control of pitch-yaw channels, that is essentially diagonalization of the transfer function described by the state equation. The feed-forward compensation method could be used to decouple the system. The decoupling principle is illustrated in Figure 3, the specific design method is referred to [5].

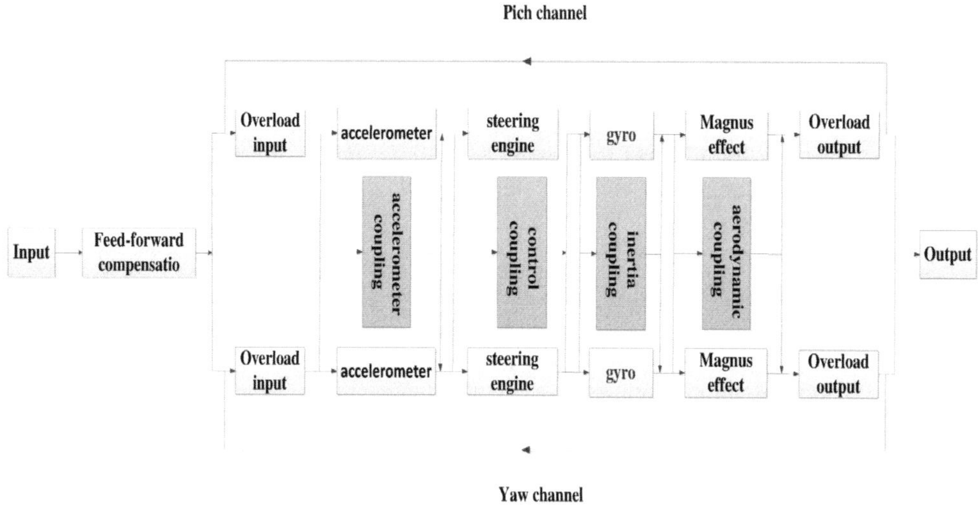

Figure 3. Pitch-yaw channels decoupling of rolling missile

3 HILS and simulation result

To verify the correctness of the scheme of multiple projectile for rolling missiles, the HILS with damping gyros of pitch-yaw channels is to be performed. The HILS is set as follows: Missile 1 uses photoelectric tracking radar mode, added initial interference whose boundary value is (-0.5, 0.5) rad/s in damping gyro. Missile 2 uses phased array radar mode, added separation interference whose boundary value is (-0.5, 0.5) rad/s in damping gyro. Missile 3 uses double radar mode, no interference. Target flight speed is 150m/s. The position of the target when the missiles launch is (13000m, 3000m, 0). Attack mode is counterpunch. The simulation result is as follows:

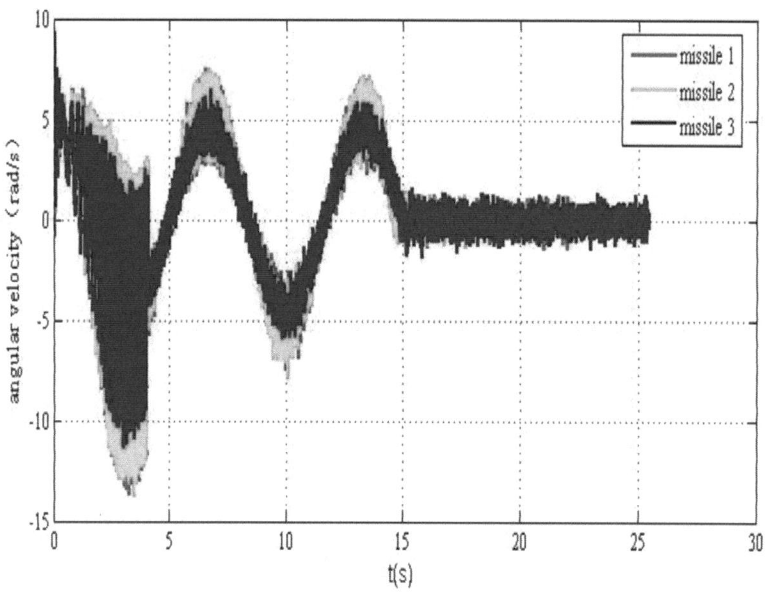

Figure 4. Data of pitch channel.

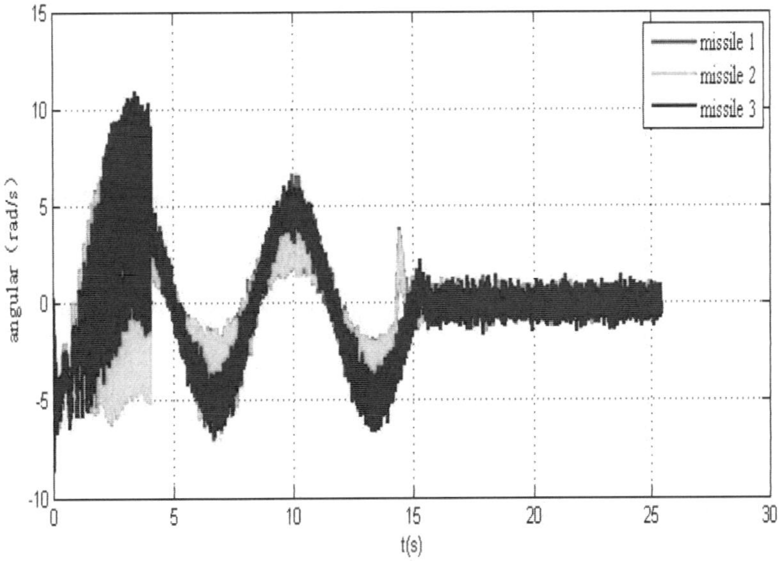

Figure 5. Data of yaw channel.

According to Fig. 4 and Fig. 5, we can see that the rolling channel and the yaw channel are decoupled. The amplitudes of the pitch channel and the yaw channel are basically same and the phases are opposite, which shows the pitch control and the yaw control have symmetry. The HILS result has no lost frames and false frames, which reflects the stability and reliability of the system. But it can be seen that there is a bigger deviation in the yaw channel than the pitch channel. At missile 2 separation time (14.5s), the yaw channel appears jumping faults .

4 Conclusion

A multiple projectiles weapon system scheme for air-defense rolling missiles is put forward in this paper in view of this problem which a missile hard to hit an target independently. The guidance model, the missile model and the stability control model are briefly analyzed. The HILS result shows that rolling channel and the yaw channel are decoupled and no lost frames and false frames in the simulation cycle, which reflects the correctness and stability of the scheme. At present, this scheme has been applied to engineering practice.

References

[1] Yu Benshui 1995 *Overall Design for Air-Defense missile.* (Beijing: Astronautic Publishing House) p 254
[2] Qian Xingfang 2011 *Missile Flight Mechanics.* (Beijing:Beijing Institute of Technology press) p 65-66
[3] Peng Guanyi 1996 *Design of Guidance and Control System for Air-Defense Missile Weapon* (Beijing: Astronautic Publishing House) p 223
[4] Chen Luojing, Liu Li, Yu Jianqiao. Transform and Coupling Analysis of Double-channel Control Rolling Missile Autopilot Loop [J].Transactions of Beijing Institute of Technology 2007(10):847-850.
[5] Chen Luojing, Liu Li, Yu Jianqiao. Decoupling Control of a Double-channel Control Rolling Missile Autopilot [J]. Transactions of Beijing Institute of Technology 2008(01):11-14.

IOP Conf. Series: Materials Science and Engineering **408** (2018) 012021 doi:10.1088/1757-899X/408/1/012021

An Automatic Aerodynamic Shape Optimisation Framework Based on DAKOTA

C C Xia[1,*], Y J Gou[1], S H Li[1], W F Chen[2] and C Shao[3]

[1] Aerospace System Engineering Shanghai, Shanghai, China
[2] School of Aerodynamics and Astronautics, Zhejiang University, Zhejiang, China
[3] Commercial Aircraft Corporation of China, Ltd. Shanghai, China

*xcc578394953@163.com

Abstract. Aerodynamic design is a must and fundamental aspect for better performance of an aircraft. This work describes a framework for aerodynamic shape optimisation (ASO) based on high-fidelity Computational Fluid Dynamics (CFD) analyses and an open source optimisation toolkit DAKOTA. The free form deformation (FFD) technique is adopted for parameterization of geometries, followed by a transfinite interpolation (TFI) method for grid deformation. CFD simulations are performed by full Navier-Stokes equations in conjunction with turbulence model. CFD, FFD and TFI modules are integrated with DAKOTA by a black-box interface, a pre-processing and a post-processing. The transonic RAE2822 airfoil is employed for demonstration of the effectiveness and capability of the developed framework. Several cases with different optimisation algorithms, design variables and turbulence models are performed. Optimum solutions with significant improvement of lift-drag ratio are presented to demonstrate the capability of the framework in this paper.

1. Introduction

The high fuel-efficiency, long-endurance and emission reduction are the persistent pursuit of modern aircraft. Aerodynamic shape optimisation (ASO) plays an important role in improving the aircrafts' performance and achieving these goals. This can be demonstrated by plenty of particular works, including the design of shock-free airfoil for a commercial airliner [1], the aspiration for better aerodynamic performance of a supersonic rocket [2], and the integration design of afterbody/nozzle with hypersonic cruise vehicle [3]. Extremely abundant researches have been focusing on related techniques in recent decades. Among the various techniques, the Computational Fluid Dynamics (CFD)-based aerodynamic optimisation [4-6] has become an area of intense interest in the past few years, thanks to the development of parallel computing, efficient optimisation algorithms and high-fidelity numerical methods.

With the increasing maturity of CFD and optimisation methods, these two techniques should be more closely connected for further improvement of design efficiency of aircraft, and hence for decreasing of complexity of design process. This may be easily achieved by the help of a robust and automatic optimisation framework. Actually, plenty of past ASO studies have been carried out by several powerful and comprehensive optimisation packages such as Isight, ModelCenter and VisualDOC. Except these commercial toolkits, some open source toolkits are also suitable for scientific and engineering applications, such as the Design Analysis for Optimisation and Terascale

Applications (DAKOTA) [7]. While the open source toolkits are usually not user-friendly, they could be more easily developed for specific problems. Besides, they are more accessible.

In this paper, a CFD-based ASO framework is developed based on the open source optimisation toolkit DAKOTA. In the optimisation flowchart of current study, geometry shapes are parameterized by free form deformation (FFD) method and grid deformation during optimisation is achieved by transfinite interpolation (TFI) method. The aerodynamic characteristics of the airfoil are obtained by solving the Reynolds-averaged Navier-Stokes (RANS) equations. Several optimisation cases are performed for the transonic RAE2822 airfoil and satisfactory results are obtained.

2. Development of the framework

2.1. Parametric modeling

The free form deformation method developed by Sederberg & Parry [8] is used for parameterization in this work. Its original goal was to deform solid surfaces in a free-form manner by moving the control point of the control volume surrounding the object. To be precise, for a given hexahedral control volume, which has a local coordinate system, any point X in it can be described as:

$$\boldsymbol{X} = \boldsymbol{X}_0 + s\boldsymbol{S} + t\boldsymbol{T} + u\boldsymbol{U} \tag{1}$$

where X0 is the origin of control volume and local coordinates s, t and u can be easily solved.

The number of control points will be (l+1)(m+1)(n+1) once the control volume is divided into l, m and n sections in S, T and U directions separately.

Evidently, the global coordinates of the control points can be written as:

$$\boldsymbol{P}_{ijk} = \boldsymbol{X}_0 + \frac{i}{l}\boldsymbol{S} + \frac{j}{m}\boldsymbol{T} + \frac{k}{n}\boldsymbol{U} \tag{2}$$

where i=0,1,…,l, j=0,1,…,m, k=0,1,…,n.

If the control volume is constructed by a trivariate Bernstein polynomial, the global coordinate of an arbitrary point after deformation can be expressed as:

$$\boldsymbol{X}_{new} = \sum_{i=0}^{l}\sum_{j=0}^{m}\sum_{k=0}^{n} B_l^i(s)B_m^j(t)B_n^k(u)\,\boldsymbol{P}_{ijk} \tag{3}$$

where the Bernstein polynomial is defined as:

$$B_l^i(s) = C_l^i(1-s)^{l-i}s^i = \frac{l!}{i!(l-i)!}(1-s)^{l-i}s^i \tag{4}$$

The more detailed operations of FFD can be found in the reference. In this paper, the displacements of the specified control points are regarded as design variables.

2.2. Grid deformation

Once the surface of the geometry changes, the grid should be deformed or regenerated for CFD simulation. Regeneration of the grid might be time-consuming for many cases and deformation is currently the first choice. Since the grid has to be updated at each optimisation step, the robustness and efficiency are of great importance to the method. The transfinite interpolation method used throughout this paper is an algebraic method [9]. A large number of works have shown that satisfactory quality can be maintained by this simple method. The method computes the volume grid points from previously computed displacements of points on the block faces, which are supported by the FFD technique in this work. Taking the RAE2822 airfoil for example, the airfoil shapes, FFD control volumes and meshes before and after deformation are shown in figure 1. Design variables are defined as the displacements in the vertical direction of the six control points between the leading and trailing edges. Both uniformity and orthogonality are maintained after deformation, as can be seen from the figures.

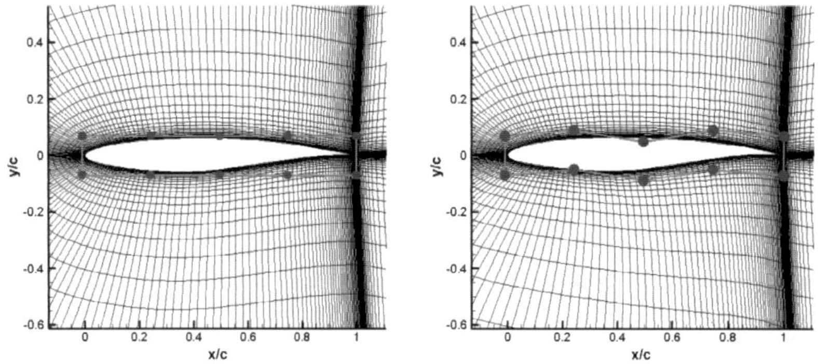

Figure 1. Illustration of the airfoil shape and mesh before (left) and after (right) deformation.

2.3. CFD method

The flow field and aerodynamic coefficients of the airfoil for the current work are obtained by CFD simulation. The three-dimensional full Navier-Stokes equations for steady, compressible flow are selected as the governing equations and can be given as:

$$\frac{\partial \boldsymbol{Q}}{\partial t}+\frac{\partial}{\partial x}(\boldsymbol{E}+\boldsymbol{E}_v)+\frac{\partial}{\partial y}(\boldsymbol{F}+\boldsymbol{F}_v)+\frac{\partial}{\partial z}(\boldsymbol{G}+\boldsymbol{G}_v)=\boldsymbol{S} \quad (5)$$

where $\boldsymbol{Q}$ denotes the conservative variable vector, $\boldsymbol{E}$, $\boldsymbol{F}$, $\boldsymbol{G}$ are inviscid flux vectors, $\boldsymbol{Ev}$, $\boldsymbol{Fv}$, $\boldsymbol{Gv}$ are viscid flux vectors, $\boldsymbol{S}$ denotes the source term vector.

Turbulence model is adopted for calculation of the turbulent viscosity. Since the turbulence model is still an aspect with large uncertainty in CFD simulation. Four commonly used turbulence models (k-ε, k-ω, S-A and SST [10]) are selected for comparison of the effect on optimisation results. Second-order accuracy is obtained by Total Variation Diminishing (TVD) scheme and steady solution is achieved by implicit method. Validation of the CFD method is performed by comparison of simulation result and experimental data. The case considered has a Mach number of 0.734 at an angle of attack of 2.79°, and the Reynolds number is about 6.3 million [11]. Adiabatic viscous wall is set for the simulation, and the free-stream turbulence intensity and temperature are 2% and 255.56K separately. Parallel computing is achieved by Message Passing Interface (MPI) to enhance the efficiency of the calculation. As shown in figure 2, good agreement is achieved between the CFD and experimental data, which demonstrates that the numerical simulation method is accurate for prediction of aerodynamic forces.

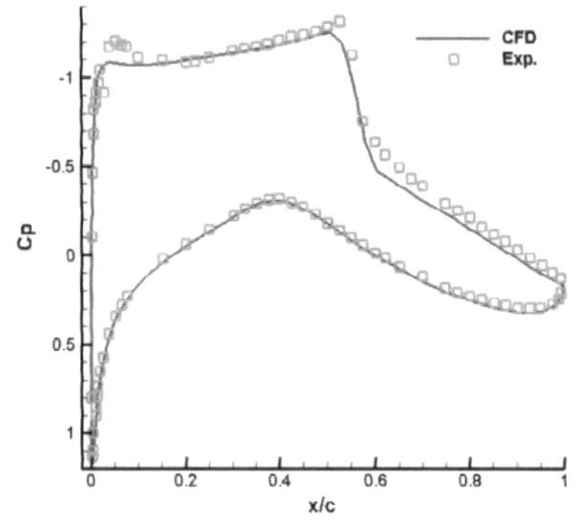

Figure 2. Pressure coefficient of the RAE2822 airfoil (SA model).

2.4. Optimization toolkit

The optimisation framework in this work is driven by the design toolkit DAKOTA, developed by Sandia National Laboratories. The toolkit has been successfully applied to several ASO problems, such as the optimisation of a generic rocket shape[12], the helicopter rotors[13], and the hypersonic aeroshell[14]. It contains various techniques for design and optimisation, including sensitivity analysis, parameter estimation, uncertainty quantification and optimisation (gradient and non-gradient methods). Only the optimisation capability of the toolkit is utilized for the moment.

DAKOTA offers a flexible and robust interface for user's simulation codes and optimisation techniques. For simplicity and generality, connectivity between DAKOTA and optimisation modules in this work are achieved through an easy-to-use black-box interface. Flowchart of the optimisation is shown in figure 3. When an input file containing the relevant information of the problem is supplied, the automatic optimisation process starts and a parameter file with new variables is generated by DAKOTA. A FORTRAN code is then invoked to transform the parameter file to an input file requested by grid deformation. Once the CFD computation is finished, its results are extracted and transformed to the format accepted by DAKOTA, followed by new iterations until the completion of optimisation.

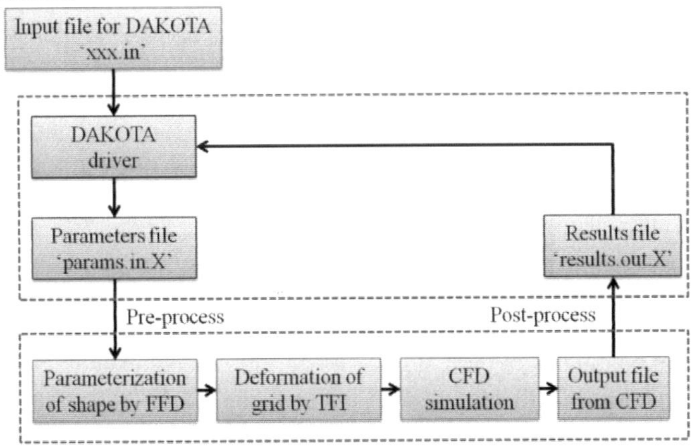

Figure 3. Flowchart of the optimization framework based on DAKOTA.

For comparison, three derivative-free global optimisation algorithms in DAKOTA are adopted, including Evolutionary Algorithms (EA), DIvision of RECTangles (DIRECT) and Efficient Global Optimisation (EGO).

The DIRECT algorithm has proven to be an effective global optimisation method for engineering applications. It is derivative free and can balance local search in promising regions with global search in unexplored regions of the design space. By subdividing the space of feasible design points, iterations can be generated in the neighbourhood of the global optimum.

The EA algorithm is a more commonly used global optimisation method based on the theory of survival of the fittest. Its key idea is that the best points in the population are allowed to survive while others are eliminated in each generation. The basic steps of EA are as follows.

1) Population initialization: a set of design points is randomly created.
2) Fitness evaluation: function evaluations are performed for each individual.
3) Parents selection: pairs of individuals are selected based on fitness value.
4) Crossover operation: apply crossover with a chosen probability for the parents.
5) Mutation operation: apply mutation with a chosen probability for the new individuals.
6) New individuals evaluation: apply function evaluations for the newly generated individuals.
7) New population determination: apply replacement to determine the new population.
8) Convergence check: return to step 2 if the loop is not converged.

Compared to DIRECT and EA, the EGO is a surrogate-based technique used for global search. In EGO, the Gaussian process (GP) model is used in each iteration for approximation of the objective function. The key idea of EGO is to maximize the Expected Improvement Function (EIF) to lead to the optimal solutions. Any new training point, which could be expected to provide better solution than

IOP Conf. Series: Materials Science and Engineering **408** (2018) 012021 doi:10.1088/1757-899X/408/1/012021

the current best solution will be selected and added to the GP model. The EGO has the following basic steps.

1) Determine a set of sample points and evaluate the objective function.
2) Construct the initial Gaussian process model.
3) Determine the new point that maximizes the EIF.
4) Evaluate the objective function at the point determined by step 3.
5) Update the Gaussian process model using the new point.
6) Return to step 3 until the loop is converged.

3. Results and discussion

3.1. Study of different optimization algorithms

Optimization cases are firstly carried out for study of effect of optimisation algorithms. The integrated framework is applied for the airfoil with the free-stream conditions from the previous validation case. For simplicity, a total of six variables are considered (shown in figure 1) and the S-A turbulence model is chosen. The original airfoil has a lift coefficient of 0.7759 and a drag coefficient of 0.01909, and the corresponding lift-drag ratio is 40.64. In all of the optimisation cases, maximization of the lift-drag ratio is adopted as the objective with the constraint that the final drag coefficient should less than its original value.

Summary of optimisation cases with different algorithms are listed in table 1. The maximum function evaluations (*Ne*) are set to 300 and 1000 for the DIRECT and EA algorithms respectively. As shown in the table, approximately 40% of increase of lift-drag ratio (denoted as *K*) has been achieved by all of the optimisation loops. Among the cases, results from the three global algorithms are very close to each other and are supposed to be the neighbours of real global optimum. The evolution-based EA algorithm has the highest objective value. However, the surrogate-based EGO algorithm is quite efficient since it requires much fewer CFD evaluations, and hence is more suitable for this kind of CPU-intensive optimisation problem.

Table 1. Summary of different cases with different optimization algorithms.

Algorithms	*Cl*	*Cd*	*K*	*Ne*
DIRECT	0.7985	0.01358	58.8	303
EA	0.8000	0.01355	59.0	1002
EGO	0.7915	0.01367	57.9	39

The iteration history of DIRECT and EA algorithms are presented in figure 4 (iteration history of the EGO algorithm is not available). It is clearly that the optimisation loops are converged since the maximal objective functions become stable after a number of iterations. Notably, the DIRECT algorithm finds its optima faster than that of the EA algorithm, and its convergence history is more oscillating because of its nature of division in the design space.

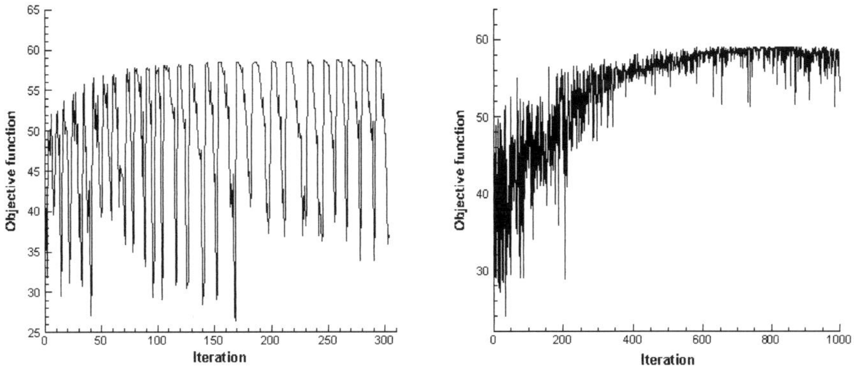

Figure 4. Convergence history of the DIRECT (left) and the EA (right) algorithm.

The comparison between the original and the optimized airfoil shapes is presented in figure 5, along with the distribution of pressure coefficient from different shapes. The optimal shapes and their pressure distribution have minor differences, especially between the results from DIRECT and EA. It can be clearly seen that the shock wave on the upper surface of optimized airfoils has been greatly eliminated. In other words, the intensive adverse pressure gradient has been significantly reduced, which results in the substantial decrease of drag force. Figure 6 shows the contours of pressure coefficient of the original and optimized (EA algorithm) airfoils. Again, the benefit of optimisation can be observed from the flow field of the shock-free airfoil.

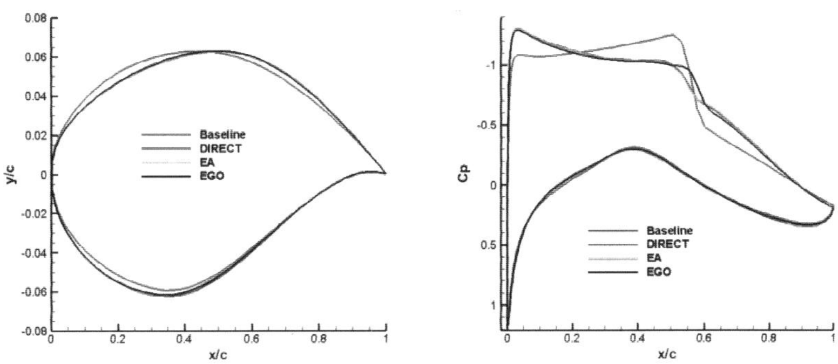

Figure 5. Airfoil shapes (left) and pressure distributions (right) .

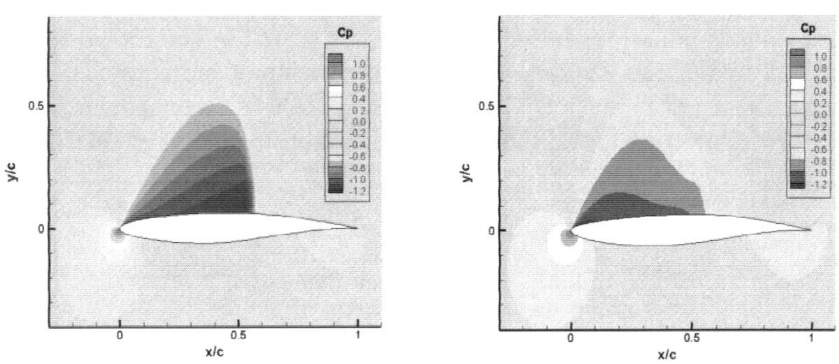

Figure 6. Pressure coefficient contour of the original (left) and optimized (right) airfoil (EA algorithm).

3.2. Study of number of design variables

Optimization cases are performed in this section for study of effect of variable number on the results. By adding control points for the control volume of FFD, numbers of design variable from 4 to 12 are compared. All optimisation loops are executed using the S-A turbulence model and efficient EGO algorithm.

Optimisations are also performed for maximization of the lift-drag ratio with the constraint of drag coefficient. Table 2 summarizes the results from different cases with different design variables. It can be observed that all of the optimized shapes have lower drag coefficient than the baseline, which means that the constraint is strictly satisfied in each of the optimisation loop. Noticeably, the objective value changes little when the optimisation has a number of design variables over 10. With the increase of design variables, the geometry shapes can be elaborately depicted, with the expansion of the design space of the optimisation problem. However, the corresponding computation resource required could be significantly increased, especially for the global optimisation using evolutionary algorithms directly. The optimal value of design variable should be a compromise of fidelity and efficiency.

Table 2. Summary of different cases with different design variables.

Number of design variables	K	Constraint (Cd-Cd_{ini})
4	55.4	-0.005747
6	57.9	-0.005426
8	60.0	-0.005079
10	61.8	-0.004740
12	61.7	-0.004898

3.3. Study of turbulence models

The final results of ASO are directly affected by the accuracy of objective function, which is obtained by CFD simulations in this work. Meanwhile, turbulence model is still one of the tough factors with large uncertainty on fluid modelling. This section explores the effect of turbulence model on the optimisation of airfoil, and all optimisation loops are executed using the efficient EGO algorithm with 10 design variables. Turbulence models considered for the same airfoil shape are the popular one equation S-A model, two-equation SST, k-ω and k-ε models.

The initial and optimized aerodynamic performance of airfoils from different turbulence models are summarized in table 3. We can see that significant improvement has been achieved by the four optimisation cases. The optimal result from SST model has the highest lift-drag ratio, while the k-ω model has the largest improvement. It is not easy to determine which model is most suitable or accurate for the case at present. However, it is clearly that the uncertainty of turbulence model has innegligible effect on the optimisation result, which has also been demonstrated in similar work [15]. Figure 7 shows the optimized airfoil shapes and the corresponding distribution of pressure coefficient from different cases. Shock reduction of different turbulence models can be clearly seen from the figure. The pressure coefficient distribution of the optimized configuration from SST model is the smoothest compared to other shapes.

Table 3. Summary of different cases with different turbulent models.

	Cl_{ini}	Cd_{ini}	K_{ini}	K_{opt}	Improvement of K
S-A	0.7759	0.01909	40.6	61.8	52.2%
SST	0.7565	0.01795	42.1	62.3	48.0%
k-ω	0.8264	0.02134	38.7	60.8	57.1%
k-ε	0.7887	0.01968	40.1	61.3	52.9%

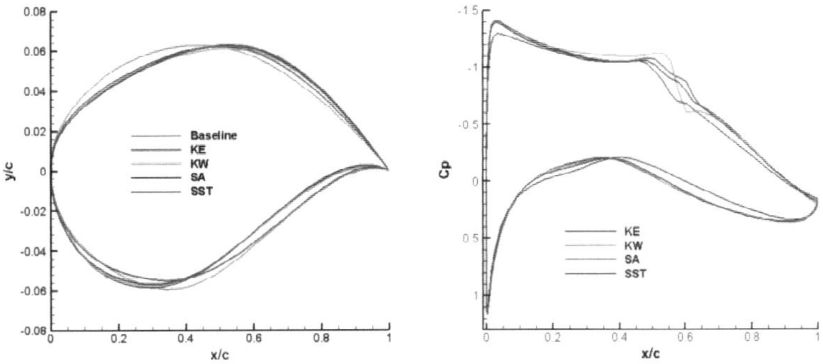

Figure 7. Optimized airfoil shapes (left) and pressure coefficients (right) from different turbulence models.

4. Conclusion

The presented work focuses on the development and demonstration of a DAKOTA-based aerodynamic shape optimisation framework. In the optimisation loop, geometry parameterization is achieved by free form deformation technique and grid deformation is carried out by a fast transfinite interpolation method. Objective functions of interest are obtained by the high-fidelity CFD simulations. Effectiveness of the framework is assessed by optimisation of the typical transonic airfoil RAE2822. Results from optimisation cases show that significant improvement of aerodynamic performance is achieved. The automatic and efficient optimisation loop can facilitate the aerodynamic design problem effectively.

For the optimisation cases considered in this work, we can see that the global method such as evolutionary algorithm is more likely to find the global optimum, while it is more time-consuming. As a compromise, the surrogate-based global method should be more appropriate for practical time-consuming engineering problems. The number of design variables and the choice of turbulence model are also important factors which should be considered carefully in the aerodynamic design process. The framework of this study could be extended to more complex aircraft configurations and even multi-disciplinary problems. Further attention will be paid to efficiency and interactivity of the optimisation loop, along with the practicability of parameterization and grid generation modules.

Acknowledgments

This research is sponsored by the Natural Science Foundation of Shanghai (Grant No. 17ZR1430100) and the State Key Development Program for Basic Research of China (Grant No. 2014CB340201).

References

[1] Chen X and Agarwal R K 2013 *Proc. Inst. Mech. Eng., Part G: J. Aerosp. Eng.* **228** 1654-67
[2] Deepak N R, Ray T and Boyce R R 2008 *J. Spacecraft Rockets* **45** 428-37
[3] Gao T, Cui K, Wang X, Hu S, Yang G and Ren L 2012 *Chin. Sci. Bull* **57** 849-57
[4] Epstein B and Peigin S 2007 *Comput. Fluids* **36** 1399-414
[5] Jahangirian A and Shahrokhi A 2011 *Comput. Fluids* **46** 270-76
[6] Lyu Z and Martins J 2013 *Proc. of the 21st AIAA Computational Fluid Dynamics Conference* San Diego
[7] Adams B M, Ebeida M S, Eldred M S, Jakeman J D, Swiler L P, Stephens J A, Vigil D M, Weldey T M, Bohnhoff W J, Dalbey K R, et al. 2016 *DAKOTA version 6.5 user's manual* SAND 2014-4633
[8] Sederberg T W and Parry S R 1986 *Proc. of the 13th Annual Conference on Computer Graphics and Interactive Techniques* New York
[9] Thompson J F, Soni B K and Weatherill N P 1998 *Handbook of Grid Generation* (CRC Press)
[10] Wilcox D C 2006 *Turbulence modeling for CFD* (DCW Industries)
[11] Cook P H, Mcdonald M A and Firmin F C P 1979 *AGARD report*, AR 138

[12] Kim J E, Rao V N, Koomullil R P, Ross D H, Soni B K and Shih A M 2009 *Math. Comput. Simul* **79** 2373-84
[13] Imiela M 2012 *Aerosp. Sci. Technol.* **23** 2-16
[14] Neville A G, Candler G V 2015 *J. of Spacecraft Rockets* **52** 76-88
[15] Guo C, Kipouros T, Shapiro E, Savvaris A 2012 *Proc. of the 53rd AIAA/ASME/ASCE/AHS/ASC Structures, Structural Dynamics and Materials Conference* Honolulu

Numerical Study on Configuration of Scramjet Combustor

Wenjia Yang [1,a], Juntao Fu [1], Xinpeng Ma [1] and Ruikang Xing [1]

[1]Air Force Engineering University, Xi'an 710051, China

[a]1158285514@qq.com

Abstract. The numerical simulation has been carried out to further investigate the cold flow characteristics for mixing efficiency and flame stability of the composite structure of pylon and cavity in the supersonic combustor and optimize the supersonic combustor flow field. It is found that the pylon can greatly improve the penetration depth of the jet and enhance the blending, and it will not bring the greater total pressure loss. Cavity can promote the diffusion of ethylene and improve the penetration depth of fuel, but also cause greater total pressure loss. The combination of the pylon and cavity can make good use of the advantage of the each single device and the geometry dimension and the combination mode of the pylon and cavity can be further analyzed from to improve the performance of the scramjet engine in next step.

1. Introduction

The supersonic combustor is one of the most important part of the scramjet engine[1-4]. In the supersonic combustor, a series of physicochemical changes such as mixing and combustion should be completed in limited space, very short time and complex flow state, and the performance loss of scramjet engine should also be reduced as much as possible. This is a very challenging task, and it is also the core problem to be solved in the design of scramjet engine. To improve ignition and flame stability, Russian Federal Central Space Agency took the lead in the successfully application of cavity flame stabilizer in scramjet combustor. Since then, the device has attracted the attention of researchers [5].Cavity has excellent flame-stabilizing ability, but because of its direct injection on the combustor wall, the fuel and its combustion products mostly stay in the cavity and near the combustion chamber wall [6-8]. This is not only unfavorable to the mixing of fuel and air, but can not make full use of the combustor space, but also increases the heat resistance requirements of the combustor material.

The reasonable structure design before the cavity can enhance the flame stability while enhancing the fuel flow mixing and improving the comprehensive performance of the combustor[9-10].To this end, domestic and foreign scholars proposed hybrid enhancement devices such as slopes, strut, and tabs[11-12]. Recently, pylon, a more promising structure, has been proposed on the basis of the strut[13-17].The test and simulation results show that all the hybrid augmentation devices have their own advantages in enhancing mixing, improving penetration depth, reducing total pressure loss and stabilizing flame and so on, but also have their own deficiencies. A more ideal combustion chamber fuel injection scheme may be obtained by a reasonable combination of different devices.

In this paper, the numerical simulation method has been carried out to analyze the influence of the pylon and cavity structure on the flow field of the combustor and its mechanism. The qualitative and quantitative research is conducted from the aspects of fuel distribution, mixing efficiency, and total pressure loss to analyze the advantage of the combination of pylon and cavity, which will provide an important reference for the design improvement of the scramjet engine [18].

IOP Conf. Series: Materials Science and Engineering **408** (2018) 012022 doi:10.1088/1757-899X/408/1/012022

2.Calculation setting and grid partition

To deeply study the influence of the combination of pylon and cavity on the flow field of the scramjet combustor, three cases are given in reference of the literature [13] and the literature [15]. See Table 1 for details. The "√" indicates that the hybrid enhancement device is applied in the case.Figure.1 shows the basic conFigureuration of the combustor where cavity and pylon are combined and the combined distance is L_C=6d, where d=1.4mm. A rectangular cross section free channel conFigureuration is adopted in the supersonic combustor to eliminate the interference of other factors. It can be seen from the Figure that the height of the combustion chamber is 25.4mm, the width is 24mm, the total length is 180mm, the pylon heigh is 5.6mm, and it is 35mm from the inlet of the combustor. The depth of cavity is D=14mm, the length is L=42mm, and the back edge angle is 45 degrees. The injector uses the square with side length a=1.24mm to facilitate the modeling and calculation (the area is equivalent to a circular hole with a diameter of d = 1.4), and the distance between the pylon and injector is 2d.

Table 1. Calculation setting

case	case1	case2	case3
cavity	√		√
pylon		√	√

The Mach number in the case is Ma=2, total pressure P_t=850KPa, static pressure p=108KPa, temperature T_t=300K, the mass fraction of the composition of the flow is α_{O_2}=23.2%, α_{N_2}=76.8%. The total pressure of ethylene injection is P_t=20000KPa, static pressure p=1400KPa, temperature T_t=1200K. The Fluent software was used to simulate the cases, SST $\kappa-\omega$ turbulence model was adopted in the numerical simulation, and the non-slip walls were used. The ICEM is used to structurally mesh all the computational domains, and mesh refinement is performed near the pylon, injector and cavity wall. The first grid node is 1×10-6m from the wall, and the number of grids in each model is about 3 million. Figure .2 shows the grid of the case3 scramjet combustor.

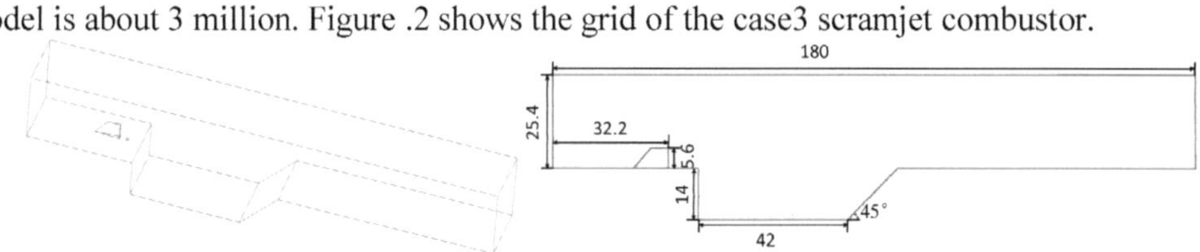

Figure 1. Schematic diagram of combustor conFigureuration (Unit: mm)

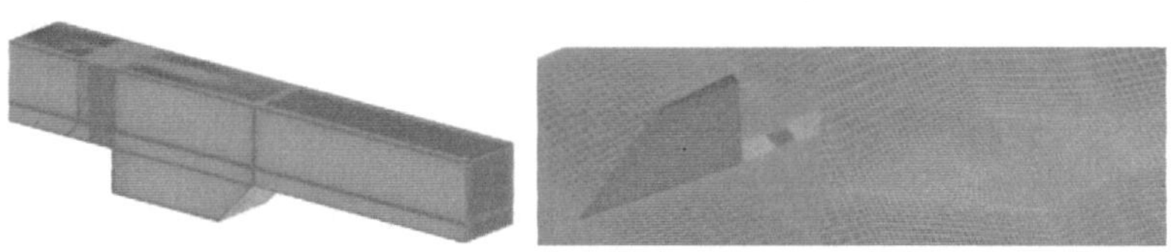

Figure 2. Schematic diagram of the combustor grid

3.Result analysis

3.1.Effect of pylon on performance of scramjet combustor

Figure.3 shows the pressure contours of the symmetrical section of the combustor. The former is the case1 pressure contour, which produces strong shock waves near the injector, while the latter is case3 pressure contours, and the shock wave is mainly generated on the upper edge of the pylon. With the flow movement, both shock waves are refracted through the wall and form shock waves strings in the combustor. Because the shock wave can cause the decrease of the flow velocity, it will lead to the loss of the total pressure. It can be found that the structure of pylon changes the position of the shock wave, and the shock intensity of case3 is slightly lower than that of the former. The shock waves distribution at the rear end of the combustor also hardly affected, and the shock intensity of case3 was slightly larger than that of case1. It is worth noting that case 3 can produce a larger low pressure zone behind the pylon, which will bring more obvious advantages to improving jet injection height and accelerating the mixing between fuel and incoming flow.

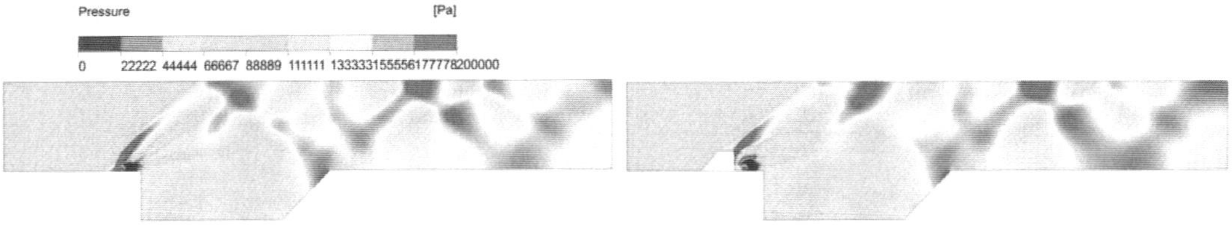

Figure 3. Pressure contours

Figure.4 is a plot of the equivalence ratio of the ethylene composition $0.4 \le \Phi \le 5.5$ and the $\Phi = 0.2$ isoline on the section X = 25d, 30d, 35d, 40d, 45d, 50d, 60d, and 70d for cases 1 and 3, respectively. The size of the contours represents the area of the combustible mixture. Outside the contour is the equivalence ratio line, and the equivalence ratio line contains the area that characterizes the area of the flow field mixing area. As can be seen in the Figure, the shape of the latter mixing zone is even narrower and longer in the same slice. The former still has ethylene with an equivalence greater than 0.2 at the X=70d section, and the latter at the X=45d section, the ethylene isoline no longer intersects the lower wall. It can be concluded that the ethylene injection height is significantly higher than the former, and the pylon effectively weakens the lateral diffusion of the fuel. At the X=35d section, the latter does not have an ethylene component with an equivalence ratio greater than 5.5, but the former also has a large area of high-concentration ethylene components, suggesting that the fin structure accelerates ethylene diffusion. In summary, the pylon can significantly lift the height of the ethylene component in the combustor, increase the penetration depth of the jet, increase the diffusion rate of ethylene, promote the blending of ethylene and incoming streams, and enhance the blending effect. It can not only accelerate the mixing of the fuel with the flow, improve the degree of mixing, thus effectively reducing the length of the combustion chamber, at the same time, also can make the mixture away from the combustion chamber wall, reduce the heat of combustion chamber wall, reducing the material heat resistance requirements.

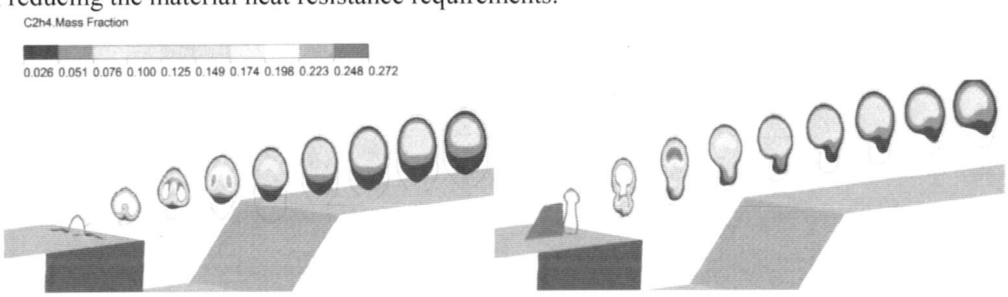

Figure 4. Ethylene mass fraction contour plot in $0.4 \le \Phi \le 5.5$ iso-clip surface and $\Phi = 0.2$ line

For further quantitative analysis of the difference in flow field characteristics of pylon combustor and no-pylon combustor, Figure.5 shows the flow combustible mixing area, plume mass center height, maximum ethylene concentration decay curve, and total pressure loss coefficient curve. From Figure 5(a) we can see that the area of the flammable mixing zone gradually increases in each case. In the case3, the combustible mixing area at the front of the combustor has a faster growth rate, while the later growth is relatively gradual. The overall growth rate of the mixed area of the case1 is not large, and the early growth rate is slow, which is lower than case3, indicating that the pylon can promote fuel blending and diffusion; When reaching near the cavity, the growth rate of combustible mixing area increased rapidly, becoming almost the same as in case 3, and maintained a high growth rate in the later period. This may be that the height of fuel in the case1 is low, and there are many high concentrations of ethylene components near the lower wall, as can be seen from the Figure 5 ethylene composition diagram. Therefore, the fuel in the case1 is better to be absorbed by the eddy current in the cavity to increase the diffusion speed.

It can be seen from Figure 5(b) that the mass center height of plume growth of no pylon case are more gently from first to last and the pylon makes jet plume mass center height began to increase rapidly from X=25d. The growth rate became fairly flat around X =50d and the final result was that the latter plume height was much higher than that of the non-pylon structure, which was consistent with the conclusions from previous qualitative analysis. From the Figure 5(c), the ethylene concentration in each case decreases with the flow movement, and the decay rate in the early stage is fast, and the late period is relatively flat. The pylon increases the rate of ethylene diffusion.

From the Figure 5(d),the total pressure loss coefficient curve shows that the two curves basically coincide in the front of the combustor. After X=70d section, the total pressure loss coefficient curve of case3 was higher than that of the case1, but the difference was not significant. Overall, the pylon can not only enhance the penetration depth of the jet, but also enhance the blending of the fuel and the incoming flow. At the same time, it will not bring about a large total pressure loss or distortion of the flow field.

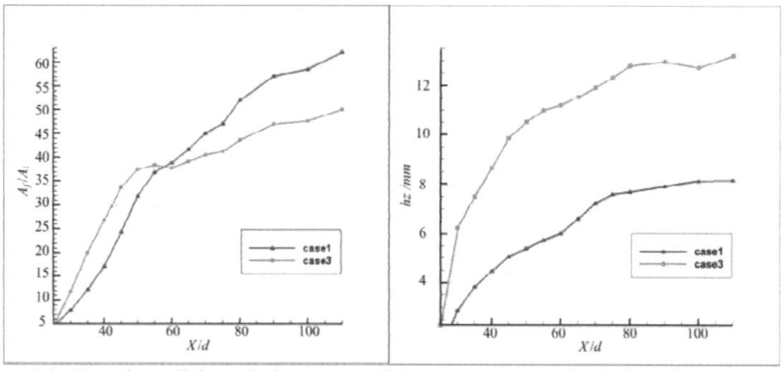

(a) Combustible mixing area (b) Mass center height of plume

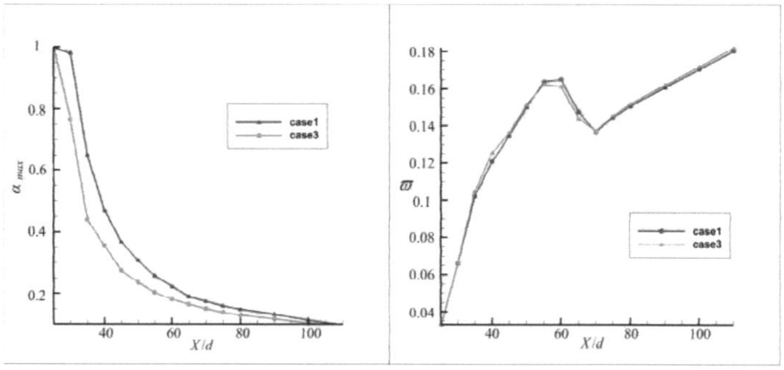

(c) Maximum ethylene concentration decay curve (d) Total pressure loss coefficient

Figure 5. Different numerical examples of flow field characteristic parameters

3.2.Influence of cavity on the performance of the scramjet combustor

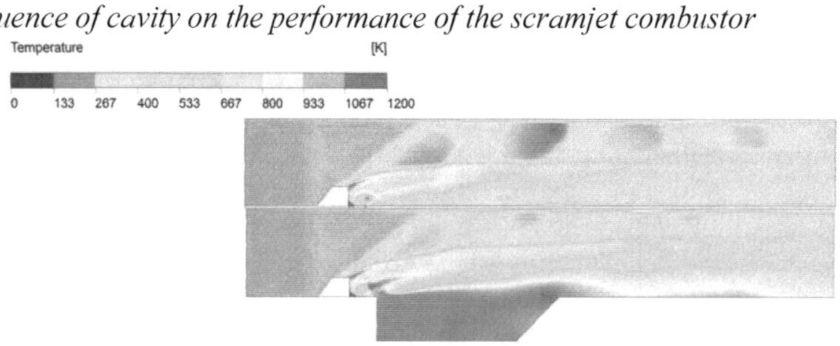

Figure 6. Temperature distribution contours

Figure.6 is a temperature distribution contour of the symmetrical cross-section of the scramjets of case 2 and case 3 under cold flow conditions. It can be seen from the Figure that the temperature inside the chamber of case 3 is significantly higher than that of other areas, which means that the cavity can generate a high-temperature zone by reflow. At the same time, it can be found that the temperature distribution in case 3 is relatively uniform, and there are multiple large-scale low-temperature zones in the case2 combustor. The high temperature area can not only effectively promote the blending of fuel and realize the ignition of the combustion chamber, but also achieve the effect of stabilizing the flame during combustion, which means that the cavity structure can not only promote fuel mixing, but also enhance the flame stabilization ability of the combustion chamber.

Figure.7 Pressure contours

Figure.7 shows the pressure distribution on the symmetrical section of the combustion chamber. It can be observed that shock waves are generated on the upper edge of the pylon and the shock waves are generated by the wall surface refraction. The comparison shows that the range of the low pressure zone behind the pylon is not much different, but in the cross section X=45d, case3 is a high pressure zone with a relatively small range, and case2 is a band-shaped high pressure zone connecting the upper and lower walls, which may have a great influence on the diffusion of fuel. In the case3, a large-scale shock wave is also generated on the rear wall of the cavity and acts together with the shock wave generated at the pylon. After the wall surface refracts, a shock wave string with a stronger intensity is generated in the combustion chamber, which will inevitably result in greater total pressure loss.

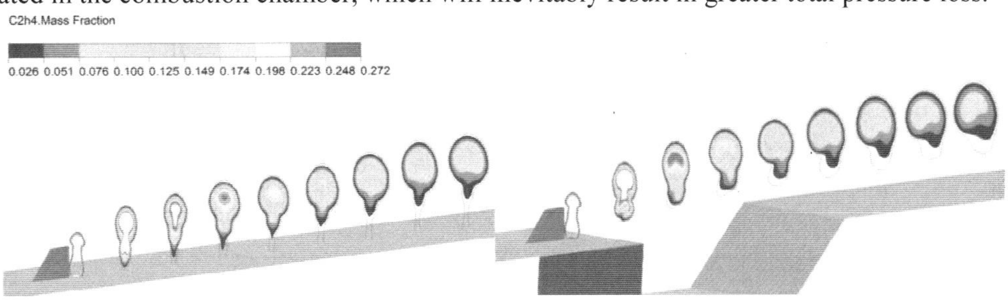

Figure 8. Ethylene mass fraction contour plot in $0.4 \le \Phi \le 5.5$ iso-clip surface and $\Phi = 0.2$ line

To further analyze the influence of the cavity on the performance of the combustion chamber, Figure.8 is a plot of the equivalence ratio of the ethylene composition $0.4 \le \Phi \le 5.5$and the isoline $\Phi = 0.2$ for cases 2 and 3, respectively. The distribution of fuels in all cases is in the form of a "gourd-shaped". In comparison with Figure 5, it may be due to the pylon. With the gradual transformation of flow field into circle, it can make full use of the channel space and increase the uniformity of fuel distribution in the combustion chamber, which is more conducive to ignition and combustion. On the X=35d section, case3 has no more ethylene components whose equivalence ratio is greater than 5.5, while case2 still has large area of high concentration ethylene components, which indicates that the cavity structure promotes the diffusion of ethylene. Ethylene with an equivalence ratio greater than 0.2 is always present on the lower wall of the combustor, and the ethylene isoline of case 3 no longer intersects with the lower wall surface after injection. It can be known that the cavities can significantly raise the height of lean components in the combustion chamber. It is worth noting that the lower part of the case 2 component chart is more tapered, while case 3 is more slick, which may be due to the effect of retraction of the cavity.

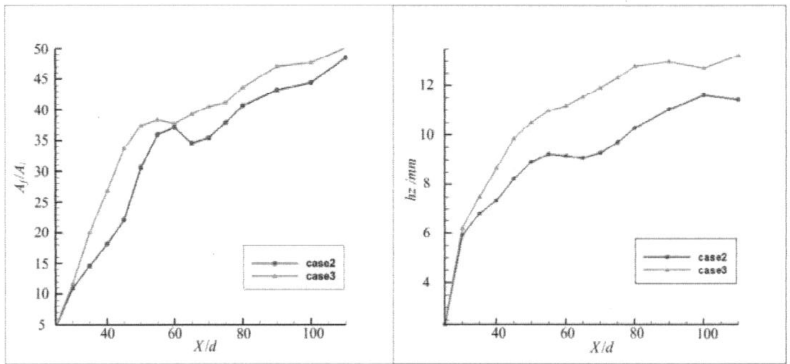

(a) Combustible mixing area (b) Mass center height of plume

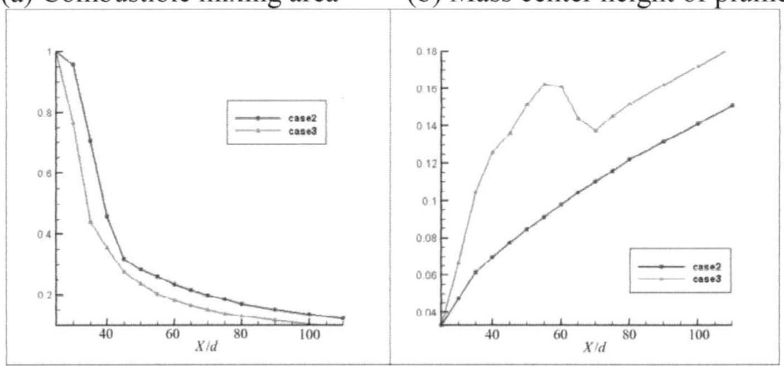

(c) Maximum ethylene concentration decay curve (d) Total pressure loss coefficient

Figure 9. Different numerical examples of flow field characteristic parameters

To quantitatively analyze the influence of cavity on jet penetration and blending, Figure.9 shows the area of the combustible mixing zone in the flow field, the plume center height, the maximum ethylene concentration decay curve, and the total pressure loss coefficient curve. As obtained by Figure 9(a), the area of the combustible mixing area is gradually increased. In the early period, the growth rate did not differ significantly. In the vicinity of cavity, the growth rate of case 3 was significantly higher than that of case 2. In the latter half of the combustion chamber, the growth rate of the combustible mixture area was not much different, indicating that cavity promoted the diffusion of ethylene component. This may be due to the suction effect of the cavity, so the former ethylene blending effect is poor, consistent with the previous qualitative analysis conclusions. As we can see from Figure 9(b), with the flow movement, the height of the mass center of the plume is increased gradually. The initial growth rate of the two case is consistent. Case 3 was significantly higher than case 2 in the vicinity of the cavity, and remained almost constant in the later period. This is similar to the change trend of the combustible mixing area, indicating that the cavity increases the penetration depth of the jet.

It can be seen from Figure 9(c) that the ethylene concentration in each case decreases with the flow. The decay rate is fast in the early stage and relatively gentle in the later stage. The addition of cavity structure increases the rate of ethylene diffusion. Figure 9(d) shows that the trends of case 2 and case 3 in the total pressure loss curve of the rear section of the combustion chamber are almost the same. The total pressure loss near the concave cavity suddenly increases, which is obviously larger than that of the former. It may be caused by the strong shock wave generated by the back wall of the concave cavity, which is consistent with the conclusion obtained from the previous analysis of pressure cloud diagram. In conclusion, cavity has a significant effect on improving the penetration depth and mixing of the combustion chamber. Next step is to further optimize length to depth ratio or the posterior wall inclination of cavity to reduce total pressure loss and improve the performance of scramjet engine.

4.Conclusion

In this paper, the qualitative and quantitative research is conducted from the aspects of fuel distribution, mixing efficiency, and total pressure loss to discuss the advantages of the combination of pylon and cavity. The conclusions are drawn as follows:

1) Pylon can not only enhance penetration depth of the jet, but also enhance the blending of fuel and incoming flow, and it does not bring about a large total pressure loss or distortion of the flow field.

2) Cavity can promote the diffusion of ethylene and increase the penetration depth of fuel, but it causes a large total pressure loss.

3) The combination of pylon and cavity can make good use of the advantage of the each single device and the geometry dimension and the combination mode of the pylon and cavity can be further analyzed from to improve the performance of the scramjet engine in next step.

References

[1] Xu Xu, Chen Bing, Xu Dajun. Principles and techniques of ramjet[M]. Beijing Aerospace University Press, 2014.2:175-200.
[2] Cai Guojun, Xu Dajun. Hypersonic Vehicle Technology [M]. Science Press, 2012.1:77-95.
[3] Yu Gang, Fan Xuejun. Supersonic combustion and hypersonic propagation [J]. Advances in Mechanics, 2013, 43(5): 449-471.
[4] Huang Wei, Yu Hui, Luo Shibin, Wang Zhenguo. Study on key technologies of shock wave induced ramjet engine[J]. Chinese Science, 2010.40(1):64-70.
[5] Roudakov A S, Schikhmann Y, Semenov V, et al. Flight testing of an axisymmetric scramjet-Russian recent advances[R]. ONERA-TAP-95-81, 1995.
[6] Vinagradov V, Kobigsky S A , Petrov M D. Experimental investigation of kerosene fuel combustion in supersonic flow[J]. Journal of Propulsion and Power, 1995, 11(1): 130-134.
[7] Ben-Yakar A, Hanson R K. Cavity flame-holders for ignition and flame stabilization in scramjets : an overview[J]. Journal of Propulsion and Power, 2001, 17(4): 8699-877.
[8] SUN Mingbo, FENF Hui, LIANG Jianhan,et al. Mixing characteristics of gaseous fuel injection upstream of a flame holding cavity in supersonic flow[J]. Journal of Propulsion Technology , 2008, 29(3): 306-311. (in chinese)
[9] ZHANG Yan, LI Chong-xiang, WEI Bao-xi. Review of Scramjet Engine Fuel Injection Schemes[J]. Aeronautic Missiles, 2014,(2): 61-67.
[10] Gao Feng, Wang Hongyu, Zhang Han. A review of the numerical analysis of the flow field in scramjet engine combustion chambers[J].Aircraft Missile,2014.1:80-84.
[11] LIU Shijie, PAN Yu, LIU Weidong. Experimental study on the combustion and flow process in a scramjet with strut injector[J]Journal of Aerospace Power, 2009.24 (1):55-59. (in chinese)
[12] Zhang Han, Wu Da, Wang Xudong. Numerical Analysis on Supersonic Combustor Using Ethylene with Cantilever Ramp Injector[J]. Aeronautical Engineering Progress, 2016.7(3):286-293.
[13] Mitchell R. Parametric Analysis of Pylon-Aided Fuel Injection in Scramjet Engines [J]. Journal of Propulsion and Power, 2013, 23(4): 24501.
[14] JIN Jin-rui, LIU Yu-ying, HONG Yan. Effects of injection fin on flow and mixing characteristics of cavity flame holder[J]. Journal of Aerospace Power, 2011, 26(12): 2716-2721.
[15] Wang Yingyang, Li Xuchang, Wang Hongyu, Wang Xudong. Numerical simulation on pylon pattern in a Supersonic Combustor [J].Journal of Solid Rocket Technology, 2016. 39(1):28-35.
[16] Aguilera C, Pang B, Winkelmann A, et al. Supersonic mixing enhancement and optimization using fin-guided fuel injection[R]. Orlando, Florida: AIAA-2010-1526, 2010.
[17] Aguilera C, Yu K H, Gupta A K. Fin-guided liquid fuel injection into mach 2.1 airflow[R]. San Diego, California: AIAA-2011-5763, 2011
[18] Zhang Wanzhou, Le Jialing, Tian Ye et al. Evaluation methods of the fuel mixing efficiency for scramjet[J].Journal of Aerospace Power,2012.27(9):1958-1966

IOP Conf. Series: Materials Science and Engineering **408** (2018) 012023 doi:10.1088/1757-899X/408/1/012023

Research on Multidisciplinary Integrated Design and Optimization System of Tactical Missile

Xiaojian Sun*, Tao Yang, Jianquan Ge, Chuangye Zhao, Qiangqiang Xu and Ye Tao

College of Aerospace Science and Engineering, National University of Defense Technology, Changsha, China

*sunxiaojian0202@163.com

Abstract. Aiming at the problem of multidisciplinary coupling design of short-range tactical missile, an overall design model with minimum takeoff mass as the optimization objective is established. Based on the integrated design method, Aerodynamics, Motor, Mass Properties, Trajectory and other disciplines are analysed and their models are established. The multidisciplinary design optimization integration framework is constructed, and a missile multidisciplinary integrated design system is established through the program encapsulation, which can realize the coupling design and the overall parameter optimization of each sub-discipline. The system can design, analyze, simulate and optimize functions both system and subsystem, provide methods for optimization of aircraft programs and parameters, and support design schemes management in different technical states. The simulation results show that the system has obvious optimization effect and practical engineering.

1. Introduction

The overall design of the missile involves the sub-disciplines of aerodynamics, motor, trajectory, control and so on. It is a complicated systematic project. It is necessary to look for the optimal design scheme on the basis of considering the coupling of each sub-discipline. For such a complex, nonlinear and coupled system, Multidisciplinary Design Optimization (MDO) has established various models and methods[1-6] for object characteristics and has gradually become the mainstream method to solve this problem [7, 8]. However, there are some problems in the multidisciplinary optimization of aircraft, which is represented by Baker [1, 2]: the multi-disciplinary design has no user interface and the human-computer interaction is less friendly, which causes great inconvenience to users and beginners.

Firstly, this paper analyzes the aerodynamic, motor, mass properties and trajectory of the missile in the initial design stage, extracts the subject-level variables, secondly considers the coupling relationship among the sub disciplines, builds the overall optimization problem model, sets up the multidisciplinary design optimization integration framework, and finally constructs the optimization program. With the help of the UI design idea of OpenRocket[9, 10], a missile design open source software, a human-friendly interface was established and a Multidisciplinary Integrated Design (MID) system for tactical missile was established.

The MID system is targeted at the overall multidisciplinary design optimization for tactical missiles, aiming to realize the distribution, digitalization, integration, automation and visualization of the overall design process of tactical missiles. The desired effect is to improve the design quality of the missile by using the system engineering idea and adopting the appropriate systematic analysis and

IOP Conf. Series: Materials Science and Engineering **408** (2018) 012023 doi:10.1088/1757-899X/408/1/012023

optimization strategy. Mid system aims at supporting the requirements of the tactical missile demonstration stage and plan design stage, and provides the function of the missile overall parameter rapid design, aerodynamic design, motor system design and trajectory design around the core warfare standard. It can design, analyze, simulate and optimize functions both system and subsystem, provide methods for optimization of aircraft programs and parameters, and support design schemes management in different technical states.

2. Short-range Tactical ballistic missile characteristic

Different from traditional ballistic missile, the coupling relationship between disciplines is more complicated and should be designed by MDO method in the overall optimization design of short-range tactical ballistic missile, which is embodied in the following aspects:

a) From the view of range, the coupling relationship between Motor-Structure/Mass-Trajectory must be paid attention to in the overall design of short-range tactical ballistic missile.

b) The trajectory characteristics of short-range tactical ballistic missiles have changed greatly with the traditional ballistic missiles, which are flying throughout the atmosphere. After the motor is finished, they can change the flight trajectory and improve the range of the missile by aerodynamic control.

c) Range is not only affected by the thrust-weight ratio, but also directly affected by the missile lift-drag ratio (L/D) characteristics and ballistic design. For missile free-gliding and orbit maneuvering flight, the larger the L/D, the smaller the loss of speed fo gliding and orbit maneuvering, the farther the range of gliding flight and the maneuver.

In addition, the trajectory control method and the design parameters also directly affect the trajectory angle, height and velocity of the shutdown point, and then affect the maneuvering range and the total range.

3. Integated design

In the integrated design, the coupling between the disciplines is complex and affects each other. Therefore, we need to establish the overall design black box. Only need to change the overall parameters, through analysis and calculation, you can get the integrated design results. This integrated design includes five disciplines: geometry, aerodynamics, motor, mass and trajectory. Figure 1 is a design variable matrix of the integrated design, showing the relationship between the various disciplines. The diagonal of the matrix is the discipline analysis module, the upper-right triangular part represents the parameter transfer between the modules, and the lower-left triangle represents the optimization process. Note that the MID system is far more complex than the diagram process. The subsystem-level module setup will be shown in the following sections.

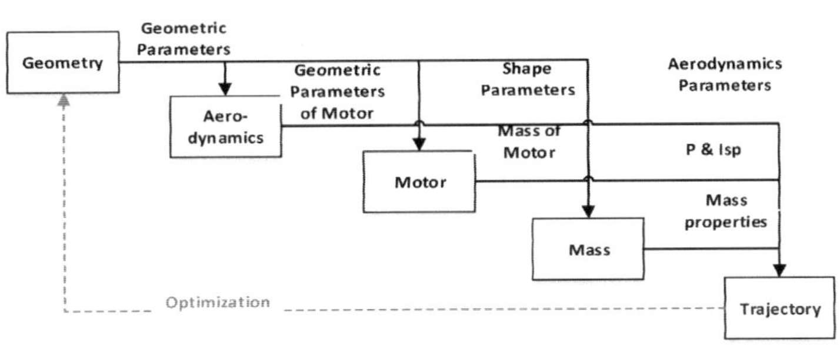

Figure 1. Relationship and optimization process of each discipline.

3.1 Geometric modeling module

The main function of the geometric module is to establish the parametric shape and provide the contour parameters to the downstream disciplines such as aerodynamics, motor and mass. MID provides a visual geometric design interface that enables the autonomous selection of missile components such as nose cones, bodies, interstage segments, and wing flaps. The parameters of each part can be modified freely, the modified model will be displayed in the interface in real time, the model can choose a variety of view modes: Face, Side, 3D sketch, 3D diagram. Mid system provides a rich variety of missile components to add models. The bodies and internal loads can be changed arbitrarily, and the center of mass(CG) and the center of pressure (CP) are calculated in real time. For example, there are 6 choices for the nose curve: Conic, Ogive, Ellipsoid, Power, Haack and Parabolic. Meanwhile, wings can set any number, shape and thickness.

In this paper, parameters which can significantly change the shape of the missile and affect the aerodynamic performance of the missile are selected, such as missile diameter D, missile slenderness ratio λ_B, one cone slenderness ratio λ_{l1}, two cone slenderness ratio λ_{l2}, and columnar length slenderness ratio λ_{lc}. Based on parametric modeling method, a series of shapes shown in Figure 2 can be obtained by changing the above parameters.

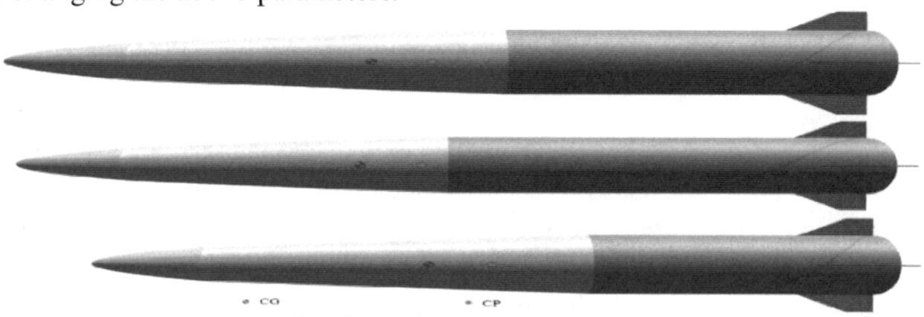

Figure 2. Example of 3D sketch based on parametric shape.

3.2 Aerodynamic modeling module

Missile Datcom 1997 is an engineering aerodynamic estimation software developed by the United States Air Force to collect flight test data of a large number of missiles in the United States, suitable for conventional layout missiles. The aerodynamics module integrates the Missile Datcom into the MID and converts the parameterized shape, the Mach number(Mach), the angle of attack(AoA) and the height(H) into the input of the Missile Datcom, makes an aerodynamics table by calculating the parameters of the lift & drag coefficient(Cl & Cd) and CP. The length of the missile and the change of the different section diameters are described by the missile's outline, the wing part needs the positioning dimension, number, type, plane shape and the mounting position of the wing.

Figure 3 shows the operation of the aerodynamics module in the MID system. Figure 3(a) is set for005.dat input file, the appearance parameters and Mach, AoA and H are set by visual operation to generate the file, the generated for005.dat and the output file for006.dat calculated by Datcom can be displayed on Figure 3(b). The module can also show the calculated aerodynamic table and draw it.

No.	Mach	AoA	H
1	0. 1	0	50, 000
2	1. 41	2. 63	50, 000
3	2. 72	5. 26	50, 000
4	4. 03	7. 89	50, 000
5	5. 34	10. 53	50, 000
6	6. 65	13. 16	50, 000
7	7. 96	15. 79	50, 000
8	9. 27	18. 42	50, 000
9	10. 58	21. 05	50, 000
10	11. 89	23. 68	50, 000
11	13. 21	26. 32	50, 000
12	14. 52	28. 95	50, 000
13	15. 83	31. 58	50, 000
14	17. 14	34. 21	50, 000
15	18. 45	36. 84	50, 000
16	19. 76	39. 47	50, 000
17	21. 07	42. 11	50, 000
18	22. 38	44. 74	50, 000
19	23. 69	47. 37	50, 000
20	25	50	50, 000

(a) Flight parameters

IOP Conf. Series: Materials Science and Engineering **408** (2018) 012023 doi:10.1088/1757-899X/408/1/012023

```
Input | Output
DIM M
DERIV DEG
$FLTCON NALPHA=20.,NMACH=20.,
        MACH=0.1,1.41,2.72,4.03,5.34,6.65,7.96,
        MACH(8)=9.27,10.58,11.89,13.21,14.52,15.83,
        MACH(14)=17.14,18.45,19.76,21.07,22.38,
        MACH(19)=23.69,25.,
        ALPHA=0.,2.63,5.26,7.89,10.53,13.16,15.79,
        ALPHA(8)=18.42,21.05,23.68,26.32,28.95,
        ALPHA(13)=31.58,34.21,36.84,39.47,42.11,
        ALPHA(18)=44.74,47.37,50.,
        ALT=50000.,50000.,50000.,50000.,50000.,
        ALT(6)=50000.,50000.,50000.,50000.,50000.,
        ALT(11)=50000.,50000.,50000.,50000.,
        ALT(15)=50000.,50000.,50000.,50000.,50000.,
        ALT(20)=50000.,
$$REFQ SREF=1.,LREF=1.,$
$AXIBOD NX=17.,
        X=0.0,0.226,0.452,0.678,0.904,1.13,1.56,1.99,2.42,2.85,
        X(11)=3.28,3.47,3.66,3.85,4.04,4.23,7.28,
        R=0.0,0.134,0.19,0.232,0.268,0.3,0.33,0.36,0.39,0.42,
        R(11)=0.45,0.46,0.47,0.48,0.49,0.5,0.5,
        TRUNC=.FALSE.,DEXIT=0.,$
$FINSET1 SSPAN=0.,0.3,CHORD=0.7,0.25,PHIF=0.,90.,180.,270.,XLE=6.58,6.580.7,
        SWEEP=0.983,STA=0.0,NPANEL=4.,$
$PROTUB NPROT=0.,
        PTYPE=
        XPROT=
        NLOC=
        LPROT=
        WPROT=
        HPROT=
        OPROT=$
SOSE
DAMP
PRINT GEOM BODY
BUILD
PLOT
NEXT CASE
```

(b) Generated for005.dat

Figure 3. Calculation process of aerodynamics module.

3.3 Motor calculation module

The function of motor calculation module is to calculate motor parameters such as motor mass, thrust and specific impulse, which is provided to trajectory module. MID embedded a variety of engine design methods: a) Produce thrust curve by manual selecting points, b) Embedded solid rocket motor interior ballistic performance calculation module, the motor can be detailed design.

The theoretical specific impulse of the solid motor is calculated by the theoretical extrapolation formula. When interpolating, the influences of nozzle expansion ratio and the pressure in the combustion chamber are considered. On the basis of thermodynamic calculation, extrapolation formula[11] can be used.

$$I_{sh} = 2408\times\sqrt{0.3189+0.4187\ln\varepsilon-0.0279\ln^2\varepsilon}\times(0.9914+0.044\ln p_c)+ \\ 522.0226p_c^{0.006965}\cdot\varepsilon^{-0.3351}-1557.6p_c^{-0.9930}\cdot\varepsilon\cdot p_a \tag{1}$$

In formula (1), ε represents the nozzle expansion ratio, p_a (MPa)represents the ambient pressure, p_e (MPa) represents the combustion chamber pressure.

The results of the extrapolation formula are not more than 0.1% relative to the thermodynamic calculation. At the same time, the actual specific impulse is also calculated by the SPP empirical formula[12] Motor mass analysis includes charge mass, shell mass and nozzle mass.

3.4 Mass calculation module

Through the parameters provided by the Geometry and Motor disciplines, the total mass is calculated and provided to the trajectory discipline. MID provides a detailed calculation of quality features. The internal components and load parts of the missile can be visually configured, such as engine, main load, parachute, etc. At the same time these components will be included calculation of the center of mass and mass properties. Figure 4 shows the visualization of the engine, warhead and missile-borne equipment by the 3D diagram. MID also shows CG and CP of the missile.

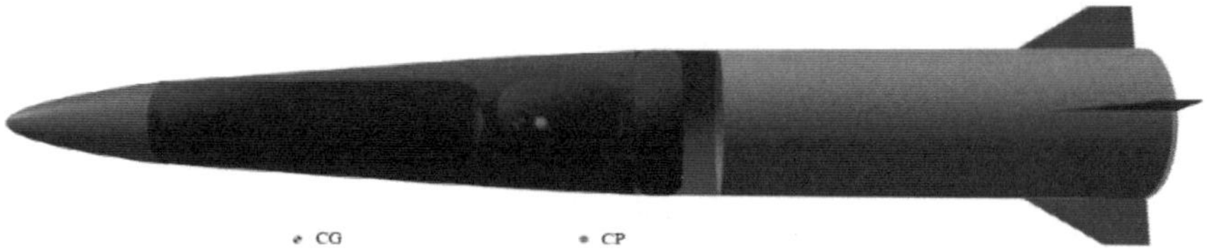

Figure 4. Visual layout of missile-borne equipment.

To simplify processing, Mass disciplines is divided into three parts: the shell mass, payload mass, motor mass.

$$m_0 = m_{shell} + m_{load} + m_{eng} \tag{2}$$

In formula (2), m_0 represents the total mass, m_{shell} represents the quality of the shell, simplified by pimping shell, m_{load} represents motor quality, including motor charge quality, shell quality and nozzle quality, m_{eng} represents payload quality, including warhead quality, missile-borne equipment quality, and other constant quality.

3.5 Trajectory calculation module

The range is calculated from the various parameters provided by the above four disciplines. MID can provide many different trajectory modes. The concrete trajectory model is realized by changing the AoA command or the engine model, AoA command and ballistic parameters are also visualized. It is supported to calculate different types of trajectory at the same time, which is easy to compare overall parameters.

A 3DOF ballistic equation without considering the Earth's rotation and flat rate is used to compute the standard ballistic trajectory[13]. The kinetic equation and kinematics equation are as follows.

$$\begin{aligned} m\frac{d^2\vec{r}}{dt^2} &= \vec{P} + \vec{R} + \vec{F}_c + m\vec{g} + \vec{F}_k' \\ \frac{d\vec{r}}{dt} &= \vec{v} \end{aligned} \tag{3}$$

In formula (3), $\vec{P}$ represents the thrust vector of the motor, $\vec{R}$ represents the aerodynamic vector, $\vec{F}_c$ represents the control force vector, $\vec{F}_k'$ represents the Gothic force vector and $m\vec{g}$ represents the gravitational vector.

3.6 Multidisciplinary integration

The construction process of the above modules has been clearly stated. Based on multidisciplinary integration technology, the above modules are integrated into MID system, and suitable optimization algorithms are used to optimize multidisciplinary design.

4. Multidisciplinary optimization example of tactical missile overall performance

Tactical missile multidisciplinary optimization is based on the establishment of the geometry, aerodynamics, mass, trajectory of the discipline model, through appropriate optimization algorithm, to obtain the result that can make the overall performance of tactical missile optimal. In the optimization of the overall performance of tactical missiles, optimization problems usually aim at the optimization of take-off mass m_0, and the size and the range R are reduced to the desired effect.

That is to take the overall parameters design of a short-range tactical missile for the study, requiring the design of a vertical launch and without control of the single-stage short-range tactical missiles, and range is not less than R_0. The design target is the minimum m_0. Missile diameter D is not optimized. The optimization problem is to ensure that the missile can reduce its m_0 and missile length under the premise of meeting certain range requirements.

4.1 Multidisciplinary Optimization Model

Multidisciplinary design model is the synthesis of discipline design model, which needs to establish the objective function, constraint condition and design variable of the whole system. Design variables play a crucial role in multidisciplinary design. The interaction of design variables among different sub-disciplines reflects the coupling and influence between different sub-disciplines.

In multidisciplinary design, design variables need to be unified, not only in the discipline-level design variables, but also in system-level design variables. These system-level design variables are coupled variables that are part of the discipline-level design variables and are responsible for interacting with other disciplines and playing a decisive role throughout the optimization process.

The principle of choosing a design variable is to express the content of the sub-discipline as completely as possible with as few variables as possible.

According to the above design variable selection principle, the design variables shown in Table 1 are selected, and the system-level design variables are arranged according to the sensitivity. The sensitivity of each design variable to m_0 is shown in Figure 5. The sensitivity of design variables indicates that the change of design variables can cause the drastic degree of change when the m_0 is the optimal target. The sensitivity of each design variable was calculated using the Full Factorial Design of Experiments, which can accurately assess the effect of the factor on the coupling effect.

Table 1. All design variables

NO.	Varible	Description	Notes
1	t	Motor working time	system-level (motor)
2	p_c	Combustion chamber pressure	system-level (motor)
3	D	Missile diameter (motor diameter)	system-level (motor, aerodynamics)
4	ε	Nozzle expansion ratio	system-level (motor)
5	λ_B	slenderness ratio	system-level(aerodynamics)
6	λ_{l2}	Two cone length ratio	system-level (aerodynamics)
7	λ_{l1}	One cone length ratio	system-level (aerodynamics)
8	λ_{LE}	Motor slenderness ratio	discipline-level(motor)
9	λ_{lc}	Column length ratio	discipline-level (aerodynamics)
10	m_{eng}	Motor mass	discipline-level (mass)
11	S_{body}	Missile surface area	discipline-level (mass)
12	m_0	Take-off mass	discipline-level (mass)
13	R	Range	discipline-level (trajectory)

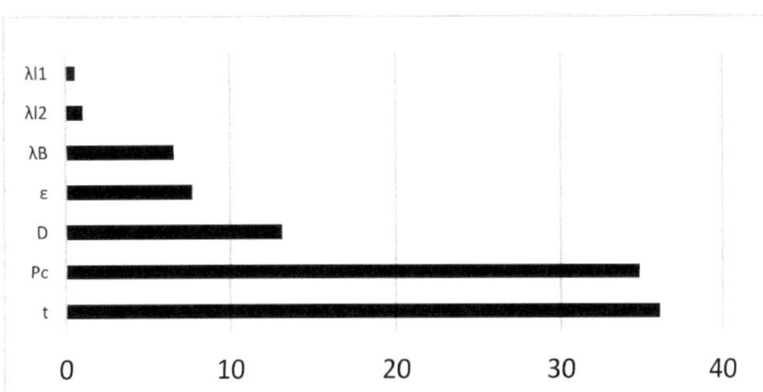

Figure 5. System-level design variable sensitivity

In this paper, in order to improve the efficiency of integrated design, part of the design parameters will be cured, such as the size of the wing, the motor shell material and charge, the missile shell material. At the same time, according to the sensitivity analysis of each design variable, 13 design variables shown in Table 1 are selected finally, of which the first 7 are optimized as system-level design variables and the latter 6 discipline-level design variables are derived from system-level design variables, which play a binding role in the process of optimization. Through the solidification of the design parameters and these design variables, the integrated design of the missile's overall parameters of the aerodynamic, motor, mass, trajectory and other disciplines can be described.

Due to the limitations of practical problems, design variables need to select a certain range of values, variables that beyond the scope of the value of the design should be discarded.

As a multiple constrained single target optimization problem, the optimization problem can be described as follows:

$$\begin{aligned}
&min\ m_0(X) \\
&X=(t,p_c,D,\varepsilon,\lambda_B,\lambda_{l1},\lambda_{l2}) \\
&s.t.\quad R\geq R_0, \\
&\qquad \lambda_{lc\min}\leq\lambda_{lc}\leq\lambda_{lc\max}, \\
&\qquad \lambda_{LE\min}\leq\lambda_{LE}\leq\lambda_{LE\max}.
\end{aligned} \tag{4}$$

In formula (4), R_0 represents the minimum range constraint, $\lambda_{lc\min}$ and $\lambda_{lc\max}$ represent the upper and lower bounds of column length ratio, $\lambda_{LE\min}$ and $\lambda_{LE\max}$ represent the upper and lower bounds of motor slenderness ratio.

4.2 Optimization Methods

MID system has several methods in place for users to optimize. In this paper, Gradient Optimizer (the gradient optimization method) is adopted to optimize. Gradient optimization method has the characteristics of small calculation, short optimization time and good calculation effect. Although gradient optimization does not necessarily converge to the global optimum, in engineering practice, the local optimal point has already satisfy the actual engineering requirements.

4.3 Optimization Examples and results

A missile's baseline shape for $D=1000mm$, $\lambda_B=6.3$, $m_0=4176kg$, $R=360km$. Specific constraints for $R\geq 350km$, $0.35\leq\lambda_{lc}\leq 0.65$, $2\leq\lambda_{LE}\leq 5$.

Optimized with MID system, spents a shorter time of 45min,. m_0 is reduced by 22.25% under the precondition of satisfying the design requirement, and the optimization effect is remarkable.

Table 2 lists the optimization results of the missile's overall parameters. It can be seen that m_0 decreases significantly under the precondition of satisfying various constraints. Due to λ_B , λ_{LE} and λ_{lc} decreases, m_{shell} and m_{eng} both are reduced, and m_{load} as constants remain unchanged, so m_0 decreases. At the same time, decrease of m_{eng} leads to t decrease, p_c and ε both increase and direct result is F (Thrust) increased to meet the predetermined requirement of range. According to the above analysis, the optimization result is reasonable and usable.

Table 2. Optimization results of missile overall parameters

Varible	Initial Value	Value Range	Optimal
$m_0\ /\ kg$	4176	-	3416
$R\ /\ km$	360	-	352
λ_B	6.3	[4,10]	6.1229
λ_{l1}	0.087	[0.05,0.15]	0.0847
λ_{l1}	0.389	[0.3,0.5]	0.4255
λ_{LE}	2.3	[2,5]	2
$p_c\ /\ kPa$	6	[4,10]	7.8435
$t\ /\ s$	57	[40,80]	41
ε	4	[3,10]	10
$F\ /\ kN$	109.5	-	130.1

MID system can visualize the comparison of the results before and after the optimization. Figure 6 graphically shows the shape change of the missile before and after optimization. You can see clearly that λ_B and λ_{lc} are reduced.

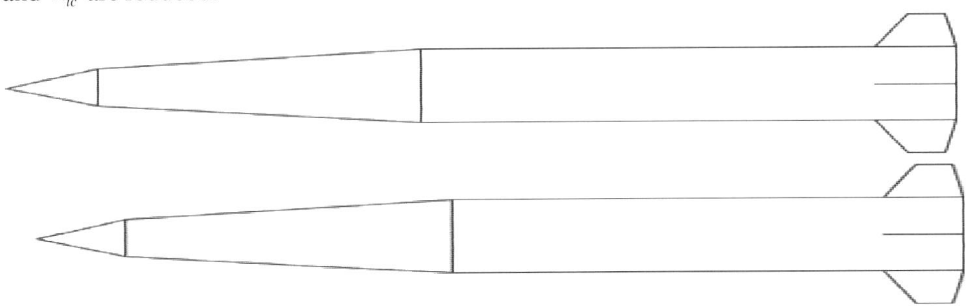

Figure 6. Comparison of missile profiles before and after optimization

5. Conclusions

In this paper, MID system with human-machine friendly operation interface is built with the help of MDO idea and integrated design method. Through the overall preliminary design and optimization of single-stage solid-motor short-range tactical ballistic missiles, MID system shows its rich operability and engineering practicability. The overall preliminary design takes less time and achieves the desired optimization effect. It shows that the integrated design model and MID system are effective under certain conditions, and can be used in the overall preliminary design of engineering practice. Mid, as an effective tool in the overall preliminary design stage of tactical missile, will greatly reduce the R&D time, improve the design efficiency, and help the missile designers to meet the demand of war standard and improve the design quality of the missile.

To further improve the generality of mid, the next research work mainly includes: refining the aircraft parametric shape, increasing the optimization algorithm, integrating more discipline design modules. To solve the problem of the current overall optimization module.

Acknowledgments

The authors wish to acknowledge the rest member of MID System team, whose contribution work is the base of this research.

References

[1] M. L. Baker, M. J. Munson, and K. Y. Alston, "Integrated Hypersonic aeromechanics tool (IHAT),"2002 *9th AIAA/ISSMO Symposium on Multidisciplinary Analysis and Optimization, Atlanta, GA*,.

[2] M. L. Baker, M. J. Munson, G. W. Hoppus, and K. Y. Alston, 2004 "The Integrated hypersonic aeromechanics tool (IHAT), Build 4,"*10th AIAA/ISSMO Multidisciplinary Analysis and Optimization Conference, Albany, New York*.

[3] Gong C L, Gu L X, and Yuan J P 2009 "Main Model Technology for Multidisciplinary Design Optimization of Aircraft,"*Journal of Astronautics*. **30** 914-918, ,.

[4] Xu W, Sun P Z, and Xia Z X, 2008 "Integration Design and Optimization for Boost-Glider Missile,"*Journal of Solid Rocket Technology*, **31** 317-320,.

[5] Lestage R, Lesage F, Stowe R A, and Hamel N, 2010 "Multidisciplinary integrated analysis for engagement simulation of hypersonic weapons,"*Defence R&D Canada-Valcartier*.

[6] Committee A M D O T, 1991 ,"Current state of the art on multidisciplinary design optimization (MDO),"*AIAA White Paper*.

[7] Carty A, 2002 "An approach to multidisciplinary design, analysis & optimization for rapid conceptual design,"*9th AIAA/ISSMO Symposium on Multidisciplinary Analysis and Optimization*,, 4-6.

[8] Giesing J P and Barthelemy J-F M, 1998 "A summary of industry MDO applications and needs,"*AIAA White Paper.*
[9] Niskanen S, 2009 "Development of an Open Source model rocket simulation software,"*Helsinki University of Technology*.
[10] Niskanen S, 2013 "OpenRocket technical documentation,"*Development of an Open Source model rocket simulation software* 11-13.
[11] Yang J, Chen R, and Zhao X, 2010 "Integal Optimization Design for Tactical Rocket and It's Srm by Utilizing the Synthesis of Internal Ballistics and Trajectory," *Journal of Astronautics* **20** 0.
[12] Fang G Y, 1990 "Overall Design Optimization of Solid Rocket Motor,"*Journal of Solid Rocket Technology,* 1-8,.
[13] Chen K J, Liu L H, and Meng H Y, 2014 *Remote Rocket Flight Dynamics and Guidance*.Guo fang gong ye chu ban she.

IOP Conf. Series: Materials Science and Engineering **408** (2018) 012024 doi:10.1088/1757-899X/408/1/012024

Effect Analysis of the Hangar Rear Edge Curvature on the Ship Airwake

Jinling Wang*, Guangwen Jiang and Xiaocheng Wang

Systems Engineering Research Institute, CSSC, Beijing 100094, China

*bellincsscl6@163.com

Abstract. In order to examine the effects of hangar rear edge curvature on ship airwake, the SFS2 ship configuration is modified by changing rear vertical surface of hangar into curved surface. Airwake with different configurations subject to free-stream of 25.7m/s at headwind are simulated using commercial CFD solver ANSYS FLUENT, and the traditional Reynolds-average Navier-Stokes k-ε two equation turbulence model is applied. A very good agreement of experiment results in wind tunnel with predictions from the CFD demonstrate that both the FLUENT code and the meshes generated by ICEM perform well in simulating ship airwake. The comparison of different hangar rear edge curvatures shows that the curvature modifications improved airwake over the flight deck region of SFS2 by reducing mean downwash and turbulence intensity. 90° is the best angle corresponding to SFS2 hangar rear edge curved arc, which will reduce pilot workload during launch and recovery operations.

1. Introduction

The superstructure and the hangar of ship usually contain sharp corners, the airflow is blocked and creates flow separation, vortex shedding and attachment when crossing these structures. The blockage and viscosity are the driving force for flow separation and vortex shedding. For rear-flight deck ship with hangar, hangar shape is the key structure that affects the airwake in the helicopter operating area. By changing hangar shape, the pressure changes in the helicopter operating area can be slowly and smoothly, avoid the strong vortex and shear layer caused by the speed jump, so as to optimize the ship design. In 2002, Kumar[1] conducted subscale wind tunnel experiments to assess the effectiveness of passive control involving a porous surface and cavity underneath for reducing the level of surface pressure fluctuations in reattaching flows, and it demonstrated the design can significantly reduce the peak surface pressure fluctuations in the reattachment zone, a maximum reduction of as much as 35% has been observed; in 2003, Tai[2] modified the SFS ship configuration by adding a rounded bow section at the front end to avoid complete separation along the side walls, the concluding remarked that the complete separation along the hull surface is avoided; in 2005, Shafer[3] explored both active and passive flow control techniques on airwake improvement in the final decent of helicopter onto the flight deck, results show that it would cause a reduction in unsteadiness in the landing region of 6.6% at 0° wind-over-deck (WOD) and 8.3% at 20° WOD by injecting air through the porous surfaces; in 2010, Forrest[4] found that it could decrease the adverse pressure gradient and the turbulence at the hangar edge by installing a 30° flap at the hangar edge of starboard side or applying a chamfer with an angle of 30°; in 2013, the midshipman LaSalle[5] of the United States Naval Academy examined the effects of passive flow control techniques on the ship airwake of YP, and the investigation showed that the particular passive flow control fence produce a less favorable ship airwake for helicopter launch and recovery operations due to an increase in shear, turbulent kinetic energy density, and mixing within the helicopter landing region.

The work presented in this paper relates to a number of curved geometric modifications which have made to the vertical hangar rear edge of SFS2 in an attempt to alter the airwake characteristics over the flight deck for headwind. Airwake for different modifications are measured with computational fluid dynamics (CFD) simulations to provide analysis for effects of the curvatures on the SFS2 airwake.

2. Aerodynamic Design and Meshing

As shown in figure 1, the rear vertical surface of the hangar is modified to curved surface. The top view of 13 different kinds of curvatures of hangar rear edge is shown in figure 2. Different curvatures are expressed by the center angle α corresponding to the arc. The α ranges from 0° to 180°, and the interval is 15°, actually when α>150° the mesh quality around the arc degraded severely, which will result in calculation divergence, so the upper limit of α in this paper is 150°.

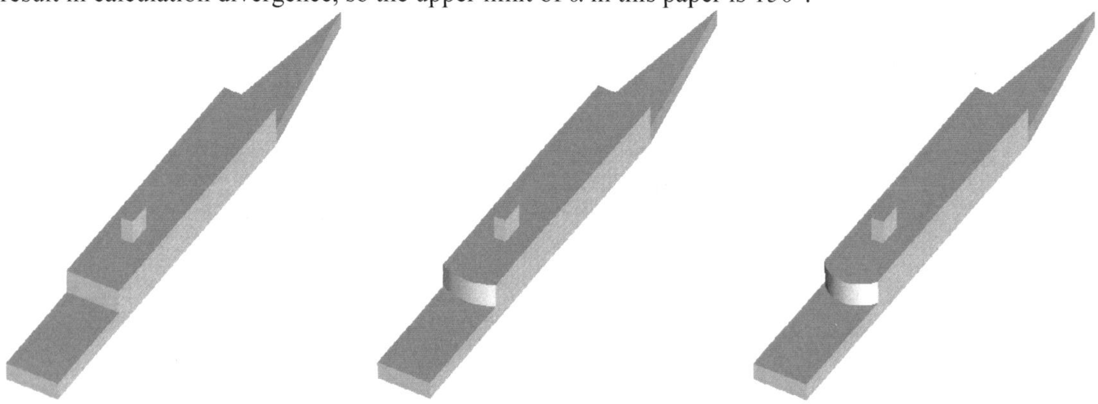

(a) α=0° (b) α=60° (c) α=120°

Figure 1. Schematic of the different curvatures of hangar rear surface for SFS2

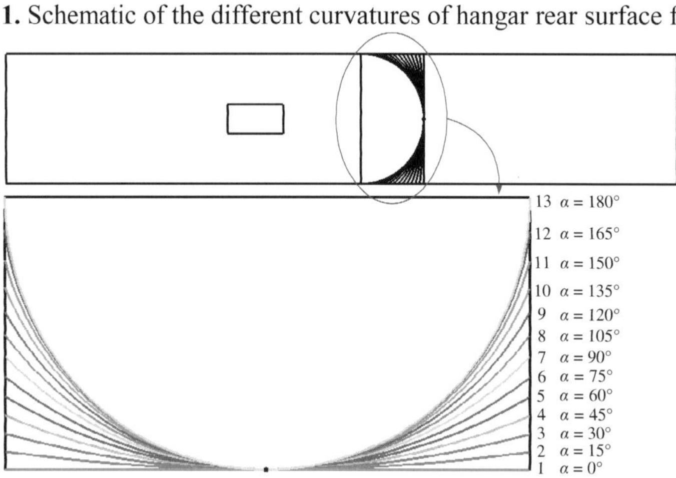

Figure 2. Top view of different curvatures of hangar rear edge for SFS2

The SFS2 models were imported into the ANSYS ICEM mesh generation software to generate structured meshes. Original SFS2 and the modified SFS2 structured mesh are shown in figure 3, after verifying the grid independence, both the original and modified SFS2 mesh number used for numerical computation is 1.89×10^7.

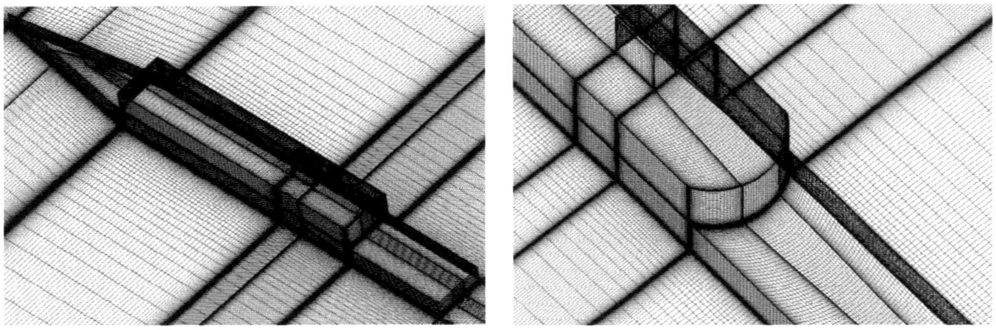

(a) mesh for original SFS2 mesh (b) mesh for modified SFS2 mesh

Figure 3. Top view of mesh distribution for different SFS2 models

3. Flow Solver and Boundary Conditions

The steady-state airwake of SFS2 is simulated using commercial CFD solver ANSYS FLUENT, and the traditional Reynolds-average Navier-Stokes k-ε two equation turbulence model is applied.

The boundary conditions for the FLUENT solver are:

(1) free-stream of 25.7m/s for upstream;

(2) pressure far field for downstream;

(3) viscous nonslip flow at the surface of the SFS2 ship;

(4) frictionless surface at stationary water surface.

4. Method Validation

The validation study was performed to gain confidence in the accuracy of the CFD modelling by comparing the experiment results obtained from the Chinese Aeronautical Establishment (CAE). The experiment was conducted in the 4.5m×3.5m low-speed wind tunnel as shown in figure 4. Particle Image Velocimetry (PIV) was used to obtain the mean velocities over the SFS2 deck. In order to simulate uniform flow condition, the 1:60 scale SFS2 model was mounted on a plate about 1.45m above the wind tunnel floor. The outlet wind speed of wind tunnel is 25.7m/s.

Figure 4. SFS2 model in the test section of the wind tunnel

Figure 6 shows a comparison of CFD and wind tunnel experiment velocity. The comparison contains the x-velocity and z-velocity of L_S plotted in figure 5. These two data sources are referred to in figure legends as 'CFD' and 'EXP' respectively. CFD and wind tunnel experiment predict the same wake pattern, with the CFD velocities a factor of approximately 0.1 lower than the wind tunnel results outside the separation, while inside the separation the velocities profiles almost overlap. The overlapped zero points of A and B indicate that the CFD and wind tunnel predict the same separation. In summary, the comparison of the detailed CFD results with data obtained from wind tunnel experiment show excellent agreement. This agreement holds not only for the velocity profiles but also for the velocity magnitude. These results demonstrate that both the FLUENT code and the meshes generated by ICEM perform well in simulating ship airwake.

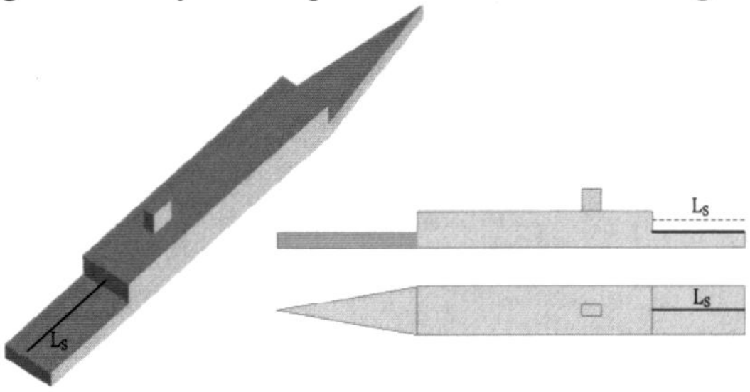

Figure 5. Coordinate and location of L_S

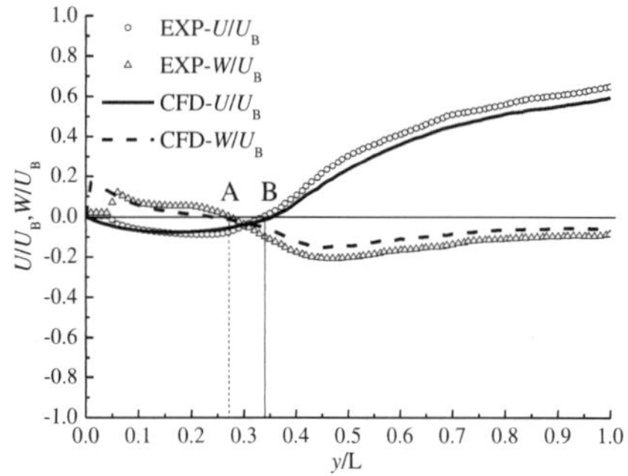

Figure 6. Comparison of CFD and experiment for SFS2

5. Results and comparison

The early publications of Lee and Zan [6,7] already revealed that both the turbulence intensity and the downwash mainly affect the helicopter in the ship airwake, the turbulence intensity mainly affect the pilot workload, and the downwash affect the rotor performance. This paper will analyse the airwake focused on the parameters mentioned above.

The z-component velocity contours on SFS2 symmetry plane for different α are shown in figure 7, with the increasing of α, the downwash around the flight deck hangar increases gradually, while the upwash shows the opposite trend; for the airwake around the rotor(the red line above the flight deck), both the downwash magnitude and the affect range reduced. When α is equal to 90°, the downwash velocity contour of 2 m/s disappears; and when α is greater than 120°, upwash appears around the rotor centre. The streamlines on SFS2 symmetry plane for different α shown in figure 8 indicate that with the increasing of α, the recirculation behind the hangar gets smaller and smaller; when α is equal to 90°, the helicopter almost escape the effect of the recirculation. Due to the limitation of mesh quality, it does not simulate the conditions when α is bigger than 150°, while based on the trend of existing data, it can be inferred that as α continue to increase, both the downwash and the recirculation zone will further decrease. It is beneficial to improve the stability of helicopter in the ship airwake.

Figure 9 gives the downwash and the turbulence intensity contours predicted by CFD for the rotor disk of a Westland Lynx helicopter sitting on the flight deck for different α. The results show that as α increases, the airwake changes obviously, both the downwash and the turbulence intensity decrease sharply. Similar to the results in figure 7, when α is greater than 105° , upwash appears around the rotor centre. Which may cause a rollover danger for the helicopter located in airwake with both upwash and downwash flow. So it can be conclude that α should be less than 105° .

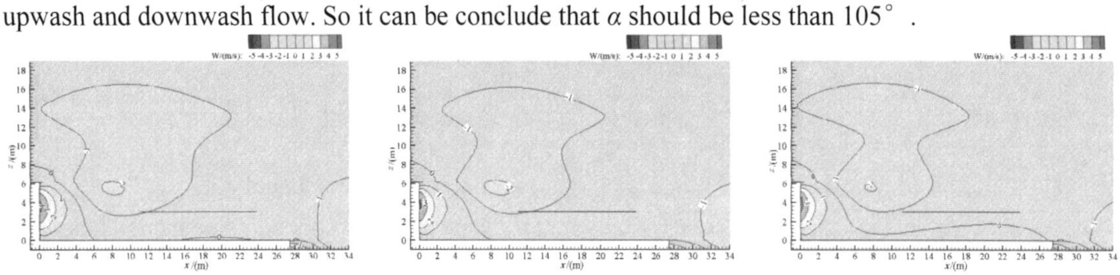

(a) α=0° (b) α=30° (c) α=60°

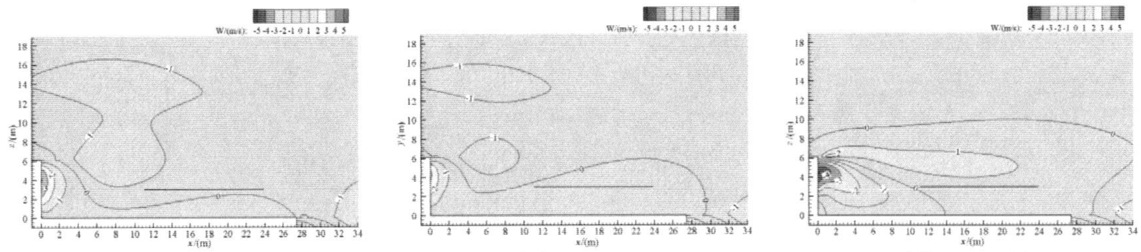

(d) α=90° (e) α=120° (f) α=150°

Figure 7. z-component velocity contours at SFS2 symmetry plane of different α for headwind

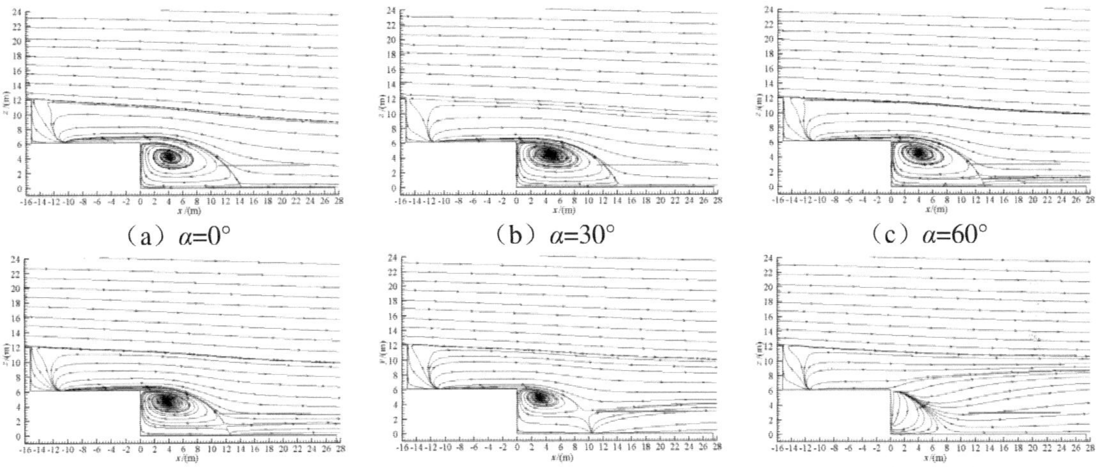

(a) α=0° (b) α=30° (c) α=60°

(d) α=90° (e) α=120° (f) α=150°

Figure 8. Streamlines at SFS2 center deck of different degrees of circular hangar transition for headwind

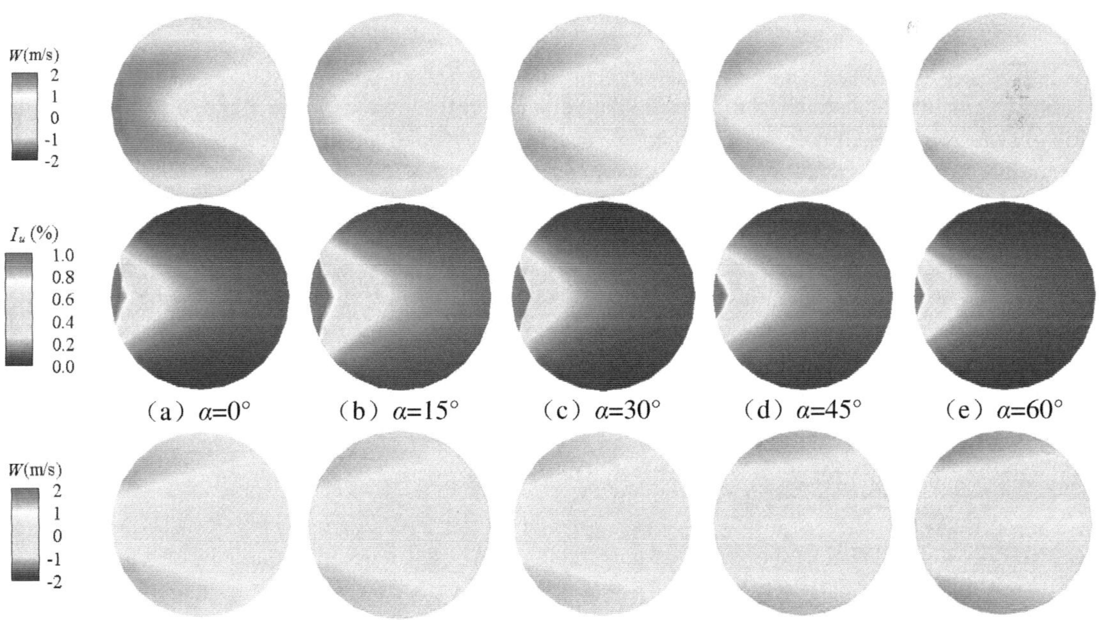

(a) α=0° (b) α=15° (c) α=30° (d) α=45° (e) α=60°

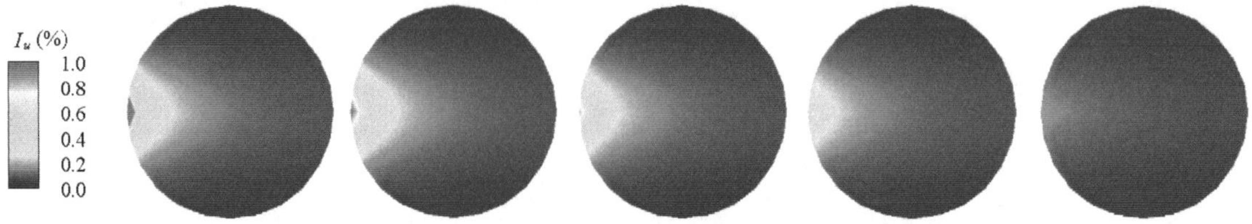

(f) α=75° (g) α=90° (h) α=105° (i) α=120° (j) α=135°

Figure 9. z-component velocity, pressure and turbulence intensity contours at Westland Lynx rotor plane of SFS2 different degrees of circular hangar transition for headwind

The z-component velocity profiles on the vertical line passing ITDP with different α shown in Figure 10 indicate that, it only exists upwash above the ITDP(Ideal touchdown point) when α is less than 30°; while when α is greater than 30°, helicopter will be subjected to the downwash firstly and then the upwash during the landing process. And as α increases, the upwash velocity increases while the downwash velocity decrease in opposite.

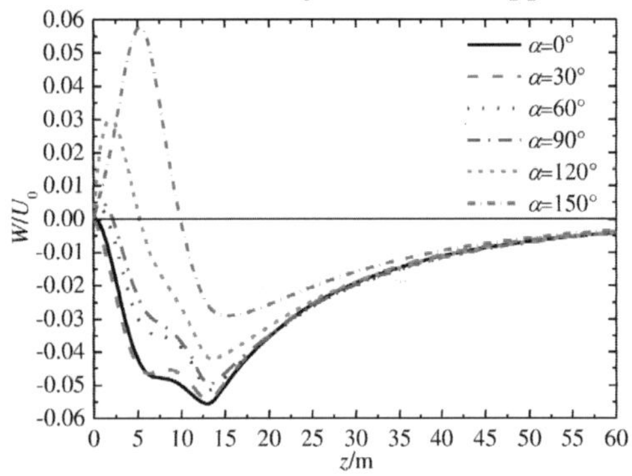

Figure 10. z-component velocity on the vertical line passing ITDP of SFS2 different degrees of circular hangar transition for headwind

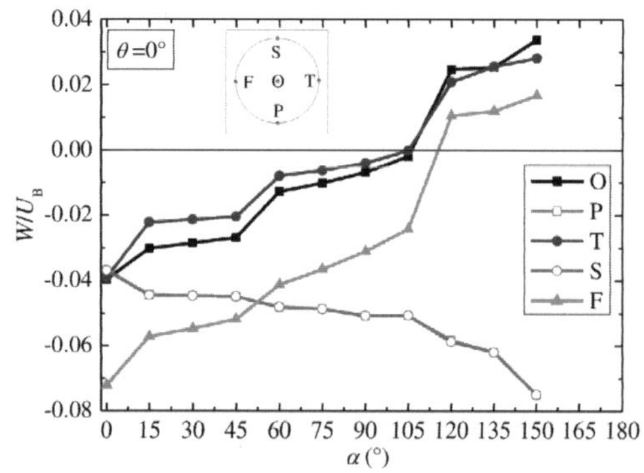

Figure 11. z-component velocity at point O,P,T,S and F of SFS2 different degrees of circular hangar transition for headwind

Figure 11 gives the z-component velocity profiles of the key points labelled as O, F, T, P and S with different α. Compared to the contours the velocity profiles show clearer trend, as α increases, the downwash velocity at points O, F and T decreases gradually, and when α=105°, the velocity magnitude drop to the minimum, then it turns to upwash and increases as α increases on. For the symmetrical points P and S, the downwash velocity profiles coincide completely, and the velocity magnitude increases as α increase from 0° to 90°, while from 90° to 105° the velocity magnitude keep as a constant, and when α is equal to 105° the downwash velocity gradient increases significantly.

The numerical simulation results provide details of the flow around SFS2 hangar rear edge curvatures. Analysis demonstrates that by changing the curvature of SFS2 hangar can significantly optimize the ship airwake around the helicopter, while when $\alpha$is greater than 90°, it may cause a rollover danger with both upwash and downwash load on the rotor and fuselage, and the downwash velocity gradient increases significantly when $\alpha$is greater than 105°, so 90° is the best angle corresponding to the SFS2 hangar rear edge curvature arc.

6. Conclusions

A methodology for testing the effect of ship hangar rear edge curvature on ship airwake has been presented, featuring steady CFD ship airwakes validated with PIV wind tunnel experiment. A very good agreement of the mean flow measured in experiments with predictions from the CFD data is demonstrated. The comparison of different hangar rear edge curvatures simulation results showed that the curvature modifications improved airwake over the flight deck region of SFS2 by reducing mean downwash and turbulence intensity in the local flow. With increasing of α (The centre angle

IOP Conf. Series: Materials Science and Engineering **408** (2018) 012024 doi:10.1088/1757-899X/408/1/012024

corresponding to the arc of curvature), the improvement effect is more and more obvious when α is less than 90°; the curvature modification more than 90° could potentially lead to the rollover of helicopter with both upwash and downwash load on the rotor and fuselage, and the downwash velocity gradient increases significantly when α>105°, thus α=90° is the best angle corresponding to the SFS2 hangar rear edge curvature arc, which will reduce pilot workload during launch and recovery operations.

For future investigations, the PIV should be applied to validate the effects of curvature modification on SFS2 ship airwake.

References

[1] Kumar R Viswanath PR. Passive control of surface pressure fluctuations in reattaching flows[J]. The Aeronautical Journal, 2002, 49(2): 149-159.
[2] Tai T C. Airwake Simulation of Modified TTCP/SFS Ship[C]//NATO RTO Symposium on Vortex Flow and High Angle of Attack. Loen, 2003: 1-13.
[3] Shafer D M. Active and Passive Flow Control over The Flight Deck of Small Naval Vessels[D]. Virginia: Virginia Polytechnic Institute and State University, 2005.
[4] Forrest J S, Kääriä C H, Owen I. Determining the Impact of Hangar-Edge Modifications on Ship-Helicopter Operations using Offline and Piloted Helicopter Flight Simulation[C]//American Helicopter Society 66th Annual Forum. Phoenix, 2010: 1-17.
[5] LaSalle N R. Study of Passive Flow Control for Ship Air Wakes[R]. Annapolis, 2013.
[6] Lee R G, Zan S J. Unsteady Aerodynamic Loading on a Helicopter Fuselage in a Ship Airwake [J]. Journal of the American Helicopter Society, 2004, 49(2): 149-159.
[7] Lee R G, Zan S J. Wind Tunnel Testing of a Helicopter Fuselage and Rotor in a Ship Airwake[J]. Journal of the American Helicopter Society, 2005, 50(4): 326-337.

Research on Cooperative Planning Technique of Gliding Trajectory for Multi-aircraft

Qiangqiang Xu*, Jianquan Ge, Tao Yang, Xiaojian Sun and Guoqiang Li

College of Aerospace Science and Engineering, National University of Defense Technology, Changsha, China

*xqqnudt@126.com

Abstract. In order to make multi-aircraft cooperating with each other and reaching the specified target area at the same time, the gliding trajectory cooperating planning is studied. The trajectory planning model is established. For a single aircraft, online trajectory generation based on the height vs velocity (H-V) profile is achieved, and a feedback linearization tracking guidance law is designed. For multi-aircraft, a multi-aircraft trajectory planning strategy based on time coordination is designed. Multi-aircraft trajectory cooperating planning under two different operational backgrounds is simulated. Simulation results show that the algorithm can achieve multi-aircraft trajectory cooperating planning, and multi-aircraft can reach the target area at the same time.

1. Introduction

With the strengthening of the aerial defence system, traditional single-aircraft is facing severe challenges in implementing penetrating and striking target. To achieve effective penetration and improve the effectiveness of cooperative operations, the US military took the lead in putting forward the concept of cooperative operations in the mid-1970s and achieved a series of research results.

At present, the main contents of multi-aircraft cooperative combat research include attack time coordination [1-5], attack angle coordination [6], attack time and attack angle coordination [7-11] and multi-aircraft formation control. In general, there are three deficiencies. First, most of the research objects are low-velocity aircraft, and they are considered to fly under uniform or uniform acceleration. There is a lack of research on cooperative operations for hypersonic vehicles. Second, it does not consider the effect of complex battlefield environment such as heat flow, dynamic pressure and overload. Third, the cooperative operation distance is close and the guidance is mostly terminal cooperative guidance.

In this paper, the cooperative engagement of multi-hypersonic gliding aircrafts are used as the study object. In order to make multi-aircraft reaching the specified target assembly area at the same time, the gliding trajectory cooperative planning is studied. Firstly, the trajectory planning model is established. Secondly, the trajectory of a single aircraft is generated online based on the HV flight profile. A nonlinear tracking law using feedback linearization method is used to track the standard trajectory. Finally, a multi-aircraft trajectory cooperative planning strategy is proposed. The simulation results are carried out in two different operational backgrounds. The illustrative examples show that the method is effective to solve the trajectory cooperative planning problem.

2. Dynamic equation

IOP Conf. Series: Materials Science and Engineering **408** (2018) 012025 doi:10.1088/1757-899X/408/1/012025

In order to simplify the calculation, the influence of the flattening rate is ignored and the three-degree-of-freedom motion equation of the aircraft is established as follows [12].

$$\begin{cases} \dot{r}=V\sin\theta \\ \dot{\lambda}=\dfrac{V\cos\theta\sin\sigma}{r\cos\phi} \\ \dot{\phi}=\dfrac{V\cos\theta\cos\sigma}{r} \\ \dot{V}=\dfrac{-D}{m}-g\sin\theta+\omega^2 r\cos\phi\left(\sin\theta\cos\phi-\cos\theta\sin\phi\cos\sigma\right) \\ \dot{\theta}=\dfrac{1}{V}\left(\dfrac{L}{m}\cos\nu-g\cos\theta+\dfrac{V^2}{r}\cos\theta+2\omega V\cos\phi\sin\sigma+\omega^2 r\cos\phi(\cos\theta\cos\phi+\sin\theta\sin\phi\cos\sigma)\right) \\ \dot{\sigma}=\dfrac{1}{V}\left(\dfrac{L\sin\nu}{m\cos\theta}+\dfrac{V^2}{r}\cos\theta\sin\sigma\tan\phi-2\omega V\left(\tan\theta\cos\phi\cos\sigma-\sin\phi\right)+\dfrac{\omega^2 r}{\cos\theta}\sin\phi\cos\phi\sin\sigma\right) \end{cases} \tag{1}$$

Where m is the mass of the aircraft, g is the gravity acceleration, V is the velocity; θ is the flight path angle; σ is the heading angle; r is the radial distance from the earth center to the aircraft. And $r=R_e+h$, where R_e is earth's average radius, h is the height of the aircraft from the local level; λ is the geographical longitude; ϕ is the geographical latitude; ν is the bank angle; $\omega_e=7.292\times10^{-5}$ rad/s is the earth rotation. L and D are the aerodynamic drag force and lift force respectively. Their specific expression are given by

$$\begin{cases} L=0.5\rho V^2 C_L S \\ D=0.5\rho V^2 C_D S \end{cases} \tag{2}$$

Where S_M is the reference area. The atmospheric density $\rho=\rho_0\exp\left(-\beta\left(r-r_0\right)\right)$, where ρ_0 =1.225kg/m^3 and β =1/7100. The drag coefficient C_D and lift coefficient C_L are related to the shape and flight state of the aircraft.

3. H-V flight profile generation and tracking law using feedback linearization

3.1. H-V flight profile generation

In general, the aircraft reentry process can be divided into three phases: the initial descent phase, the gliding phase, and the terminal guidance phase. The gliding trajectory is designed to two segments which can be obtained by cubic polynomial planning, namely

$$H_0=c_1+c_2V+c_3V^2+c_4V^3 \tag{3}$$

Where c_1, c_2, c_3, c_4 are the undetermined coefficients. The gliding trajectory is divided into two phases which are connected by control point. By changing the height and velocity of the control point, the shape of the two curves can be changed so that the trajectory of the entire gliding trajectory can meet the process constraints and task requirements and reach the target area.

The initial descent phase of the aircraft adopts the constant angle of attack and zero-bank angle to control. At the end of the descent phase, the pull-up point parameters including height H_d and speed V_d can be obtained. According to the terminal guidance requirements, the push-down point parameters including height H_f and speed V_f can be pre-determined. The speed of control point can be taken as $V_c=(V_d+V_f)/2$, and the height H_c must meet the constraints of the reentry corridor. The specific value is determined by the range to be flying. According to the parameters of the pull-up point, the control point, and that the change rate of the height vs the speed is equal to zero, the undetermined coefficient of the curve can be determined. Similarly, the curve form between the control point and the push-down point can be solved.

In the case of a small lateral manoeuvring range of the aircraft, the range is mainly determined by the longitudinal range. The longitudinal range can be determined based on the H-V flight profile. Without considering the earth rotation, the equation of longitudinal motion is simplified as following.

$$\begin{cases} \dot{r} = V\sin\theta \\ \dot{V} = -\dfrac{D}{m} - g\sin\theta \\ \dot{\theta} = \dfrac{1}{V}\left[\dfrac{L\cos\nu}{m} + (\dfrac{V^2}{r} - g)\cos\theta\right] \\ \dot{R} = V\cos\theta \end{cases} \tag{4}$$

Where R represents the range.

$$\frac{dR}{dV} = \frac{V\cos\theta}{-D/m - g\sin\theta} \approx -\frac{mV}{D} = -\frac{2m}{C_D \rho VS} \tag{5}$$

When the H-V flight profile is determined, the range can be estimated by integrating the formula above. Generally, the specific height H_c of the control point can be determined by the Newton iteration algorithm according to the range mission. At the same time, since the speed V of the control point is determined and the height H_c needs to meet the reentry corridor constraints, the height of the control point can also be determined by traversing the search. That is, when the speed of the control point is known, the height interval is determined according to the reentry corridor constraints. The search step length is set. Then, the control point height H_c is searched and determined.

3.2. Tracking law using feedback linearization

In order to track the H-V curve, a feedback linearization is used to design the tracking controller. According to the dynamic equation in longitudinal plane, the state variables are taken as $[r, V, \theta]$, and the control variable is taken as longitudinal lift.

$$u = Y\cos\nu \tag{6}$$

The output of the control system is

$$y = H \tag{7}$$

According to the above transformation, the H-V profile tracking problem is converted into a single-input single-output problem with input as u and output as H. The system equation can be expressed as

$$\begin{cases} \dot{y} = \dot{H} = \dot{r} = V\sin\theta \\ \ddot{y} = \ddot{H} = \dot{V}\sin\theta + V\dot{\theta}\cos\theta \end{cases} \tag{8}$$

The simplified dynamic equation is brought into the above system equation. Then

$$\ddot{y} = a(x) + b(x)u \tag{9}$$

Where $a(x) = (-\dfrac{X}{m} - g\sin\theta)\sin\theta + (\dfrac{V^2}{r} - \dfrac{\mu}{r^2})\cos^2\theta$, $b(x) = \dfrac{\cos\theta}{m}$.

Let $u=\frac{w-a(x)}{b(x)}$, the input-output relationship will become a simple linear double-integral relationship.

$$\ddot{y}=w \tag{10}$$

In order to ensure the stability of tracking, the error must meet the following requirements at the same time.

$$\delta\ddot{H}+2\zeta\omega\delta\dot{H}+\omega^2\delta H=0 \tag{11}$$

Where $\delta H=H-H_0, \delta\dot{H}=\dot{H}-\dot{H}_0, \delta\ddot{H}=\ddot{H}-\ddot{H}_0$, H_0 is the reference height, H is the current flight height.

Therefore, the control system input is given by

$$u=\frac{\ddot{H}_0-2\zeta\omega\delta\dot{H}-\omega^2\delta H-a(x)}{b(x)} \tag{12}$$

As can be seen from the relationship between input and output of the control system, if the angle of attack or bank angle is determined, then the other can be obtained. Assuming that the angle of attack profile is determined, the bank angle is calculated by

$$\nu_c=\arccos(\frac{u}{Y}) \tag{13}$$

For the control of the lateral trajectory, the method of designing the heading angle error corridor is usually used.

The height - velocity curve obtained by tracking is shown in figure 1.

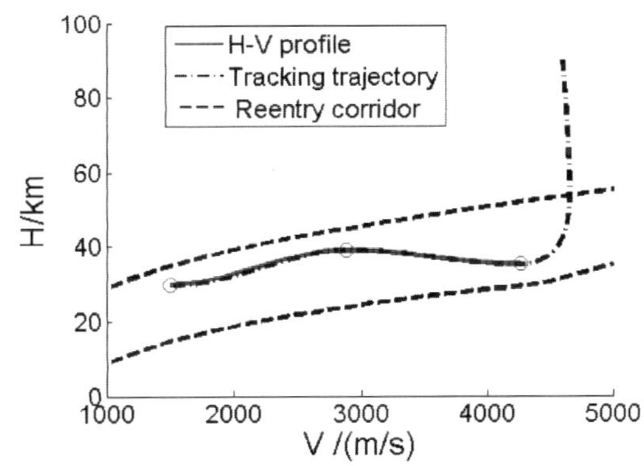

Figure 1. Velocity vs height curve.

4. Trajectory planning based on time coordination

In order to cope with the complex battlefield environment in the future and make multiple aircraft reach the attack target area at the same time, the following cooperative path planning process is designed on the basis of online trajectory generation method.

First, estimate the remaining flight time required for each aircraft to reach the target. The remaining flight time is available from the current time t and the terminal time t_f and given as $t_{go}=t_f-t$;

Second, determine whether cooperative is needed. Calculate the difference among the remaining flight time of each aircraft, and compare it with the pre-given time difference ε. If $\left|t_{go}^i-t_{to}^j\right|>\varepsilon$, the trajectory needs cooperative planning; otherwise, it does not need;

Third, do cooperative planning. Considering that the flight process is one no-power reentry and gliding phase, it is difficult to achieve acceleration for the aircraft. So the aircraft with the larger remaining flight time is used as a reference, and the remaining flight trajectories are regenerated and tracked online.

Without loss of generality, the trajectory online generation and cooperative planning process of two aircrafts can be shown in figure 2.

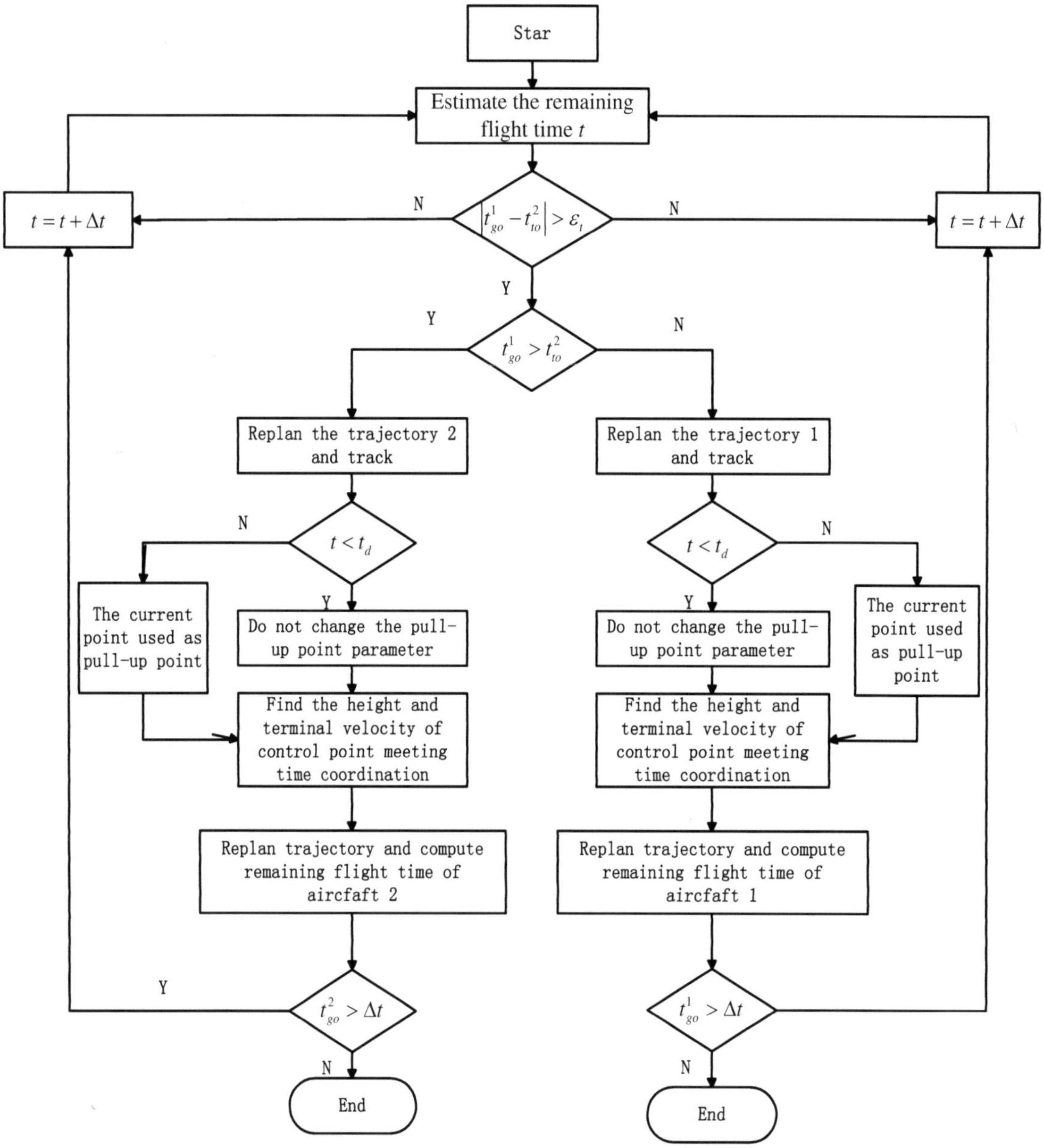

Figure 2. Flow chart of online generation and cooperative planning.

In the figure, it means that the cooperation is performed at this time Δt, which is the coordination time interval. t_d is time of the pull-up point.

IOP Conf. Series: Materials Science and Engineering **408** (2018) 012025 doi:10.1088/1757-899X/408/1/012025

5. Experiments and results

5.1. Launch at the same place but the different time

Assuming that the latitude and longitude of the launch point are set (0°E, 0°N). The initial velocity is 4600m/s, the height is 90km, the flight path angle is -6.5°, and the heading angle is 45°. Take the target latitude and longitude latitude as (15°E, 10°N). The push-down point satisfies the following constraints: the height is 30km, the velocity is between 1200m/s and 1800m/s, and the radius of the target area is 50km.

The time interval between the launch of the two aircrafts is 60s. The time-coordination interval of the two aircrafts is $\Delta t = 100s$. The allowable time difference is $\varepsilon_t = 1s$. The simulation results are shown in figure 3.

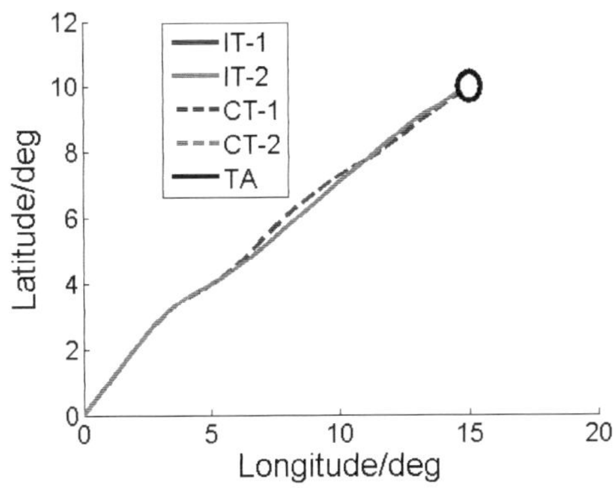

(a) Longitude-latitude curve.

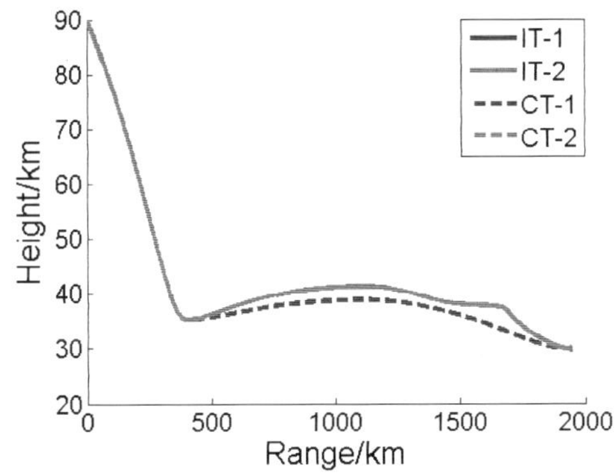

(b) Range-height curve

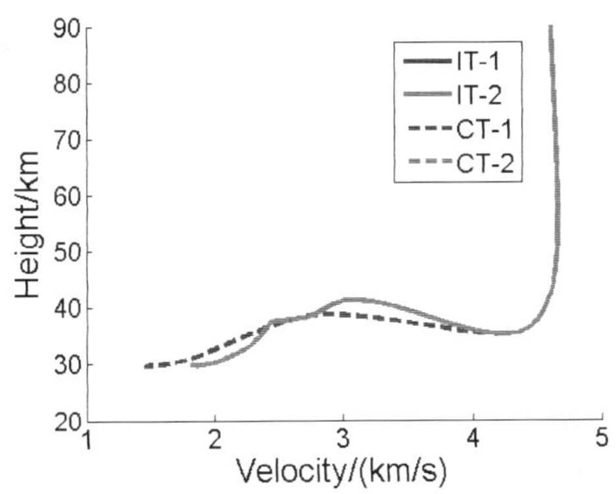

(c) Velocity-height curve

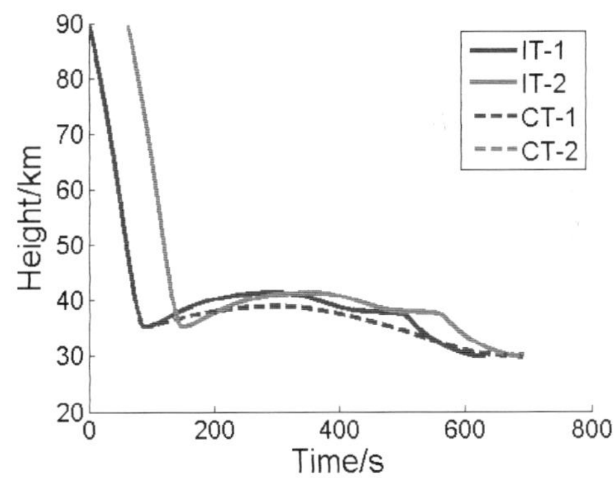

(d) Time-height curve

Figure 3. Results for launching at the same place but the different time.

In figure 3, the denotations IT-1, IT-2, CT-1, CT-2 represent initial trajectory of aircraft 1, initial trajectory of aircraft 2, cooperative trajectory of aircraft 1 and cooperative trajectory of aircraft 2 respectively. The denotation TP represents the target area.

As can be seen from the figure 3, the initial state and the terminal state are the same before the two aircrafts cooperating. There is only a gap in time, so the trajectories obtained by online generation and tracking guidance are parallel to each other. On account of that the trajectory of aircraft 2 does not need to be replanned, the IT-2 and CT-2 are coincident. The number of trajectories seen from the figure is three.

From figure 3a, it can be seen that after the cooperative trajectory planning, the two aircrafts can still reach the preset target area. But the flying height are different which can be seen from figure 3b.

IOP Conf. Series: Materials Science and Engineering **408** (2018) 012025 doi:10.1088/1757-899X/408/1/012025

Since the aircraft 1 is launched 60s earlier than 2, the height of the trajectory gliding phase of the aircraft 1 should be reduced as a whole during the cooperative planning in order to achieve the time coordination. At the same time, it can be seen from the terminal state that the two aircraft are consistent in range and have the same height, and satisfy the given terminal state constraints. From figure 3c, it can be seen that after the cooperative planning, the gliding height is reduced. The velocity decreases from 1800 m/s before the coordination to 1446 m/s. From figure 3d, it can be seen that after the cooperative planning, the two aircraft trajectories can reach the predetermined altitude at the same time, which meets the time coordination requirement.

5.2. Launch at the same time but the different places

In order to strike the same target whose latitude and longitude is (15°E, 10°N), two aircrafts are launched from two positions whose latitudes and longitudes are (0°E, 0°N) and (0°E, 5°N) respectively. The starting height is 90km and the initial velocity is 4600m/s. The flight path angle is -6.5°, and the heading angle is 45°. The two aircrafts are required to reach the designated target area at the same time, and the push-down point should satisfy the constraints, including that the height is 30km, the velocity is between 1200m/s and 1800m/s, and the range to be flying is 50km. The coordination time interval is taken as $\Delta t = 100s$, and the allowable time difference is $\varepsilon_t = 1s$. The simulation results are shown in figure 4.

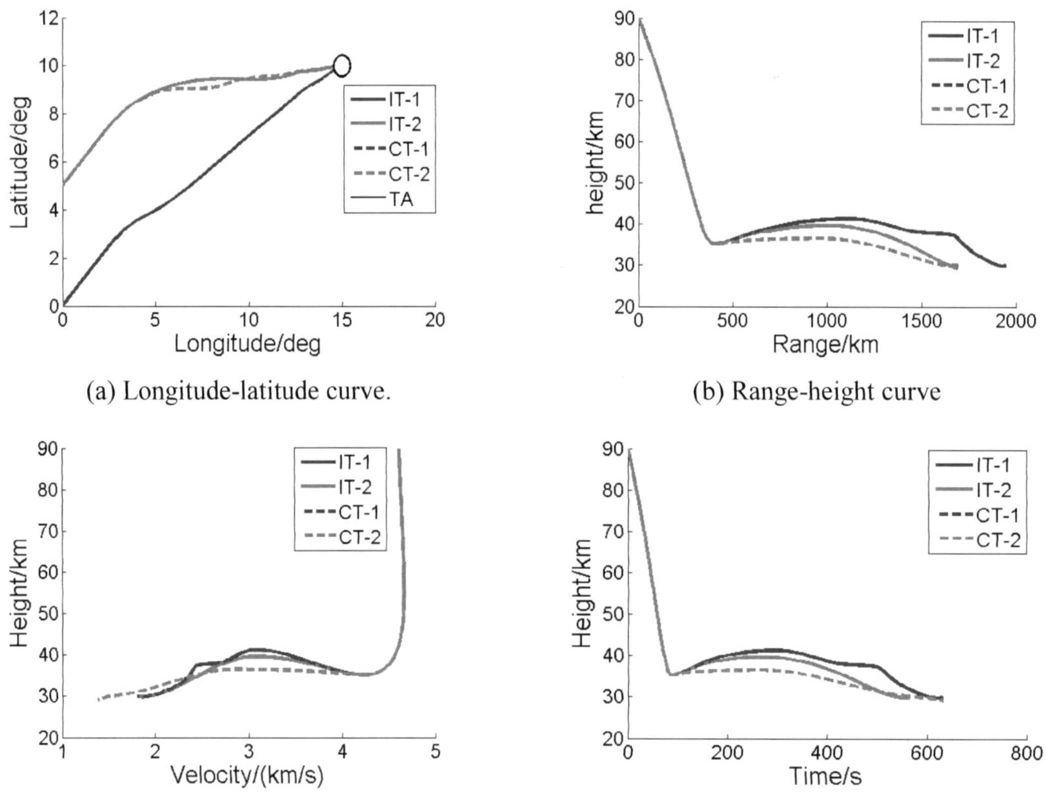

(a) Longitude-latitude curve.

(b) Range-height curve

(c) Velocity-height curve

(d) Time-height curve

Figure 4. Results for launching at the same time but the different place.

From the simulation results, it can be seen that before the cooperative is performed, the flight time of initial trajectory 1 time is 631.7 s, and the initial trajectory 2 is 564 s. The aircraft 2 reaches the target area earlier than aircraft 1. In order to make the two aircraft reach the target area at the same time, the trajectory of the aircraft 2 needs to be replanned online. Therefore, the deceleration of aircraft 2 is controlled until reaching the target area simultaneously with the aircraft 1. On account of that the trajectory of aircraft 1 does not need to be replanned, the IT-1 and CT-1 are coincident. The number of trajectories seen from the figure is three. From the flight time information of the two aircrafts, it can be seen that the time coordination control strategy successfully makes the two aircraft reach the target area at the same time. Finally, the two aircrafts complete the coordinated combat mission.

From the simulation results obtained in the above two different situations, we can see that the cooperative trajectory planning method can enable multiple aircrafts to reach the target area at the same time, which layes the foundation for the terminal guidance.

6. Conclusions

With the continuous development of hypersonic technology and the strengthening of the aerial defence system, multi-aircraft cooperative combat will become a new combat style. In view of the deficiencies in the current research, this paper does research on cooperative planning technology of gliding phase for multi-aircraft. Based on the online trajectory generation and a feedback linearization tracking guidance law for a single aircraft, a multi-aircraft trajectory planning strategy based on time coordination is designed under time coordination condition. Two cases are simulated and verified under different operational backgrounds. The simulation results show that the designed trajectory cooperative planning strategy can effectively achieve the multi-aircraft reaching the target area at the same time. The flight process meets the constraints of the reentry corridor. The trajectory satisfies the terminal constraint, which can lay the foundation for the subsequent terminal guidance.

7. Acknowledgments

The authors wish to acknowledge the editor and the anonymous referees for their suggestions and comments which have resulted useful for improving the paper and the two foundations, including Aeronautical Science Foundation of China (No. 2016ZA88003) and CALT201702.

References

[1] Dhananjay N and Dhananjay N 2014 Accurate time-to-go estimation for proportional navigation guidance *Journal of Guidance Control & Dynamics.* **37** 1378-1383.

[2] Jeon I S, Lee J I and Tahk M J 2010 Homing guidance law for cooperative attack of multiple missiles *Journal of Guidance Control & Dynamics* **33** 275-280

[3] Sun X J, Zhou R and Hou D L 2014 Consensus of leader-followers system of multi-missile with time-delays and switching topologies *Vac Optik - International Journal for Light and Electron Optics.* **125** 1202-1208

[4] Zhang L M, Sun M W and Chen Z Q 2014 Receding horizon trajectory optimization with terminal impact specifications *Mathematical Problems in Engineering* 693-697

[5] Zhang Y A, Wang X L and Wu H L 2014 Impact time control guidance law with field of view constraint *Aerospace Science & Technology* **39** 361-369

[6] Song J H, Song S M and Xu S L 2016 Cooperative guidance law for multiple missiles with impact angle constraints *Journal of Chinese Inertial Technology* **24** 554-560

[7] Zhang C Y, Song J M and Hou B. 2016 Cooperative guidance law with impact angle and impact time constraints for networked missiles *Acta Armamentarii* **37** 431-438

[8] Fang Y, Ma K and Chen Y 2014 Cooperative guidance laws with constraints on impact time and terminal angle *Journal of System Simutation* **26** 2434-2441

[9] Jung B and Kim Y 2006 Guidance laws for anti-shipping missiles using impact angle and impact time *AIAA Guidance, Navigation, and Control Conferenc. Sci. Technol, Keystone, Colorado*

[10] Kumar S R and Ghose D 2015 Impact Time and Angle Control Guidance *AIAA Guidance, Navigation, and Control Conference*

[11] Fang Y, Ma Y and Chen Y 2014 Cooperative guidance laws with constraints on impact time and terminal angle *Journal of System Simulation* **26** 2434-2441

[12] Vinh N X 1981 *Optimal trajectories in atmosphere flight* (New York: Elsevier Scientific Publishing Company)

IOP Conf. Series: Materials Science and Engineering **408** (2018) 012026 doi:10.1088/1757-899X/408/1/012026

Research on Application of Digital Assembly Technology based on MBD in Spacecraft Field

Xiaohui Song[1,2,*], Yi Lu[1], Zhibin Liu[1], Chunsheng Yang[1], Pei Wang[1] and Feng Xue[1]

[1]Beijing Institute of Spacecraft Environment Engineering, Beijing, China
[2]Beijing Engineering Research Center of the Intelligent Assembly Technology and Equipment for Aerospace Product, Beijing, china

[*]songhsiaohui@126.com

Abstract. With the development of technology based on MBD, three-dimensional model has already been the sole basis during spacecraft assembly process, three-dimensional process design technology on MBD has all-around been applied in spacecraft assembly field. On the base of the five key technology which include digital assembly process design, visualization of three-dimensional process, simulation verification of three-dimensional assembly process, three-dimensional work instruction browsing and digital inspection, spacecraft assembly has realized the transformation from two-dimensional process design system to three-dimensional process design system. The digital assembly technology on MBD has thoroughly changed the spacecraft assembly flow. Through the implement of three-dimensional process design technology, the process design results are transferred to the workshop assembly site in the form of three-dimensional assembly instruction, improving the working environment of the production site, the level of spacecraft assembly is improved.

1. Introduction

MBD means a definition based on model, is a kind of method of using integrate three-dimensional entity model to fully express the definitional information of product[1]. Based on the model of three-dimensional product, MBD assembled characters like size marking, tolerance demand, technology information to describe and share the product to fulfill the direct delivering demand of digitization producing information under the situation of cancelling two-dimensional drawing engineering definition. This technology co-defined PMI (3D Product Manufacturing Information) and three-dimensional information into the three-dimensional digital model of product. Also this technology fulfilled the highly integration CAD and CAM (process, assemble, measure and test) by replacing two-dimensional design pattern with three-dimensional dimensioning model as the producing base of product. In the technological system of MBD, MBD data set contains several information from different sections, including design, technology and manufacture[2]. The schematic diagram of MBD data structure is shown in Figure 1.

Three-dimensional digital technology that is based on MBD had applied to many of the general assembly fields of China's spacecraft with the improvement and perfect of the CAD/CAM technology. It has substantially promoted the development of digital assembly technique, improved the current production status of manually assemble spacecraft and enhanced productivity, reduced coordination links and developed the assembly technique of spacecraft.

This article illustrates the digital assembly technique and appliance of spacecraft based on the technical background of MBD.

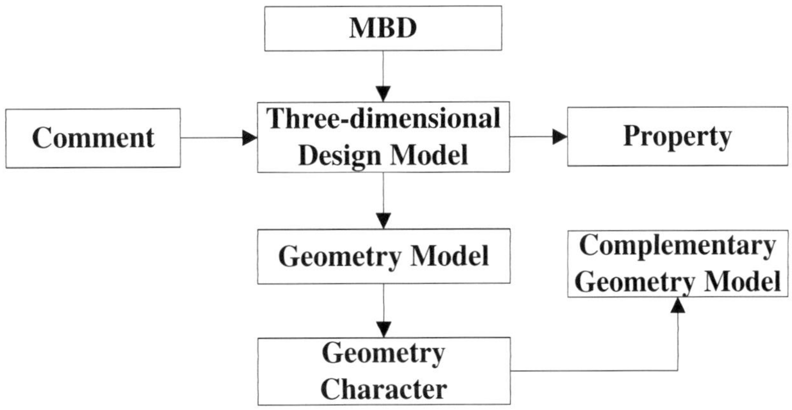

Figure 1. MBD data structure.

2. Digital assembly process design

Process design is the bridge that connects design and produce. Within the traditional two-dimensional environment, process design relies on the two-dimensional engineering drawing provided by design, carry out works like process design and matching table, establishing process cards and signing technological process in CAPP system. Then it would be distributed to production unit, combined with files like the standard code of practice onsite and the executional record of the MES system and direct production personnel to carry out operation in the end.

The assembly process technology based on MBD refers to the high-efficient and rational assembly process design system that based on light weight three-dimensional model, utilized MBD data information, digital and structural method to define and present the whole production process[3]. The system established a satisfactory three-dimensional digital spacecraft prototype through setting up the MBD digital definition standard, adopting three-dimensional modelling to define digital products. Under the instruction of MBD process design, technologists would develop their design and establish three-dimensional digital assembly process model directly according to the three-dimensional entity model and conduct digital analogue simulation of the assembly process through digital virtual assembly environment. The virtual assembly verification of the assembly process is conducted to ensure a preciseness of the process the same time as the design process, before the entity assembly process, and the authorized assembly design is issued for field use and entity assemble. The three-dimensional process diagram and multimedia animation data of the assembly operational process is produced during the process of digital analogue simulation of the assembly process, combined with assembly process to establish digital assembly design data and provide evidence for digital assembly process field.

Three-dimensional structural design system of the general assembly of spacecraft is a deep-developed and specialized system with a featured character of the general assembly of spacecraft, which is based on the "Team centre" system of Siemens. Three-dimensional structural process design is the structural combination of process digital prototype, process structure, enclosed piece, technological resources, technological material and enforcement record. The main function of this system includes receiving design data, such as three-dimensional design data, realizing a technological prototype which faces general assembly under three-digital model; realizing a multiple and full connection between model machine, craft, process, step, materials and resources; realizing the fusion of process planning and checking executive logging; realizing the bi-directional, seamless and real-time integration with MES system; multimediatizing the process; unifying the resource of data of the formal process, temporary process and process planning; breaking through the data link between process and the site; establishing a full life circle data from design , process model machine, three-digital technology to AIT site and a data net that connects them all. The logic diagram of aircraft general assemble structural process design system is shown in Figure 2.

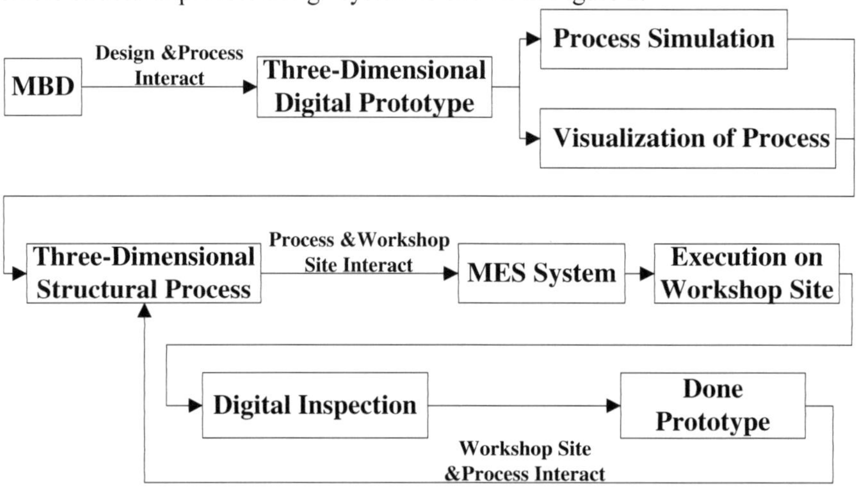

Figure 2. Working logic of three-dimensional structured process planning system for spacecraft assembly

3. Digital assembly process design

Assembly process planning is to determine the assembly plan by analyzing the structure and characteristics of the spacecraft. In a three-dimensional environment, the shape and entity model of the product are present in front of the craft, and the technicians can directly measure and analyze the process separation surface, the sequence diagram of the command assembly, the main zero component assembly datum and positioning methods, the main coordination parts and methods, the assembly work position, the assembly sequence and parts of the parts. The docking scheme is carried out and the corresponding process and equipment design work is carried out at the same time. The planning process of assembly process is divided into four modules: resource module input module, assembly process planning module, assembly process simulation module and assembly process file output module[4].

3.1. Resource module input module

The assembly process model is designed and simulated in three-dimensional environment based on the product MBD data set and the product EBOM table extracted from the data information. The product information is extracted from the EBOM form to form a product structure tree. On this basis, the assembly process plan is formulated, the assembly requirements are made and the assembly unit are divided and refined to "work step" level, therefore formed the assembly process structure tree and related to the product and manufacturing resources. The assembly process planning and simulation optimization, the design and assembly process, and the design of the three-dimensional process documents are distributed to the relevant personnel in parallel, and the assembly process model with various information of the product assembly is formed.

3.2. Assembly process planning module

According to the product structure, the assembly process structure and process flow are divided, and the workload of the assembly unit is evenly distributed by taking workload of each unit into consideration.

3.3. Assembly process simulation module

The rationality of the process planning, the parallelism of operation procedures and the correctness of product and tooling design are verified by simulation.

3.4. Assembly process file output module

In the digital process planning environment, using the product structure tree and manufacturing resources, and also using the process template as base, the output of different process documents is carried out on different process nodes through program, and a three-dimensional lightweight model (three-dimensional picture and three-dimensional dynamic painting) is formed through program integration to form a lightweight three-dimensional process file.

4. Visualization of three-dimensional process

The visualization technology of assembly process is to transfer the product design information, manufacturing resource information and process design information to the workshop site in the form of digital quantity, and the method is displayed[5]. Digital product, digital process model and its three-dimensional AO (Assembly Order, that is assembly instruction) has taken the place of traditional two-dimensional engineering drawings, process design and two-dimensional assembly instructions completely, which improved the effect of technical training for workers. It allows the instruction of workers to carry out field production vividly and intuitively, and set up the assembly for field production. It is necessary to visualize an application system facing field production.

Laying network in assembly workshop, setting up digital application terminal equipment at production site, transferring three dimensional data to assembly operation site, using multimedia assembly process information, three-dimensional product data and three dimensional tooling data, and guiding spaceflight assembly work in digital environment.

Because of the large number of spacecraft assembly components, in order to quickly simulate the process of product assembly, so that to make a higher requirement to the performance of computer on the one hand, and to make a requirement of carrying out a lightweight display on the other hand. According to the assembly datum and process of the product, lightweight product MBD model (consistent with the data set designed by the design department) and assembly operation animation made by 3DVIA Composer software to realize the lightweight index of the three-dimensional assembly process regulations. So that to lay a solid foundation for the design of technical specification for lightweight assembly on the one hand. And to enable the surface assembly technology to have an intuitive and vivid characteristics on the other hand, so that workers can operate more directly and improve assembly efficiency. An example of three-dimensional visualization of spacecraft assembly is shown in Figure 3.

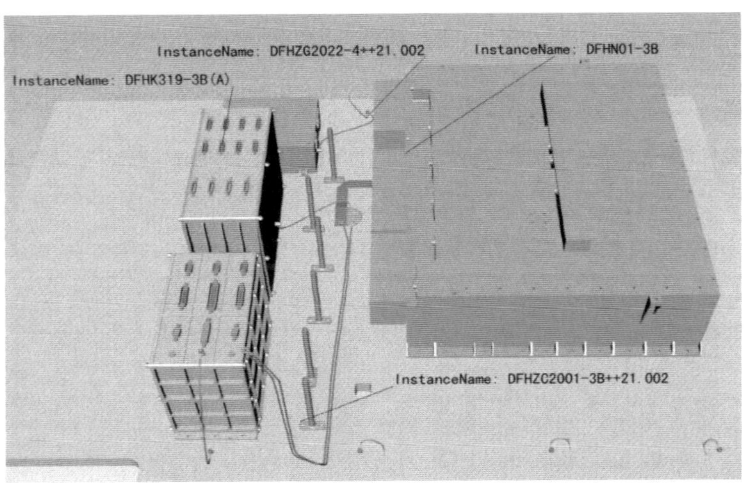

Figure 3. Application example of three-dimensional process visualization for spacecraft assembly

5. Simulation verification of three-dimensional assembly process

Assembly process simulation is an interactive method to simulate the assembly process in virtual assembly environment. It is used to analyze the rationality of the design of the assembly process model before the physical assembly, and the assembly process model is continuously optimized through the continuously discovery of the problem. Assembly process simulation verifies the assemblability of structural parts and system parts, and manual assembly simulation verifies the visibility, accessibility and operability of assembly workers, and the automatic assembly simulation verifies the interference between terminal actuator, flexible tooling and fitting parts[6]. After process certification is completed, possible modifications to corresponding process, tooling and even product design are made. The two main directions of assembly simulation are assembly sequence planning and assembly path planning.

Assembly sequence planning (ASP) is the design of assembly sequence in computer based on the product model of design phases. It is one of the key technologies for Design for Assembly (DFA), and is the core content of assembly process design. One correct assembly sequence is selected to verify whether each part is assembled unimpeded by the designed process sequence, and whether assembly time and cost are reduced. A flow diagram of the process design using a three-dimensional model is shown in Figure 4.

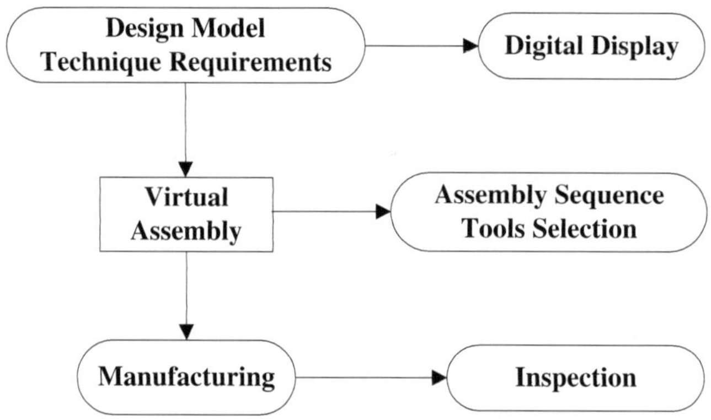

Figure 4. Flow chart of process design using three-dimensional models

Assembly path refers to the location of the product from the location of the installed parts until the parts are assembled onto the assembly body to form the trajectory of the product. The assembly path planning is based on the product structure and related technical requirements, and through a certain algorithm to find a space motion path for a part to be installed from the starting position to the designed product. An example of the application of the spacecraft assembly path planning is shown in Figure 5.

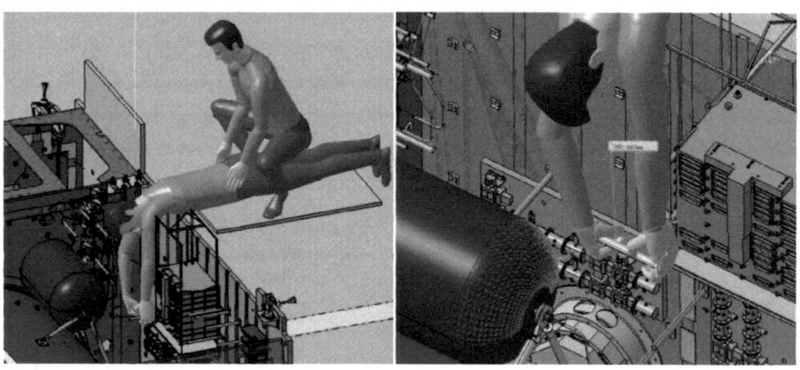

Figure 5. Application example of spacecraft assembly path planning

IOP Conf. Series: Materials Science and Engineering **408** (2018) 012026 doi:10.1088/1757-899X/408/1/012026

6. Three-dimensional work instruction browsing

The traditional AO (Assembly Order), that is the assembly instruction, is a process document used to guide production according to the requirements of engineering design and the requirements of existing technological level and quality assurance. The process instruction under three-dimensional conditions is a combination of structured process information (standard format) and three-dimensional information such as model, view and animation to form a three-dimensional assembly process (instruction) electronic data packet that can guide the work of assembly staff.

With the development of MBD technology, three-dimensional AO has become a digital process document. Which includes product design information, resource information, process design information and process animation. In the MBD manufacturing mode, three-dimensional assembly process field application system for spacecraft assembly is established, and three-dimensional product engineering data, three-dimensional tooling data, operation process graphics and operation animation are organized to complete the three-dimensional AO construction.

In the preparation stage of assembly, by watching the animation of assembly process, assembly workers can have an intuitive understanding of the whole assembly work and master the contents of assembly work, including assembly objects, assembly order and method of the parts positioning. Through digital assembly application terminal, process instructions are described in the way of three-dimensional view and three-dimensional animation, and structured text information is supplemented.

7. Digital inspection

MBD technology defines three-dimensional product manufacturing information (PMI) and three-dimensional design information together in three-dimensional numerical model of the product. Under the MBD technology environment, the inspection of products needs to be protected according to three-dimensional numerical models and relevant quality technical documents.

The spacecraft assembly data model based on MBD technology integrates three-dimensional annotation information needed for product assembly, measurement and inspection. MBD digital terminal users can extract the required data from the digital module according to their needs, and carry out the operation of digital mode measurement, marking and capture.

Three-dimensional annotation information based on MBD technology includes two kinds: geometric and non-geometric information. Geometric information refers to the annotation in addition to three-dimensional entities. Non-geometric information is mainly tagged on feature tree. Three-dimensional labeling of assembly is tagged in a newly built "component" model, and is tagged on assembly structure tree with other models. Due to the large number of spacecraft fittings, size annotation and other information provided in the MBD module do not necessarily conform to the user's needs. This requires users to select and extract useful product information actively from the number mode according to their actual needs[7].

The installation of checkpoints for spacecraft assembly is the crux of spacecraft assembly inspection. Spacecraft assembly is a complex system engineering. Many factors and links need to be coordinated in the assembly process. Therefore, the assembly inspection needs to be divided into different stages. Each stage has different inspection priorities until the whole assembly inspection is completed.

In the inspection stage of spacecraft assembly, inspection methods can be adopted: visual inspection, gauge inspection, fixture inspection and photographic examination. Different technical requirements and inspection states need to be used in different inspection and inspection methods, and the required inspection points are different. Therefore, it is necessary to ensure the scientificalness of the inspection points. The scientificalness of the test point setting is based on the familiarity and digestion of the MBD number model, the understanding of the relationship between the different parts and the relationship between the different parts and the motion relations and functions. It is a dynamic adjustment process based on the design requirements and technical documents. It may be changed according to the design and process changes at any time. Changes in relevant technical documents are made to make adjustments[8].

8. Conclusion

The general assembly of spacecraft is a comprehensive technology with high technical difficulty, high complexity and complex cooperation. At the same time, due to the high reliability requirements of spacecraft products, the requirements of assembly quality are very strict. MBD technology provides the process information for assembly site application. The process design results are transferred to the workshop assembly site in the form of three-dimensional assembly instruction, improving the working environment of the production site, improving the product quality and production efficiency. Therefore, the application of digital assembly technology in the field of spacecraft assembly will contribute to the long-term development of China's spacecraft business.

References

[1] Qiuzhong Z and Yuqing F 2008 Application of MBD on Airplane Manufacturing *AVTION MAINTENANCE & ENGINEERING* 2008(3):55-57
[2] Zhongyi M 2010 Digital Aircraft Assembly Technology Based on MBD *AERONAUTICAL MANUFACTURING TECHNOLOGY* 2010(18):42-45
[3] Jihua Z 2009 Aircraft Digitized Measurement Aided Assembly Technology and Application. *ERONAUTICAL MANUFACTURING TECHNOLOGY* 2009(24):49-52
[4] Fuzhou D & Haicheng L 2010 Research on Key Technology of MBD-Based Aviation First Article Inspection *AERONAUTICAL MANUFACTURING TECHNOLOGY* 2010(23):57-59
[5] Junyi N & Xiaofei Y 2012 Expression and Transmission of Engineering Information Under the Pattern of MBD *AERONAUTICAL MANUFACTURING TECHNOLOGY* 2012(6): 77-79
[6] Yanming Z 2012 Process Design based on MBD Technology *CAD/CAM YU ZHIZAOYE XINXIHUA* 2012(12):60-63
[7] Yuqing F 2012 Model Based Definition Technology and Its Practices *AERONAUTICAL MANUFACTURING TECHNOLOGY* 2012(6):42-47
[8] Alemanni M, Destefanis F and Vezzetti E 2011 Model based definition design in the product lifecycle management scenario *International Journal of Advanced Manufacturing Technology* 2011(1/4):1-14.

IOP Conf. Series: Materials Science and Engineering **408** (2018) 012027 doi:10.1088/1757-899X/408/1/012027

Dynamic Modelling of the Hovering Phase of a New V/STOL UAV and Verification of the PID Control Strategy

Xiaomeng Zhang* **and Weijun Wang**

Beihang University, Beijing, China

*15731168528@163.com

Abstract. This paper analyses the dynamics of a vertical/short take-off and landing (V/STOL) unmanned aerial vehicle (UAV), and obtains a nonlinear dynamics model. During the flight from vertical take-off to hovering, the roll moment, pitch moment and yaw moment are regarded as control inputs to linearize the nonlinear dynamic model to obtain the system transfer function. Finally, the control scheme of pitch angle during the flight from vertical take-off to hovering is developed, and the control scheme is verified by simulation. The simulation results reveal that the designed Proportional-Integration-Differential (PID) attitude controller can achieve stable control of the pitch attitude of the V/STOL drone during the hovering phase.

1. Introduction

Vertical/short take-off and landing (V/STOL) aircraft combines the advantages of fixed-wing aircraft and helicopter. In addition to achieving the speed of fixed-wing aircraft, V/STOL is also not subject to site constraints, which makes V/STOL aircraft of great significance in military and civil affairs.

V/STOL aircrafts are divided into two types by forms at the dynamic force: one is to achieve short-range vertical take-off and landing by deflecting thrust vector nozzles, such as "Harrier Jet"[1]; the other is through the tilting rotor, such as MV-22 "Osprey" [2]. However, aircrafts designed according to these two methods are complex in design and heavy. Therefore, these types are not suitable for unmanned aerial vehicle.

As shown in Figure 1. For the new V/STOL, the two pairs of propellers on the leading edge of the wing, combined with the help of the front propeller, produce the increasing effect of deflection slip flow of the propeller [3-4]. In this way, unconventional vertical take-off and hovering of the aircraft can be achieved. What's more, the increasing effect of deflection slip flow of the propeller can be also fully utilized by the design of the double slotted flap on the wing rudder surface. In general, the requirements of short-range take-off and landing drones, light and simple construction and low cost, are satisfied.

Figure 1. The schematic diagram of V/STOL aircraft

IOP Conf. Series: Materials Science and Engineering **408** (2018) 012027 doi:10.1088/1757-899X/408/1/012027

2. Analysis of dynamic equation

2.1. Coordinate system conversion

The earth coordinate system is attached to the surface of the earth, usually denoted by $OX_gY_gZ_g$, the usual earth coordinate system is the North-East-Earth coordinate system. That is, any point on the ground is regarded as the origin of the coordinate system, usually the position of the aircraft taking off. The body coordinate system is fixed to the aircraft. It is a dynamic coordinate system that changes continuously with position of the aircraft. The body coordinate system is generally represented by $OX_bY_bZ_b$.

According to the above setting, the transformation between the body coordinate system and the ground coordinate system is represented by the roll angle φ, the pitch angle θ, and the yaw angle ψ. By the vector transformation of Euler angles, the transformation from the body coordinate system to the earth coordinate system can be given by:

$$C_b^n = C_z(\psi)C_y(\theta)C_x(\phi) \tag{1}$$

$$C_b^n = \begin{bmatrix} \cos(\theta)\cos(\psi) & -\cos(\phi)\sin(\psi)+\sin(\phi)\sin(\theta)\cos(\psi) & \sin(\phi)\sin(\psi)+\cos(\phi)\sin(\theta)\cos(\psi) \\ \cos(\theta)\sin(\psi) & \cos(\phi)\cos(\psi)+\sin(\phi)\sin(\theta)\sin(\psi) & -\sin(\phi)\cos(\psi)+\cos(\phi)\sin(\theta)\sin(\psi) \\ -\sin(\theta) & \sin(\phi)\cos(\theta) & \cos(\phi)\cos(\theta) \end{bmatrix} \tag{2}$$

Supposing: p, q, r are the angular velocities of the aircraft in the body coordinate system, and $\dot{\varphi}$, $\dot{\theta}$ $\dot{\psi}$ are the Euler angles velocities of the aircraft in the earth coordinate system. Based on the conversion relationship, it can be expressed that:

$$\begin{cases} \dot{\phi} = p+\left(r\cos(\phi)+q\sin(\phi)\right)\tan(\theta) \\ \dot{\theta} = q\cos(\phi)-r\sin(\phi) \\ \dot{\psi} = \dfrac{1}{\cos(\theta)}\left(r\cos(\phi)+q\sin(\phi)\right) \end{cases} \tag{3}$$

2.2. Linear dynamics equations

The movement of the aircraft in space is described by 6-DOF (Six Degrees Of Freedom) model. The linear velocity vector of the aircraft in the ground coordinate system is $V_E = [\dot{x}\ \dot{y}\ \dot{z}]^T$, the angular velocity vector in the body coordinate system is $W_b = [p\ q\ r]^T$. In the hovering state, only three forces, which are the force of the combination of propeller, wing and flap, the force of front propeller and gravity, need to be considered respectively. In the body coordinate system, the schematic diagram of force analysis in the hover state is depicted in figure 2.

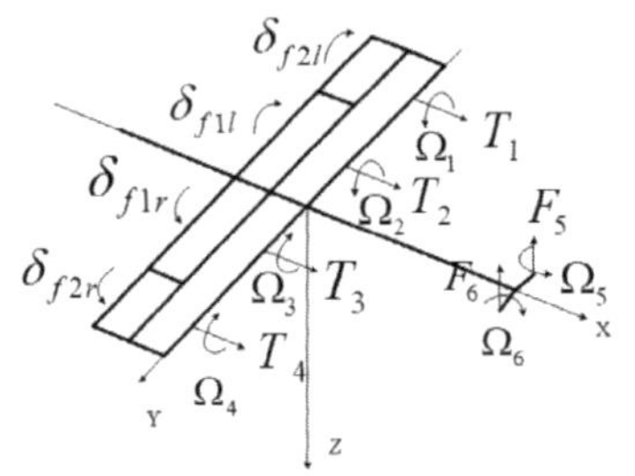

Figure 2. The schematic diagram of force analysis in the hovering state

The force of the combination of propeller, wing and flap acts on 1, 2, 3, and 4 points respectively, regarded as F_1, F_2, F_3, F_4. The propellers at the 2 and 3 positions turn inwards. The propellers at the 1 and 4 positions turn outward. The force of the forward propeller is applied at 5 and 6 points, turn outward too. The flap deflection angle is denoted by 'a 'and the aileron deflection angle is denoted by 'b '. It should be noted that in the process analysis of the present work, it is assumed that the flap and the aileron deflection angles are fixed during the hovering process, that is to say, 'a' and 'b' are constant. Equations of force analysis in the body coordinate system are as follows:

$$\begin{cases} f_x = \cos(a)(F_2+F_3)+\cos(b)(F_1+F_4) \\ f_y = 0 \\ f_z = -\sin(a)(F_2+F_3)-\sin(b)(F_1+F_4)-(F_5+F_6) \end{cases} \tag{4}$$

The joint force F_E in the earth coordinate system can be calculated by equation (5):

$$F_E = f_x\begin{bmatrix}\cos(\theta)\cos(\psi)\\ \cos(\theta)\sin(\psi)\\ -\sin(\theta)\end{bmatrix} + f_z\begin{bmatrix}\sin(\phi)\sin(\psi)+\cos(\phi)\sin(\theta)\cos(\psi)\\ -\sin(\phi)\cos(\psi)+\cos(\phi)\sin(\theta)\sin(\psi)\\ \cos(\phi)\cos(\theta)\end{bmatrix} + \begin{bmatrix}0\\0\\G\end{bmatrix} \tag{5}$$

The linear motion equation of the aircraft in the earth coordinate system can be determined from equation (6):

$$\begin{bmatrix}\ddot{x}\\ \ddot{y}\\ \ddot{z}\end{bmatrix} = \frac{f_x}{m}\begin{bmatrix}\cos(\theta)\cos(\psi)\\ \cos(\theta)\sin(\psi)\\ -\sin(\theta)\end{bmatrix} + \frac{f_z}{m}\begin{bmatrix}\sin(\phi)\sin(\psi)+\cos(\phi)\sin(\theta)\cos(\psi)\\ -\sin(\phi)\cos(\psi)+\cos(\phi)\sin(\theta)\sin(\psi)\\ \cos(\phi)\cos(\theta)\end{bmatrix} + \begin{bmatrix}0\\0\\g\end{bmatrix} \tag{6}$$

2.3. Angle kinematic equations

Under the effect of moment, the aircraft rotates around the centre of mass. At the hovering and take-off states, owing to the small mass and volume, moment of inertia of the rotor is very small, so the influence of the gyro effect can be ignored. When taking off, the aircraft is at a very low speed, so the air resistance proportional is also negligible. Therefore, only the torques generated by rolling, pitching and yawing are taken into account, which are represented by L, M, and N, respectively:

$$\begin{cases} L = \sin(a)(F_2-F_3)r_1+\sin(b)(F_1-F_4)r_2+(F_5-F_6)R \\ M = (F_5+F_6)l_1-[\sin(a)(F_2+F_3)+\sin(b)(F_1+F_4)]l_2 \\ N = \cos(a)(F_2-F_3)r_1+\cos(b)(F_1-F_4)r_2 \end{cases} \tag{7}$$

According to $\mathrm{M} = I\dot{W}_B + W_B \times IW_B$:

$$\begin{cases} L = \dot{p}I_x + qr(I_z-I_y)-(\dot{r}+pq)I_{xz} \\ M = \dot{q}I_y + pr(I_x-I_z)+(p^2-r^2)I_{xz} \\ N = \dot{r}I_z + pq(I_y-I_x)+(qr-\dot{p})I_{xz} \end{cases} \tag{8}$$

In the body coordinate system,I_x, I_y, I_z are moments of inertia in the direction of X, Y, and Z, and $I_{xz}$is the product of inertia of the axes x and z of the aircraft body. Because the mass of the aircraft is constant, each moment of inertia and inertial product are constant, so each coefficient in equation (5) can be replaced by C_1-C_9.

$$\begin{cases} \dot{p} = (c_1 r + c_2 p)q + c_3 L + c_4 N \\ \dot{q} = c_5 pr - c_6(p^2-r^2)+c_7 M \\ \dot{r} = (c_8 p - c_2 r)q + c_4 L + c_9 N \end{cases} \tag{9}$$

3. Dynamic equation modeling

Based on equations (3), (6) and (9), the nonlinear dynamics model of the aircraft can be derived.

$$\begin{cases} \ddot{x} = \dfrac{f_x\cos(\theta)\cos(\psi) + f_z\left[\sin(\phi)\sin(\psi) + \cos(\phi)\sin(\theta)\cos(\psi)\right]}{m} \\ \ddot{y} = \dfrac{f_x\cos(\theta)\sin(\psi) + f_z\left[\cos(\phi)\sin(\theta)\sin(\psi) - \sin(\phi)\cos(\psi)\right]}{m} \\ \ddot{z} = \dfrac{-f_x\sin(\theta) + f_z\cos(\phi)\cos(\theta)}{m} + g \\ \dot{p} = (c_1 r + c_2 p)q + c_3 L + c_4 N \\ \dot{q} = c_5 pr - c_6(p^2 - r^2) + c_7 M \\ \dot{r} = (c_8 p - c_2 r)q + c_4 L + c_9 N \\ \dot{\phi} = p + (r\cos(\phi) + q\sin(\phi))\tan(\theta) \\ \dot{\theta} = q\cos(\phi) - r\sin(\phi) \\ \dot{\psi} = \dfrac{1}{\cos(\theta)}(r\cos(\phi) + q\sin(\phi)) \end{cases} \tag{10}$$

Linearizing the nonlinear dynamics model of the aircraft in hovering state [5-6], the system can be defined as follows:

$$\begin{cases} \dot{X} = AX + Bu \\ Y = CX + Du \end{cases} \tag{11}$$

Where: X is a state variable, Y is the system output, u is the input vector, and the state space matrixes A, B, C, D are functions of time-varying parameters. X, Y, u are defined respectively as:$X = \left[\dot{x}, \dot{y}, \dot{z}, \dot{\phi}, \dot{\theta}, \dot{\psi}, \phi, \theta, \psi\right]^T$,$Y = [\phi, \theta, \psi]^T$,$u = [u_1, u_2, u_3]^T$

$$\begin{cases} u_1 = \dfrac{c_3}{c_4}L + N \\ u_2 = M \\ u_3 = \dfrac{c_9}{c_4}N + L \end{cases} \tag{12}$$

Laplace transform is performed on equation (11):

$$G(s) = C(sI - A)^{-1}B + D \tag{13}$$

Equilibrium computation shows that, in order to meet the requirements of vertical take-off and hovering and achieve a balance between force and moment during the vertical take-off to hovering phase, the aircraft needs to maintain an angle of 40°-60° with the ground. Afterwards, a set of trim points in the hovering state are obtained [7-8], and the pitch angle for the trim points is 60°.

After the linearization process, the transfer function can be developed.

$$G(s) = \begin{bmatrix} \dfrac{c_4}{s^2} & 0 & \dfrac{433c_4}{250s^2} \\ 0 & \dfrac{c_7}{s^2} & 0 \\ 0 & 0 & \dfrac{2c_4}{s^2} \end{bmatrix} \tag{14}$$

According to $Y(s) = G(s)u(s)$, the transfer function relationship between pitching moment and pitching angle can be expressed as:

$$\theta = \frac{c_7}{s^2}M = \frac{3.846}{s^2}M \tag{15}$$

4. Design of the PID control criterion

PID controllers have simple control structure and are easier to set in practical applications. Therefore, the PID controller is used to control the pitch angle. PID controller schematic diagram [9] is shown in Figure 3.

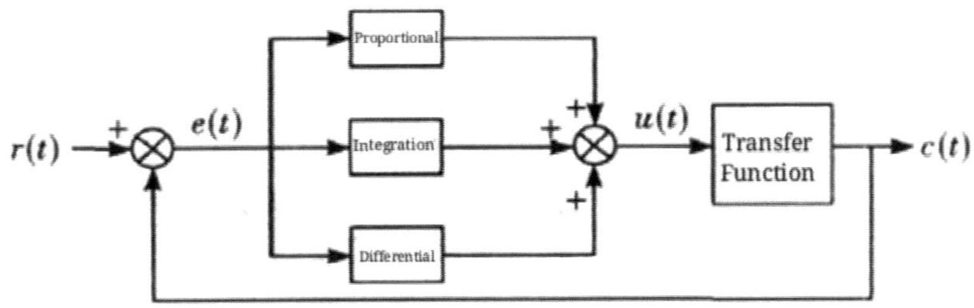

Figure 3. PID controller principle

Rise time, overshoot, settling time, and steady-state error are indicators to measure the performance of a control system. These four parameters reflect the system's responsiveness and stability, through which the performance of a system can be determined. With step response, the desirable performance specifications of the control system designed in this paper include the settling time which is less than 2 s, the overshoot which is less than 20%, and the rise time which is less than 1 s.

Based on the transfer function, the PID control strategy for pitch angle control is designed. The simulation with MATLAB/SIMULINK software verifies the correctness of the PID control strategy. Simulation block diagram is descripted in Figure 4. Parameter setting is crucial for PID control strategy. In this paper, the proportion degree method [10] is used to set the PID parameters. The parameters of the PID controller are $K_P = 1.5$, $K_I = 0.06$ and $K_D = 0.3$. In the actual simulation process, the parameters may be also adjusted by the root locus method and finally determined by the limit value range of the PID parameter in the outer-loop attitude control loop of flight control system. Simulation results are shown in Figure 5. Results indicate that the PID control strategy can basically meet the control requirements and achieve stable control of the pitch angle.

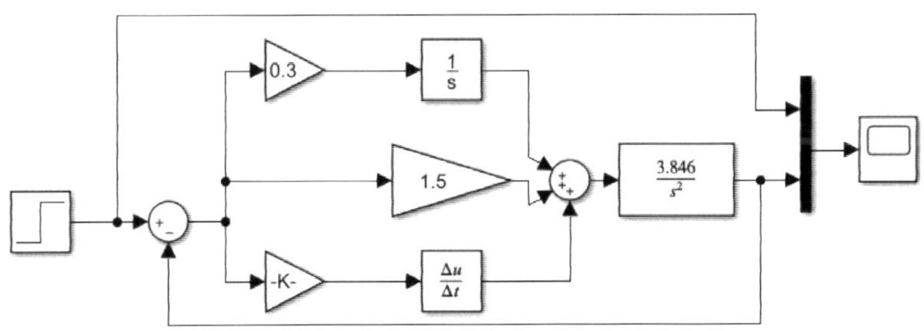

Figure 4. Pitch angle attitude simulation of the PID controller

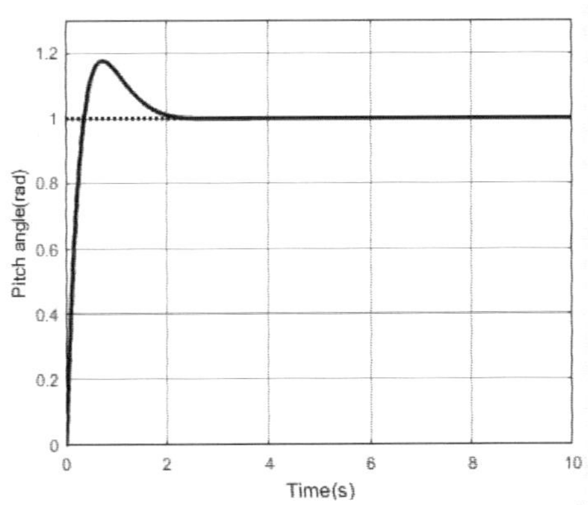

Figure 5. The pitch angle response curve under step input

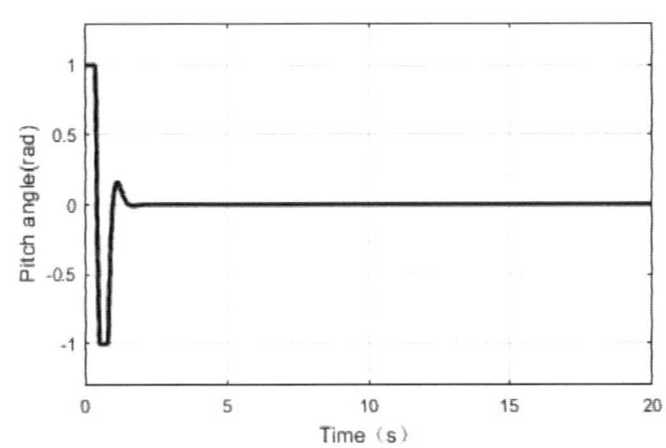

Figure 6. Changing curve of the pitch angle versus time

Finally, based on the Pixhawk flight controlling unit, the flight dynamics simulation model of the aircraft under the hovering state is constructed. The control amount of each channel for a set of trim and the PID parameters obtained by setting are input into the attitude angle control loop in the simulation model to obtain the simulation curve. As shown in Figure 6, with settling time less than 2 s, rise time less than 1 s and overshoot less than 20%, the simulation results reveal that the pitch attitude angle of the V/STOL aircraft under the hovering state is controllable and stable.

5. Conclusion

In this paper, the dynamics model built for the hovering and take-off phases of a new type of V/STOL UAV which is based on the principle of the increasing effect of deflection slip flow of the propeller aided by the forward vertical propeller is performed. Under the MATLAB/Simulink environment, the simulation model of PID controller is established and verify the control effect of PID controller. Results of the simulation suggest that this PID control system based on the vertical take-off and landing stage of the new V/STOL aircraft has good control effect and could achieve the stable control for pitch attitude.

References

[1] David L and Koldman 1981 *Introduction to V/STOL airplanes* (Iowa: Iowa State University Press)
[2] Herrero J L 2001 *Int. Society for Air Breathing Engines*
[3] Longyang Pan 2016 *Beijing Society of Theoretical and Applied Mechanics (Beijing)* vol 7 pp 60-61
[4] Kendoul F and Fantoni I and Lozano R 2005 *J. Helicopter Society Annual Forum* vol 61(1) pp 2-3
[5] WANG Qing and WANG Tong and HOU Delong 2014 *J. System Engineering and Electronic* vol 36(6) pp 1130-1136
[6] Wang Zhenchao and Peng Mingyan and Zhang Jinpeng and Yang Lingyu and Zhang Jing 2015 *J. AEO WEAPONY.* vol 1 pp 14-20
[7] ZHANG Zhi-Sheng and CHEN Huai-Min and WU Cheng-Fu and MA Song-Hui 2010 *J. Fire Control& Command Control.* vol 35(9) pp 93-97
[8] TANG Bin and HUANG Yi-Min 2007 *J. of Shenyang Institute of Aeronautical Engineering.* vol 24(1) pp 13-16
[9] Kendoul F and Fantoni I and Lozano R 2005 *Decision and Control, 2005 and 2005 European Control Conference* pp 8144-8149
[10] QIU Li and ZENG Gui and ZHU Xue and Sun Pei 2005 *J. Techniques of Automation and Application* vol 24(11) pp 28-31

The Influence of Blade Tip Loading Distribution on the Performance of a High Loading Centrifugal Impeller

Yilin Wang [1], Zefeng Li[1], Donghai Jin [1,2,*] and Xingmin Gui [1,2]

[1] School of Energy and Power Engineering, Beihang University, Beijing, China
[2] Collaborative Innovation Center For Advanced Aero-engine, Beijing, China

*jdh@buaa.edu.cn

Abstract. It has been proved that different blade loading distributions between tip and hub has a significant effect on centrifugal impeller performance, but the effect of loading difference at different location along the stream has not been fully understood. In this paper, several blades were modelled by a three-dimensional inverse design method, in which the impellers are defined by the blade loading distribution at its tip and hub and the flows are computed numerically by a commercial CFD code. Firstly, two impellers were modelled using the same distribution at the hub but giving different distributions at tip. The results show that the impeller with a smoother tip loading distribution performs better. At design flow rate, the total pressure ratio is found to be 1.8% higher and the efficiency rises 1.36% as well. Then a further research about the effect of increasing tip load at different parts along the flow path was carried out. Further analysis indicates that adding the load at the fore part of tip could increase the efficiency at design flow rate and brings more uniform impeller exit flow. However, the surge margin was narrowed by such change. Besides, adding the $v_u r$ at the aft part of tip will promote the pressure ratio, with no adverse effect on the stable operation range.

1. Introduction

Different from axial compressors, centrifugal compressors always possess a more complicated three-dimensional flow structure. Due to the narrow flow channel, large streamline curvature, and the Coriolis force, inevitable severe secondary flow happens in the impeller, which leads to a pitchwise non-uniform impeller exit flow. As a consequence, the flow in radial diffuser is always coupled with a considerable loss. Therefore, impeller design has always been a difficult part in the design process of centrifugal compressors.

At present, the inverse method of blade loading control has become a key technique in modern impeller design process. M.Zangeneh [1-3] developed a set of guidelines for the suppression of secondary flows. Therefore, a loading distribution in which the maximum loading is set in the fore part of the impeller at the shroud and the aft part at the hub has been applied in several different centrifugal pump impellers. As a result, the performance of the impeller designed using these guidelines was found to be 5 percent higher at the peak efficiency point as compared to the corresponding conventional impeller. Jiancheng Shi et al. [4] applied three kinds of loading distributions in a transonic impeller and demonstrated that the aft-loading impeller achieves the best performance. Lihua Tao [5] investigated the effects of different blade loading distributions on aerodynamic performance of a subsonic centrifugal impeller and stated that higher design efficiency can be obtained with fore-loading on the shroud and aft-loading on the hub, but the surge margin was narrowed. H F Leng [6] also made a similar conclusion in a double suction centrifugal pump. Manabu Yagi [7] thinks that the relative velocity decelerates with regards to the enhance of load at the fore part of tip, thus the flow separation at the axial to radial turning is controlled.

It is clear that the blade loading distribution has a significant influence on the performance of centrifugal impellers. Notably most of the past research objects were subsonic, low specific speed impellers with a relatively small flow coefficient and usually the loading distributions were totally changed. Actually impellers perform remarkably different along the load variation of tip even they adopt a similar kind of loading distribution. Besides, the increase of load in different parts of tip also works in different ways. In this paper, a CFD method is adopted to a high flow coefficient, high load, low specific speed transonic impeller to analyze the effect of different tip loading distribution on the performance.

2. The effect of blade loading distribution on the flow

The inverse design method based on a inviscid streamline curvature method procedure is developed by Ji Guofeng[8], where the meridional parameter $v_u r$ is specified as a design input. This parameter directly represents the amount of work done by the blades on flow, which was defined as ℓ_u:

$$\ell_u = \omega(v_{u2}r_2 - v_{u1}r_1) \tag{1}$$

Where v_u is the circumferential component of absolute speed v, r the radius. On the other side, ℓ_u could also be written as:

$$\ell_u = \frac{1}{2}(v_2^2 - v_1^2) + \frac{1}{2}(w_1^2 - w_2^2) + \frac{1}{2}(u_2^2 - u_1^2) \tag{2}$$

Where u means rotation speed, w represents the relative speed in rotating coordinate. Combined with the energy equation:

$$\ell_u = \frac{1}{2}(v_2^2 - v_1^2) + \int_1^2 \frac{1}{\rho} dp + \ell_f \tag{3}$$

It can be derived that:

$$\frac{1}{2}(w_1^2 - w_2^2) = \frac{1}{2}(u_1^2 - u_2^2) + \int_1^2 \frac{1}{\rho} dp + \ell_f \tag{4}$$

It can be considered that a higher $v_u r$ leads to an increase of v, a decrease of w, and an increase of static pressure. A higher static pressure gradient from tip to hub helps controlling the secondary flow of the blade surface, which can be easily seen from the conservation equation of the momentum in inviscid flow:

$$(\boldsymbol{W} \cdot \nabla)\boldsymbol{W} + 2\boldsymbol{\omega} \times \boldsymbol{W} = -\frac{1}{\rho}\nabla p + \nabla\left(\frac{u^2}{2}\right) \tag{5}$$

For the secondary flow within the boundary layer of the impeller, static pressure gradients and centrifugal force are the main factors driving radial migration. Adopting a higher $v_u r$ at tip than hub increases the static pressure gradient, which in turn supresses secondary flow. So by adjusting the loading distribution, a higher efficiency and more uniform exit flow are expectable, but more specific impacts still need to be discussed.

3. Computational Methodology

3.1. Impeller model

Table 1 shows some typical design parameters of this impeller. Because of a relative high load coefficient, the impeller is almost radial at its exit and the metal angle is around 5°. Besides, the relative Mach number at impeller inlet tip is about 1.12, which means the effect of shockwaves cannot be ignored.

3.2. CFD method

The commercial CFD code NUMECA FINE/Turbo was used to simulate the flow. Spalart–Allmaras model was selected for turbulence calculation. The flow model showed in figure 1 includes a radial and an axial diffuser after the impeller. Both the inlet and outlet domain were extended to ensure the convergence of the calculation. The total pressure(156486Pa) and total temperature(336.94K) are set at inlet and a static pressure are set at outlet as boundary conditions. Adiabatic non-slip wall condition was adopted.

Table 1. Typical parameters of impeller

Parameter	Value
Revolution	n=22000rpm
Flow rate	m=13.53kg/s
Total pressure ratio	Pr=4.0
Inlet tip diameter	$D_0 = 0.282m$
Outlet diameter	$D_2 = 0.4m$
Flow coefficient	$\varphi = 0.127$
Load coefficient	$\phi = 0.875$
Specific speed	0.42

The grid independence was checked and the results are showed in figure 2(solid lines represent total pressure ratio, dashed lines refer to efficiency). The number of grid is around 1.2 million, which have been proved to be meaningful since doubling the number of grid hardly changes the stage performance.

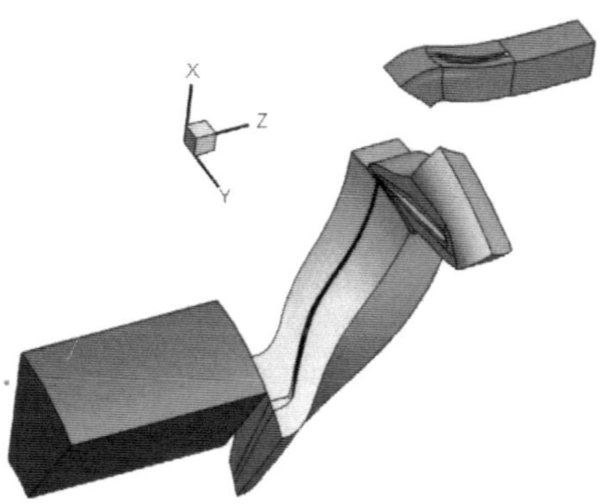

Figure 1. Numerical model

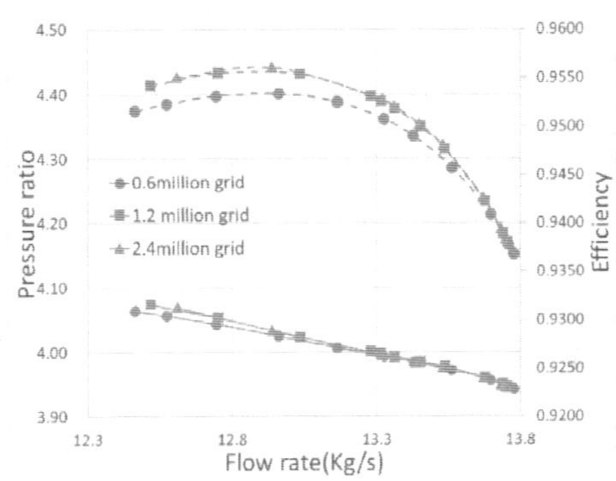

Figure 2. Grid independence validation

4. Results and discussions

4.1. Baseline impellers

Since lots of previous work has proved that an aft-loading at hub leads to a higher efficiency for the impeller, the v_ur distribution at hub were kept a aft-loaded way. But two different degrees of increased v_ur were chosen to compare the performance. However, due to the relatively high loading coefficient of the impeller, the fore-loaded pattern of v_ur may lead to forward-bent at the middle part, so all of the given distributions of v_ur are aft-loaded. The v_ur distributions of two baseline impellers and their performance are showed in figure 3 and figure 4.

As a result, the baseline2 impeller, which is based on a higher v_ur than baseline1 along the flow path, gains a 1.8% higher total pressure ratio and 1.36% higher efficiency at design flow rate. This proves that an increase of v_ur at tip than hub makes a positive influence on the performance, but whether the increase of v_ur at different streamwise location affects impeller performance the same way remains a question. Some further work has been done to address it.

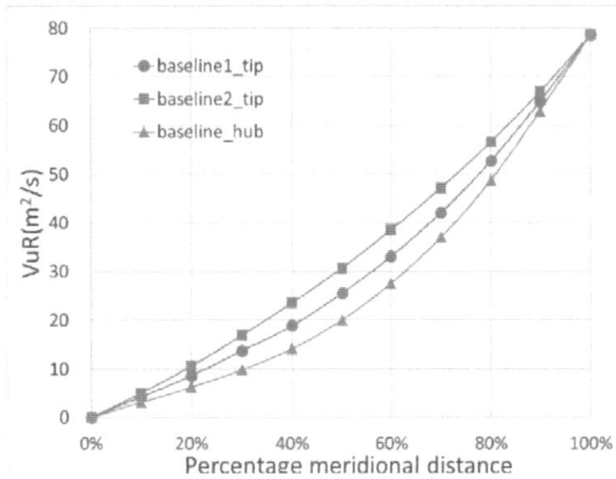

Figure 3. Blade v_ur distributions

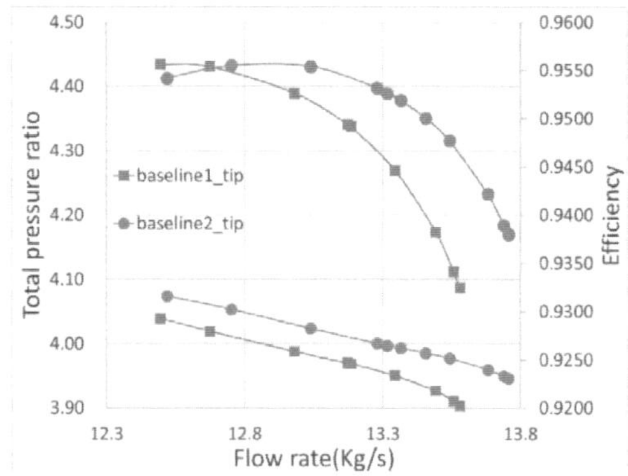

Figure 4. Impeller total pressure ratio and efficiency

4.2. *The effect of higher load at the fore part of tip*

Figure 5 shows three kinds of v_ur distributions with different degrees of increase at the fore part of tip. As shown in figure 6(solid lines represent total pressure ratio, dashed lines refer to efficiency), the total pressure ratio keeps almost the same in these impellers, while the flow rate of peak efficiency shifts as v_ur increases. A 0.6% higher efficiency can be found at the design flow rate. Efficiency at low flow rate reduces and a narrower surge margin can be found when adding the load.

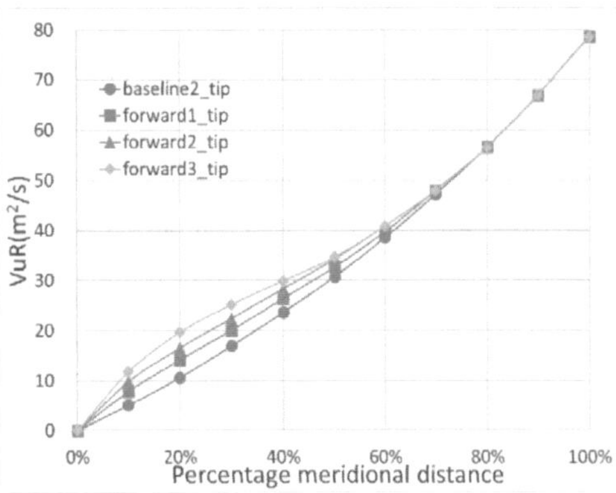

Figure 5. Blade v_ur distributions

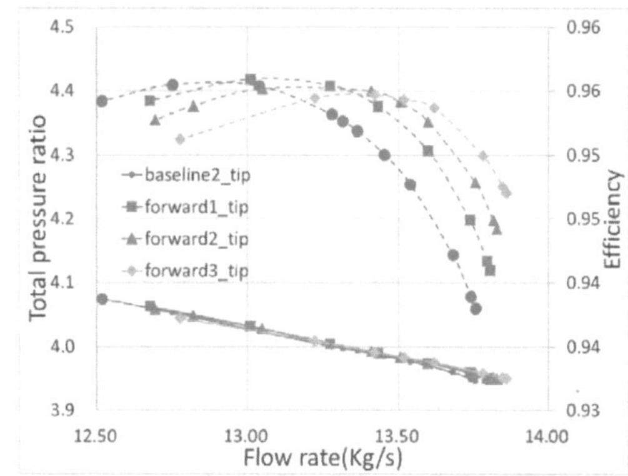

Figure 6. Total pressure ratio and efficiency

The phenomenon can be explained that by increasing the load at fore part at tip, a higher gradient of static pressure is developed from hub to tip on the suction surface of blades, thus suppressing the secondary flow.

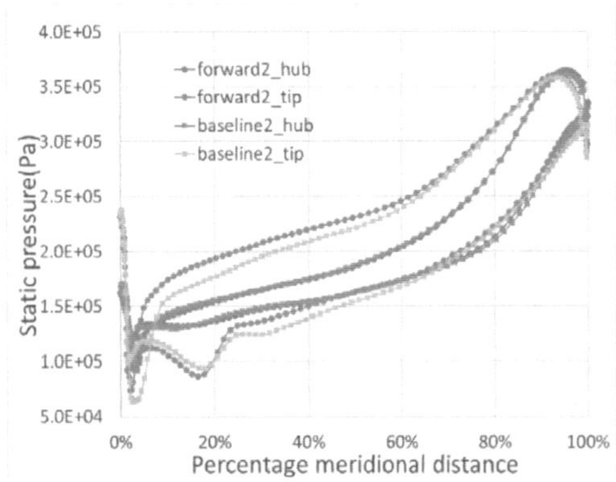

Figure 7. Blade static pressure distributions

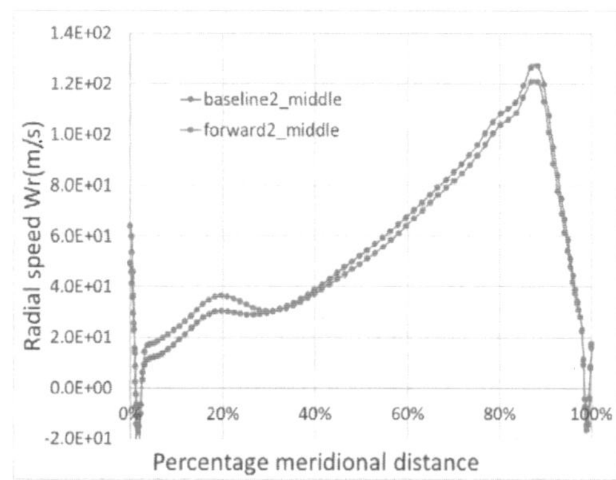

Figure 8. Suction side radial speed distributions

Figure 7 shows the static pressure distribution at 90% and 10% pitchwise for baseline2 and forward2 blades at design flow rate. The static pressure keeps the same at hub, while it rises at tip from 25% to 70% non-dimensional streamwise distance. And a decreased migration speed can be found in the same region from figure 8. The data are extracted from the first layer mesh on the suction surface of blade at 50% pitchwise height. It shows the streamwise distribution of radial component of relative velocity w_r. A positive correlation can be found between the w_r and static pressure gradient from hub to tip. At the fore part of the blade, the static pressure at tip decreases, and the w_r increases as well. However, from the 30% to 90% streamwise distance, the w_r drops with the increase of tip static pressure.

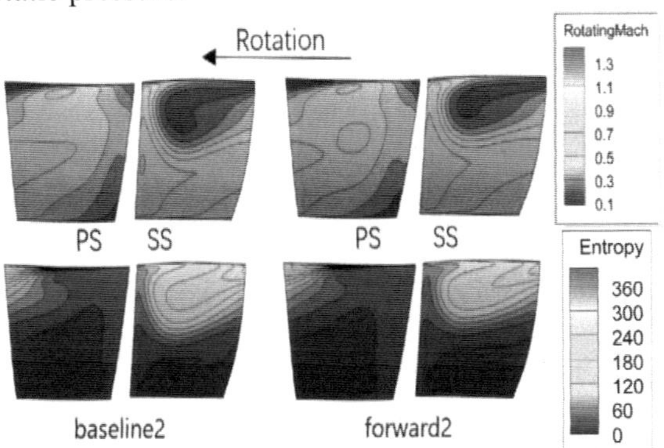

Figure 9. Mach and entropy distributions at impeller outlet

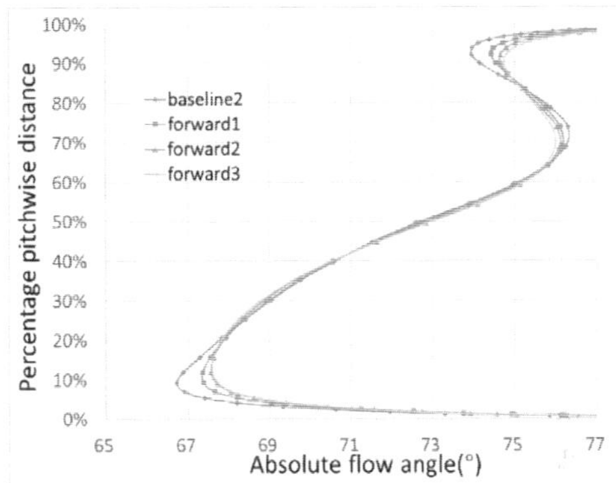

Figure 10. Absolute flow angle distribution

The effect of higher load at fore part of tip can also be proved by the relative Mach and entropy distribution at the exit of different impellers. From figure 9, in the suction surface/shroud corner, a significantly reduced low relative Mach region which are generated by secondary flow can be observed. The entropy in this region reduces as well. And a more uniform impeller exit flow condition can be found in figure 10, which shows the pitchwise distribution of absolute flow angle β for each impeller. Ignoring the relatively high β in the boundary layers close to hub and shroud, a parameter δβ is defined as the difference between the maximum β at around 70% pitchwise distance and the minimum β at around 10% pitchwise distance to present the homogeneity of the exit flow. The δβ reduced 1.2°(relatively 12.6%)in forward2 compared with baseline2.

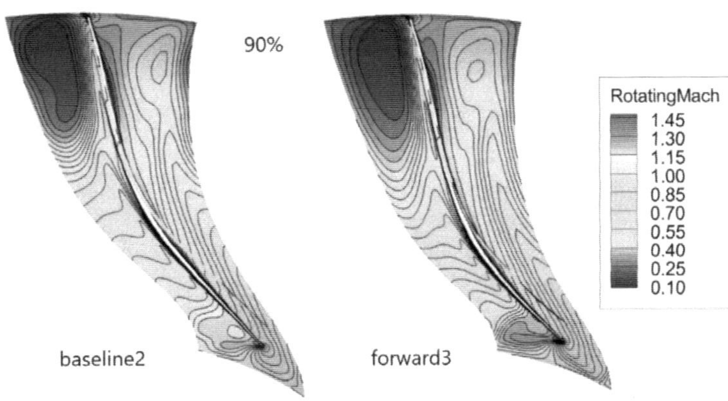

Figure 11. Relative Mach distribution at 90% pitchwise height

On the other side, the increasing load also lead to a large blade curvature at the fore part at tip, where a stronger shock wave is generated. Flow behind this shock are more likely to be separated because of the interaction between shock waves and boundary layer, especially when the flow rate is relatively low. So the surge margin is narrowed by a higher load at the fore part of tip. Figure 11 shows the relative Mach distribution of 90% pitchwise for baseline2 and forward3 impellers at a same flow rate near surge. A larger region of low relative Mach is generated as increasing the load.

4.3. The effect of higher load at the aft part of tip

In this part, three kinds of v_ur distributions with different degrees of increase at the aft part of tip were carried out to obtain the effect. The v_ur distributions and impeller performance are presented in figure 12 and figure 13(solid lines represent total pressure ratio, dashed lines refer to efficiency).

It can be concluded that the total pressure ratio rises when adding the load at the aft part of tip. At the same time, the efficiency curvature becomes gentle. The efficiency rises at high flow rate and slightly declines at low flow rate. Comparing the backward3 with baseline1 impeller, the total pressure ratio rises by 2.3%, the efficiency rises by 0.5% at design flow rate. Notably the flow rate near surge stays almost the same.

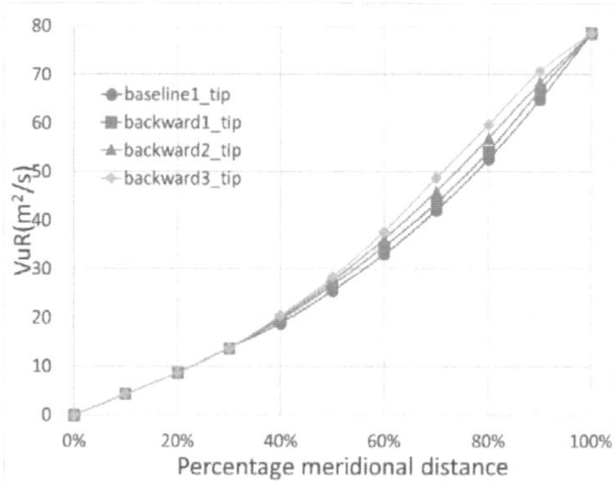

Figure 12. Blade v_ur distributions

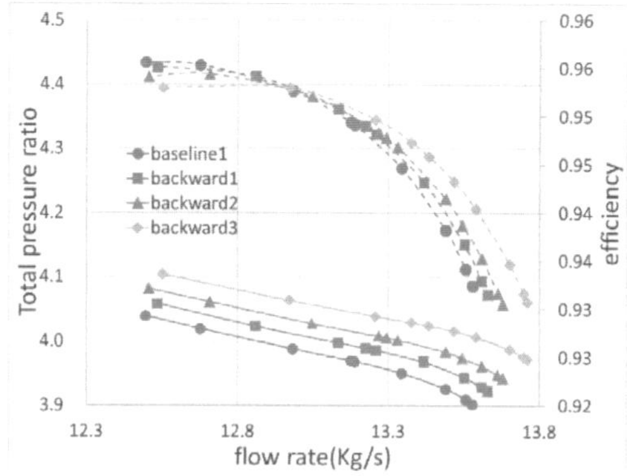

Figure 13. Total pressure ratio and efficiency

Figure 14 shows the static pressure distribution at 90% and 10% pitchwise height, and figure 15 shows the w_r distribution at 50% pitchwise height on the suction side for baseline1 and backward3 impellers. Only a slightly increment of static pressure on the suction side of the blade can be found in backward3 impeller compared to baseline1 and the change of w_r still shows an agreement with the change of static pressure. A more uniform outlet flow is obtained because the $\delta\beta$ was reduced by 0.55°. However, the increment of static pressure on the pressure side is higher than it on the suction side. That means a higher total load on the backward3 impeller, which explains the higher total pressure ratio and absolute flow angle at its exit.

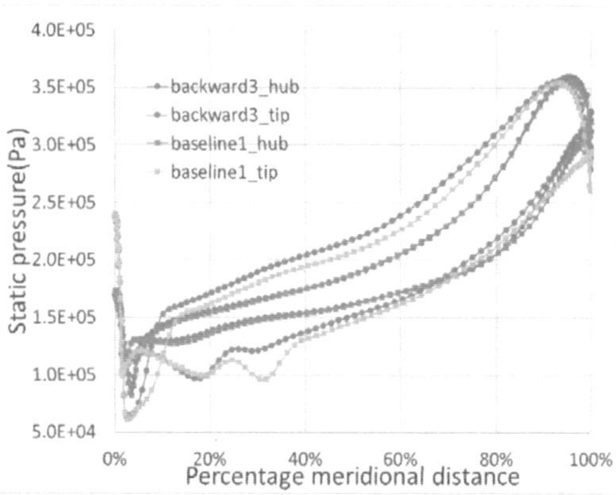

Figure 14. Blade static pressure distributions

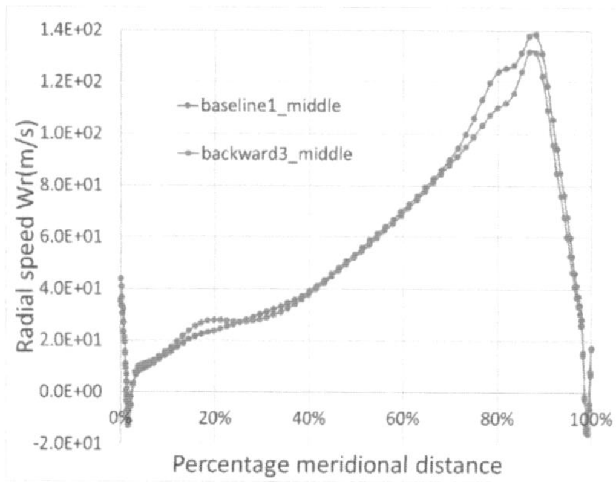

Figure 15. Suction side radial speed distributions

The relative Mach and entropy distribution are also showed in figure 16, and the absolute flow angle at blade exit is presented in figure 17. Since the reduction of high-entropy region is relatively small, the improvement in efficiency can be caused by both the reduction of secondary flow loss and the increment of input work.

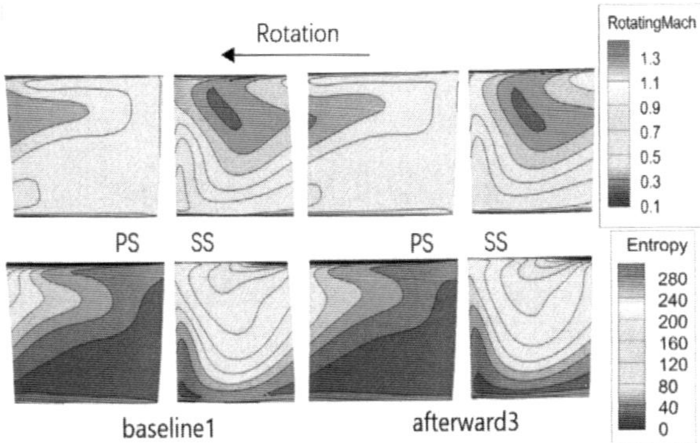

Figure 16. Mach and entropy distributions at impeller outlet

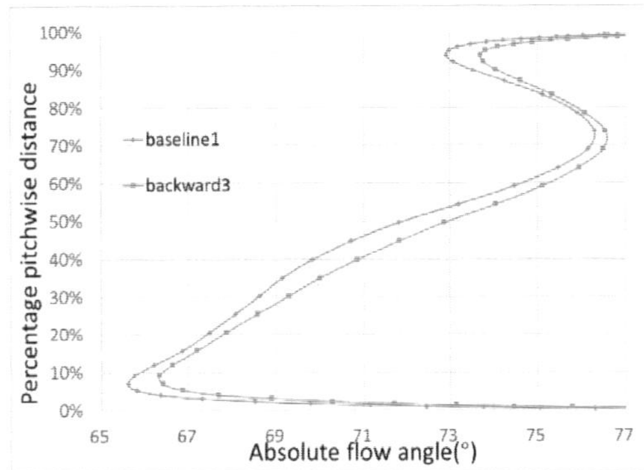

Figure 17. Absolute flow angle distribution

5. Conclusion

In this paper, the numerical calculation of several impellers designed by an inverse method was carried out to investigate the effect of blade tip loading distribution. It was proved that adopting a higher v_ur at tip than hub promotes both the pressure ratio and efficiency of the impeller. but adding v_ur at different streamwise locations at tip affects the impeller performance in diverse ways. From the research, a positive correlation between secondary flow intensity and static pressure gradient from hub to tip on blade surface was established. The numerical calculation result indicates that higher load at the front part promotes the efficiency at design flow rate and improves the uniformity at outlet. However, it decreases the efficiency at low flow rate as well as the surge margin. Higher load at the aft part promotes the total pressure ratio, with no adverse effect on the stable operation range.

References

[1] Zangeneh M, Goto A and Harada H 1998 *J. Journal of Turbomachinery.* **120** 723
[2] Zangeneh M 1998 *Int. Conf. ASME 1998 International Gas Turbine and Aeroengine Congress and Exhibition* vol 1 (Stockholm, Sweden) p V001T01A122
[3] Zangeneh M and Schleer M 2004 *J. Journal of Turbomachinery.* **126** 669
[4] Shi J C, Gao X and Liu B J 2008 *J. Journal of propulsion technology.* **29** 458
[5] Tao L H, Wang J, Liu Y and Liu X 2015 *J. Compressor, Blower & Fan Technology.* **6** 13
[6] Leng H F, Wang F J and Zhang Z C 2013 *Conf. Ser. Mater. Sci. Eng.* **52** 032008
[7] Yagi M, Kishibe T and Shibata T 2008 *Conf. ASME Turbo Expo 2008: Power for Land, Sea, and Air vol 6 (Berlin, Germany)* pp 1639-1648
[8] Ji G F, Yin M X and Gui X M. 2010 *J. Journal of Aerospace Power.* **25** 557

The Ballistic Design of Intercepting the Ultra-low Altitude Target

Xinpeng Ma[1,*], Ruikang Xing[1], Xinghao Ran[1] and Hairui Zhang[1]

[1]Air Force Engineering University, Xi'an, 710051, China

[*]18591970952@163.com

Abstract. According to the multipath effect problem that the missile intercepts the ultra-low altitude target, when the grazing angle near the Brewster Angle, the reflection coefficient of clutter reach the minimum, and the precision of guidance is higher. To ensure that the missile can intercept the target effectively under the vertical launch condition, a super-low-altitude intercept trajectory model is established. Then an improved proportional guidance law satisfying the Brewster Angle constraint is developed, and the overall trajectory of the missile from launch to strike target finally is designed. The Simulation show that the trajectory designed is relatively straight and meets the requirements. From the perspective of trajectory design, a way to solve the problem of the seeker detection have been explored.

1. Introduction

When the air defense missile intercepts the ultra-low altitude target, it will be affected by the specular reflection signal due to the influence of terrain or the density of ground and sea clutter environment, then causing a multi-path effect[1-2]. It will result in the radar seeker effectively failing to track and recognize the target so that it seriously affects the operational performance of the missile. With the rapid development of the ultra-low-level flying technology, it will be particularly important to study anti-aircraft missile which has the anti-ultra-low altitude penetration capabilities.

The angle between the connecting line of the missile and the mirror target and the horizontal plane is called the grazing angle [3]. The study shows that when the grazing angle reaches a certain angle, the density reflection coefficient of clutter environment is the smallest, and the angle is called the Brewster Angle [4-5]. Therefore, it can reduce the influence of ground image interference that using a vertically polarized radar illumination signal and making the rubbing angle is close to the Brewster Angle by ballistic design and guidance law optimization during the missile attack target and is feasible for intercepting ultra-low altitude targets. Based on this idea, the paper designs a general ballistic scheme including the initial stage (the program turning) and the middle stage (the guided law compensation) and the terminal (proportional guidance law), then carries out Matlab simulation verification and analysis and makes a conclusion.

2. Multipath effect problem description and the ultra-low altitude interception model

2.1. Multipath Effect Problem Description

Figure 1 shows: F is the ground clutter reflection point; β is the grazing angle; P_1 is the path which target signal return, and P_2 is the path which the density of clutter environment reflects. When the target is flying at low altitude, the ground clutter reflection intensity is very high (sometimes even beyond the target body), so the seeker may attack the mirror target as real target.

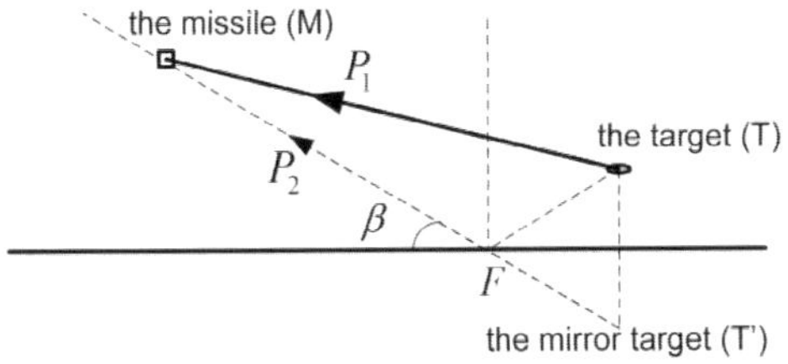

Figure 1. The geometry of missile, target and mirror

The reflection coefficient of the ground clutter varies with the change of the grazing angle. When the grazing angle reaches a certain angle, the reflection coefficient is the minimum. When this reflection coefficient is the minimum, the corresponding grazing angle is called the Brewster Angle. But the size of the Brewster angle is not the same in different environments (grassland or sea). Taking a constant environment as an example [6], the law of the reflection coefficient with the change of the grazing angle is shown in Figure 2.

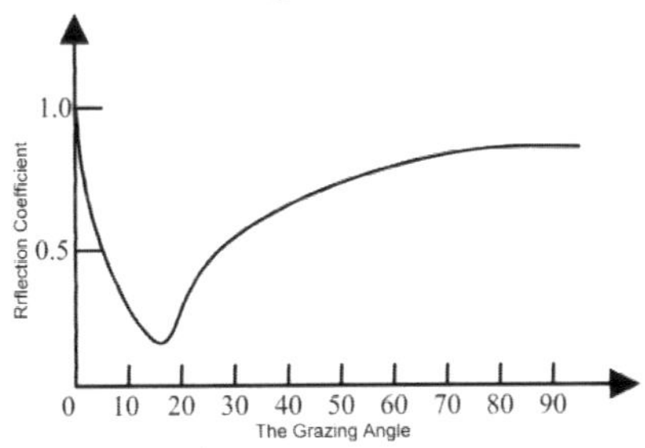

Figure 2. The relation between reflection coefficient and erasure angle

2.2. *Approximate calculation of the grazing angle*

From the point of the intercepting need for the ultra-low altitude target, we must ensure that the grazing angle of the middle trajectory is constrained near the Brewster Angle. However, actually it is difficult for the missile to measure the size of the grazing angle in real time, and it is more easily realized through the angle of sight between the missile and the target from the perspective of the guidance law. Therefore, the difference between the sightline angle and the grazing angle is analyzed at first [7].

For example, when the missile is flying the flight altitude of 150 meters, the values of the sight line angle are calculated with different distance at the same time the grazing angle is 10 degrees.

Table 1. The error between the erasure angle and the line of sight angle

distance/km	LOS/°	erasure angle /°	error/°
5	6.64	10	3.36
10	8.32	10	1.68
15	8.88	10	1.12
20	9.16	10	0.84
30	9.44	10	0.56
40	9.58	10	0.42
50	9.67	10	0.33
60	9.72	10	0.28

From Table 1, it can be seen that the difference between the sight angle of the true target and the sight angle of the mirror target relative to the missile intercepting an ultra-low altitude target is very small for the flight trajectory, when the distance between the missile and the target is far away. When the target approaches, the difference increases. It is easier to implement the correction for the angle of sight in the design process of the guidance law although there is a certain error between the two angles.

Therefore, we can implement the correction of the sight angle or the line of sight angular velocity and constrain it near the Brewster Angle to improve missile guidance accuracy when designing the mid-trajectory guidance law.

2.3. The ultra-low altitude interception model

The establishment of the ultra-low altitude interception model is as follows. The three-dimensional space can be divided into the vertical pitch plane and the transverse turning plane. For the convenience of analysis, the vertical pitch plane can only be studied. As is shown in Figure 3: M is the missile; v_m is the speed of the missile; θ_m is the trajectory inclination angle of the missile; a_m is the normal acceleration of the missile; T is the target; v_t is the speed of the target; θ_t is the dead reckoning angle of the target; a_t is the normal acceleration of the target; R is the distance from the missile to the target; q is the angle of sight between the missile and the target. According to the Flight Mechanics [8], the counter clockwise rotation is positive and the clockwise rotation is negative for all angles.

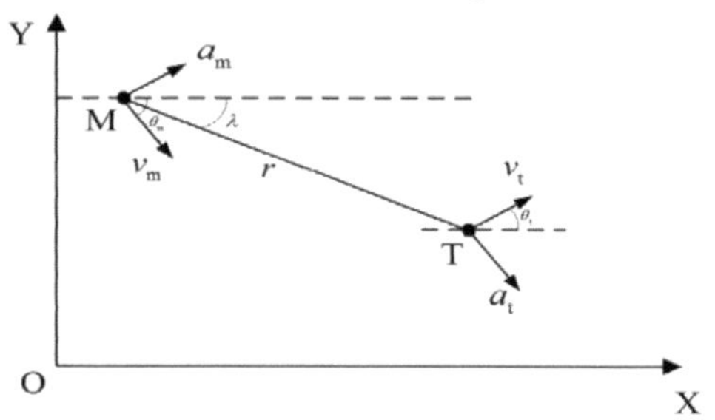

Figure 3. The motion relation between missile and target

As can be seen from Figure 3, the relative equation of motion between the missile and the target [9] is

$$\begin{cases} \dot{R} = -v_m \cos(q-\theta_m) + v_t \cos(q-\theta_t) \\ R\dot{q} = v_m \sin(q-\theta_m) - v_t \sin(q-\theta_t) \end{cases} \tag{1}$$

According to formula (1) can be drawn:

$$\ddot{q} = \frac{-2\dot{R}\dot{q}}{R} - \frac{a_m \cos(q-\theta_m)}{R} + \frac{a_t \cos(q-\theta_t)}{R} \tag{2}$$

3. Ballistic design

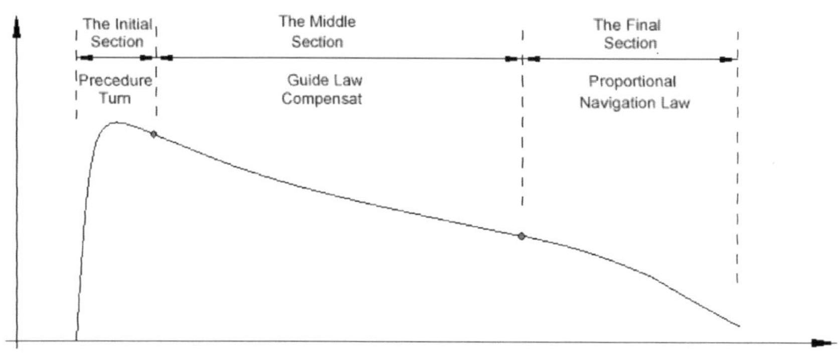

Figure 4 The ballistic scheme

The missile flies close to the ground or the surface of the water in the process of intercepting the target, and the missile can fly at a higher altitude and track the target by the radar seeker. The characteristics of the target movement determine the trajectory of interception for the ultra-low altitude targets.

The missile is launched vertically from the ground, and turns toward the target after reaches a certain altitude, then the missile enters the midcourse guidance section when the seeker intercepts the target. During the middle flight, the target information is detected by the missile seeker on the missile and the missile dives in the end. It is necessary to control the angle of sight near the Brewster Angle in the middle flight after turning.

According to the above requirements, the ballistic design scheme is shown in Figure 4.

According to the characteristics of the flight trajectory, it can be divided into the initial turn section, the middle guidance section and the end attack section [10].

3.1. The initial turn section

For the flight trajectory design for intercepting the ultra-low altitude target, the initial turning section is controlled by the program and has nothing to do with the target. After the missile is launched at a predetermined elevation angle, it reaches a certain altitude and turns toward the target. In this process, we can provide better initial conditions for pilot flight in the middle section by adopting the given design rules of the trajectory inclination angle through the reasonable parameter setting. The initial turn section is specifically divided into the following stages [11]:

(1) The booster

After launching the missile, the missile is accelerated according to the given trajectory inclination (the trajectory inclination angle is a constant value) under the influence of the booster. The working time of the booster is t_1, then the change law of the trajectory inclination angle and acceleration is:

$$\begin{cases} \theta_*(t) = \theta_0 & 0 \le t < t_1 \\ a_*(t) = a_m & 0 \le t < t_1 \end{cases} \tag{3}$$

(2)The gliding section

After the work of the booster is over, the missile keeps its speed unchanged and climbs upwards, then begins to turn toward the target after it rises to a certain height. Remember the starting turn time is t_2, then the change law of the trajectory inclination angle and acceleration is:

$$\begin{cases} \theta_*(t) = \theta_0 & t_1 \le t < t_2 \\ a_*(t) = 0 & t_1 \le t < t_2 \end{cases} \tag{4}$$

(3) The turning section

After the gliding segment was over, the missile begins to make a turn according to the predetermined procedure and eventually turned flat. The change law of the trajectory inclination angle of the turning section can be given in terms of flight time. Remember the ending time of the turning section is t_3. In order to meet the continuous change of the ballistic inclination angle and the ballistic inclination rate during the flight, the change law of the ballistic inclination of the turning section must meet the following conditions:

①At the beginning of the turning section, the ballistic inclination angle meets $\theta_*(t_1) = \theta_0$ and the change rate of the ballistic inclination angle meets $\dot{\theta}_*(t_1) = 0$;

② The change rate of the ballistic inclination meet $\dot{\theta}_*(t_2) = 0$ at the end of the turn.

3.2. The midcourse guidance section

Most of the active missiles adopt the classical proportional guidance law or the modified proportional guidance law. When the missile is guided by the proportional guidance law, the line of sight angle basically depends on the initial missile launch information, and there is no Brewster Angle constraint requirement. So it is unable to meet the Brewster Angle constraint. Therefore, based on the existing proportional guidance law, the proportional guidance law is modified by introducing the compensation amount so that the angle of sight meets the Brewster Angle constraint during the missile attack. The revised guideline can be written as[12-13]:

$$\frac{d\theta}{dt}=K\frac{dq}{dt}+x \tag{5}$$

In the formula (5), x is the compensation amount.

Assuming that at the initial time of the guidance section, the angle of sight is q_0, the change rate of the angle of sight is $\dot{q}_0$, and the angle of sight required by the Brewster Angle constraint is q_b. According to the Brewster Angle constraint requirement, the law of angular velocity change of the sight angle as shown in Figure 5 which needs to be designed. After the time t_f from the initial moment of the middle stage, the angle of sight changes from q_0 to q_b and the angle of sight angular velocity becomes smoothly 0 (namely $\dot{q}_e(t_f)=0$, $\ddot{q}_e(t_f)=0$); When $t>t_f$, keeping the line of sight unchanged ($\dot{q}_e(t)=0$).

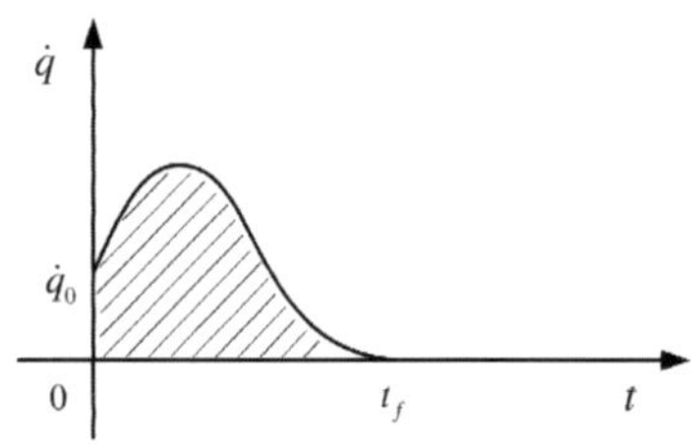

Figure 5. The line of sight angular velocity curve

From Figure 5, we can see that from time 0 to time t_f, the change law of the line-of-sight angular velocity satisfies the cubic spline curve, and its polynomial expression can be written as:

$$\dot{q}_e(\mathrm{t})=at^3+bt^2+ct+d \quad (0<t\le t_f) \tag{6}$$

Among them, a,b,c,d are the parameters to be determined. Therefore, the law of line-of-sight angle which satisfies Brewster Angle constraints can be expressed as:

$$\begin{cases}\dot{q}_e(t)=at^3+bt^2+ct+d & 0<t<t_f\\ \dot{q}_e(t)=0 & t\ge t_f\end{cases} \tag{7}$$

The derivation of the midcourse guidance law is as follows:

According to the missile-target relative motion equation, the amount of the change in the line-of-sight angular velocity caused by the compensation amount during Δt time can be expressed[14] as:

$$\begin{aligned}\dot{q}_e-\dot{q} &= [V\sin(q-\theta-x\Delta t)-V_T\sin(q-\theta_T)]/r-\\ &\quad [V\sin(q-\theta)-V_T\sin(q-\theta_T)]/r\\ &=[V\sin(q-\theta)\cos(x\Delta t)-V\cos(q-\theta)\sin(x\Delta t)/r]/r-\\ &\quad V\sin(q-\theta)/r\end{aligned} \tag{8}$$

During the missile attacks the target, because the radar seeker and the control system have higher operating frequencies, it can be assumed that the missile trajectory inc lination angle caused by the compensation amount changes a small amount in a relatively short period of time (or $x\Delta t$ is a small amount). Then:

$$sin(x\Delta t)\approx x\Delta t, \qquad cos(x\Delta t)\approx 1 \tag{9}$$

The formula (8) can be written as:

$$\begin{aligned}\dot{q}_e-\dot{q}&=[V\sin(q-\theta)-V\cos(q-\theta)x\Delta t-V\sin(q-\theta)]/r\\ &=-V\cos(q-\theta)x\Delta t/r\end{aligned} \tag{10}$$

Among them, $\dot{q}_e$ is the line-of-sight angular velocity obtained by the modified guidance law. $\dot{q}$ is the line-of-sight angular velocity obtained according to the classical proportional guidance law. The amount of compensation that can be solved from the above formula is:

$$x=-\frac{r(\dot{q}_e-\dot{q})}{V\Delta t cos(q-\theta)} \tag{11}$$

IOP Conf. Series: Materials Science and Engineering **408** (2018) 012029 doi:10.1088/1757-899X/408/1/012029

3.3. The end attack segment

After entering the end, the seeker can effectively distinguish the target, so the Brewster Angle constraint can be dismissed and the missile's own guidance law can be used to design and achieve an effective attack on the target [15]. In general, the proportional guide law is:

$$\frac{d\theta}{dt} = K\frac{dq}{dt} \tag{12}$$

4. The ballistic simulation

In accordance with the trajectory scheme of the initial corner section, and the middle section guidance section and the end attack section designed in the previous section, the missile flight process was simulated and analyzed [16-17].

Assuming that the ultra-low altitude target makes a uniform straight flight along the ground. The initial height $Y_t = 100\text{m}$, the initial position $X_t = 80km$, speed $V_t = 270m/s$, the trajectory inclination angle $\theta_t = 0^{\circ}$; the missile launches vertically from the ground, the boost acceleration $a_{m1} = 30g$, the engine working time $t_1 = 6s$, the acceleration of turning section $a_Z = g$, the starting time of turning section $t_2 = 8s$, the time at the end of the turning section $t_3 = 16s$, the trajectory inclination in the end $\theta_m = 0°$, the mid-adjustment time $t_f = 5s$, and the distance to the end $X_1 = 6km$;the Brewster's angle is 15° around [13-15].

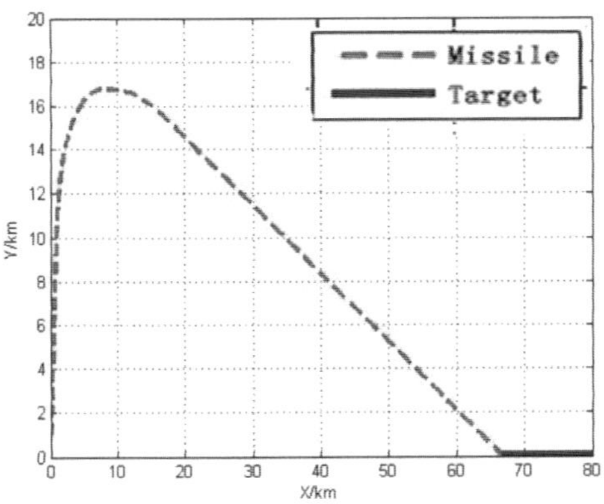

Figure 6. The trajectory of missile and target

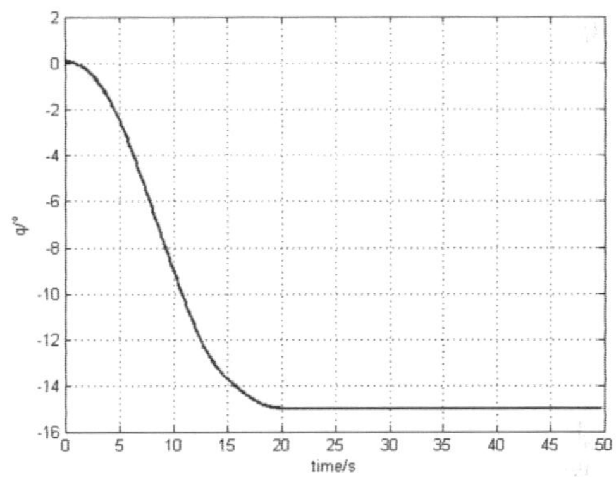

Figure 7. The variation od the view angle

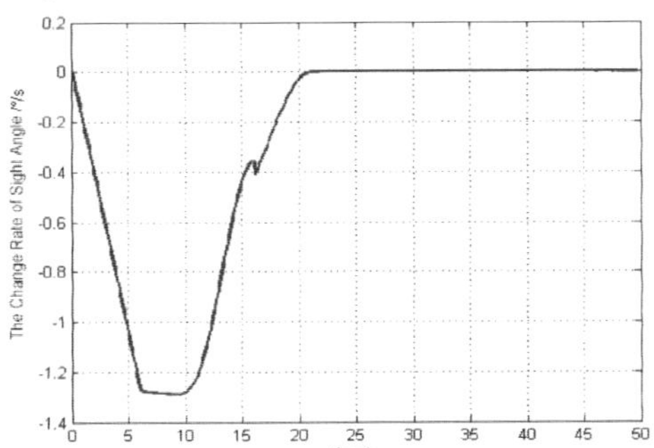

Figure 8. The variation of the sight angular rate

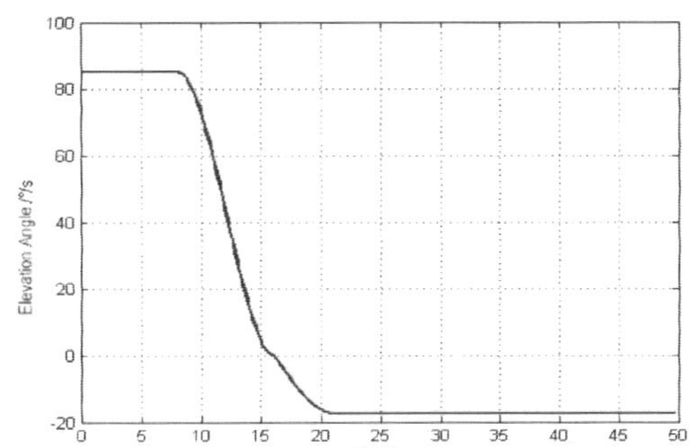

Figure 9. The variation of the angle of the trajectory

It can be seen from Figure 6 that the missile is launched vertically from the ground, then goes through a turn of the initial stage program, the midcourse guidance and the end of the attack, finally intercepts the target successfully, and the designed trajectory is relatively flat. It can greatly enhance the maneuverability of the missile, reduce the overload and improve the guidance accuracy. From

Figure 7 and 8, it can be seen that after the turn is completed, the angle of sight can be quickly approached to the Brewster Angle and remain unchanged after arrival. The change of the line-of-sight angular rate also tends to 0. So it meets the guidance requirements. It can be seen from Figure 9 that the trajectory inclination of the missile is 0° after the turning of the program. It satisfies the set value, and the velocity inclination of the ballistic trajectory remains basically unchanged.

5. Conclusion

The article discusses the problem that the missile intercepts the ultra-low altitude target under vertical launch conditions. For the multi-path problem during the interception process, the position of the Brewster Angle is determined, and the interception trajectory model is established, and the improved proportional guidance law which constrains the angle of sight near the Brewster Angle is studied to make the density of ground and sea clutter environment minimum. Then the ballistic trajectory is divided into the initial turn section, the middle guidance section and the end attack section. The full range trajectory of the missile from the launch to the final hit is designed by simulation. The results show that the designed trajectory is relatively flat and meets the requirements, and the guidance accuracy is so high that it has a strong reference value.

References

[1] WANG L J, GUO J M, HAO J. The effect of multi-path effect on radar detection of low-altitude cruise missile[J].The radar science and technology, 2010,8(1):7-11.

[2] SU C L. The super-low altitude trajectory design of the air defense missile at general point brewer[J].The groung anti-aircraft weapon, 2008,39(3):23-26.

[3] WANG G S,LI Y.A study on the guiding law that satisfies the Brewster angle constraint[J].The modern defense technology, 2014,42(3):70-74.

[4] CAI H,HU Z D,CAO Y. The overview of guidance laws with terminal angle constraints[J].The journal of aerospace, 2010,31(2):315-323.

[5] QIAN X F,LIN R X,ZHAO Y N. Missile flight mechanics[M].The Beijing university of technology press, 2000(8).

[6] ZHAO Yao, SHENG Yongzhi. Sliding mode control based guidance law with impact angle constraint[J]. Chinese Journal of Aeronautics, 2014,27(1): 145-152.

[7] WANG Hongqiang, CAO Daiqiang, WANG Xudong. The stochastic sliding mode variable structure guidance laws based on optimal control theory [J]. Journal of Control Theory and Applications, 2013, 11(1): 86-91.

[8] HAN Yanhua, XU Bo. Variable structure guidance law for attacking surface maneuver targets [J]. Journal of Systems Engineering and Electronics, 2008, 19(2): 337-341.

[9] ZHANG Y X, SUN M W, CHEN Z Q. Finite-time convergent guidance law with impact angle constraint based on sliding-mode control[J]. Nonlinear Dynamics, 2012, 7(3):619-625.

[10] XIN T D,FAN H L,YAN L. The study on buffeting problem of guidance law of sliding mode variable structure[J].The aviation weapon, 2015, 2(10): 10-13.

[11] SUN D X,TONG C M,PENG P. The broad Brewster effect of the super-low target[J].Journal of the air force engineering university, 2015, 16(4):29-32.

[12] WU H, LIU YP.A solution to the multipath effect[J].Modern radar, 2007,29(5): 26-31.

[13] LEI H M. Missile guidance and control principles[M].National defense industry press, 2006, 34-67.

[14] Dennis B. Trizna. A Model for Brewster Angle Dampingand Multipath Effects on the Microwave]Radar Sea Echo at Low Grazing Angles[J]. IEEE TRANSACTIONS ON GEOSCIENCE AND REMOTE SENSING,1997(9),VOL.35,NO.5:1232-1244.

[15] R.J. Burkholder, P. Janpugdee, and D. Colak. Development of Computational Tools for Predicting the Radar Scattering from Targets on a Rough Sea Surface[R]. ADA388046, OHIO STATE UNIV COLUMBUS ELECTROSCIENCE LAB,2001.

[16] Laura I. Thomson, Gordom R. Osinski, Wayne H. Pollard. Application of the Brewster angle to quantify the dielectric propertiers of ground ice formations[J]. Journal of Applied Geophysics, 2013(9):12-17.

[17] LIU J H. The mathematical simulation and analysis of ground-air multipath effect of active radar seeker[J].Modern defense technology, 2007(3): 82-85.

IOP Conf. Series: Materials Science and Engineering **408** (2018) 012030 doi:10.1088/1757-899X/408/1/012030

A Novel Penetration-evade Guidance Law against Two Homing Missiles

Siyu Xiong[1,*], Gang Li[1], Siyu Huang[1] and Jia Zhan[1]

[1]Air Force Engineering University, Xi'an, 710051, China

[*]xiongsiyu8313@163.com

Abstract. A novel guidance for an evader which is exposed to a threat of two homing missiles is presented. Each missile is assumed to measure solely its own line-of-sight (LOS) angle and share it with other missile. Such information sharing enables the missiles to form a triangular measuring baseline relative to the evader and to improve their estimation accuracy. However, if the separation angle between the two LOS is small enough, the observability of such double-LOS measuring approach becomes weak. Motived by this observation, the idea of the proposed concept is to bring the missile on the same LOS with target in the first stage. As for the second stage, the evader perform an appropriately timed maximum evasive maneuver. The simulation results demonstrate that the proposed guidance law can reduce the estimate and guidance accuracy of interceptors and achieve the goal of penetration.

1. Introduction

Modern high-precision interceptors can easily track and destroy our high-value ballistic missiles. Mobile escape is the best way to get rid of the interceptor. It is a direct and effective attack strategy [1].

For the one-to-one case where the enemy's defense system uses a missile to intercept, the literature [2-4] has deduced the optimal escape strategy for the target. The assumption is that the missile has all the information of the target. However, in the real environment where there is noise interference, it is difficult to obtain all the states of the target. Therefore, many estimation algorithms have been derived [5]. In order to increase the probability of interception, the multiple-to-one attack method for launching multiple missiles is very effective in increasing the probability of intercepting the success rate. It can also restrict the interception time and the interception angle [6-8]. Most tactical missiles are equipped with infrared sensors that measure the LOS angle to measure the LOS between the missile and the target. Literature [9-11] studied the methods of tracking the target and enhancing the performance of the estimation in the context of multi-missiles using purely azimuth measurements. Through cooperative detection, the tracking accuracy of the target can be enhanced. The multi-missiles share the LOS angles measured by them in real time, and the missile-target triangulation structure [12] can greatly enhance the estimated performance of the guidance system. Literature [13] proved that this information sharing estimation method has a very good improvement in interception performance. However, the estimated performance of this method largely depends on the trajectory of the missile. Regardless of the filter used, if the difference between each missile and the target LOS angle is small, the relative kinematics relationship becomes difficult to estimate, the accuracy of the estimate decreases, and the missile guided system's interception decreases. Literature [14-15], from the perspective of guided missile interception, enhances the estimation accuracy by adjusting to the appropriate LOS angle difference.

The above studies all emphasize how to enhance the estimated performance of the interceptor. For the defense side, there is less research on one-to-two or one-to-many penetration methods. This paper proposes a missile penetration prevention escape strategy from the perspective of the penetration defense missile. The guidance law is divided into two phases. In the first phase, the penetration missile is intended to reduce the estimated performance of the interceptor in the context of multi- LOS measurement. Changing the interceptor's trajectory will make them equal to the sight line angle of the penetration missile. In this case, the distance between the missiles becomes difficult to estimate, which greatly increases the survivability of the missiles. At the right moment, it switches to the second stage, during which the missiles make maximum maneuvers to escape.

2. Problem statement

The considered engagement scenario consists of two interceptors and an evader. This paper considers the issue from the perspective of the target. The two-dimensional geometric relationship shown in Figure 1. $X-O-Y$ is Cartesian inertial coordinate system. The speed, normal acceleration and flight-path angle are V, a and γ respectively; Subscript $i \in \{1,2\}$ and t means the i missile and the target respectively; (x_j, y_j), $j \in \{1,2,t\}$ denotes the position of the j th vehicle, and the distance between the target and the i th missile is denoted as r_i. The LOS of i th missile is λ_i. The difference between the two line of sight angles is defined as $\Delta\lambda \triangleq \lambda_1 - \lambda_2$, which will be used in the derivation of the guidance law.

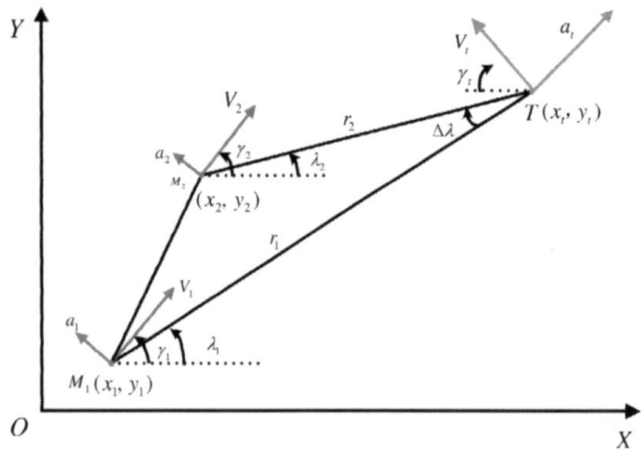

Figure 1. Planar geometric relationship

The kinematics relationship of the i th missile in the polar coordinate system (r_i, λ_i) is expressed as

$$\begin{cases} \dot{r}_i = V_{ri} \\ \dot{\lambda}_i = V_{\lambda i} / r_i \end{cases} \tag{1}$$

The speed component along and normal to the LOS are

$$\begin{cases} V_{ri} = -V_i \cos(\gamma_i - \lambda_i) - V_t \cos(\gamma_t + \lambda_i) \\ V_{\lambda i} = -V_i \sin(\gamma_i - \lambda_i) + V_t \sin(\gamma_t + \lambda_i) \end{cases} \tag{2}$$

Arbitrary-order linear dynamics of the j th vehicle is

$$\begin{cases} \dot{\boldsymbol{X}}_j = \boldsymbol{A}_j \boldsymbol{X}_j + \boldsymbol{B}_j u_j \\ a_j = \boldsymbol{C}_j \boldsymbol{X}_j + D_j u_j \,, \quad j \in \{1,2,t\} \\ \dot{\gamma}_j = a_j / V_j \end{cases} \tag{3}$$

The maximum maneuverability of the j th vehicle is

$$\left|u_j\right| \le u_{j\max}, \quad j \in \{1,2,t\} \tag{4}$$

The i th missile's interception time is t_{fi}. Under the assumption of linearization with a small angle, the interception time can be considered as fixed.

$$t_{fi} \triangleq -r_i / \dot{r}_i, \ \forall i \in \{1,2\} \tag{5}$$

The non-negative time-to-go of the i th missile is defined as

$$t_{goi} = \begin{cases} t_{fi} - t & t < t_{fi} \\ 0 & t > t_{fi} \end{cases} \quad \forall i \in \{1,2\} \tag{6}$$

When $t = t_{fi}$, $r_i(t_{fi})$ is the miss distance of the i th missile.

If at the end of the guidance stage, the deviation between the missile and the target and the respective collision triangle is very small, the linearization assumption can be established, and the linear plane geometry relationship between the missile and the target [16] is shown in Figure 2

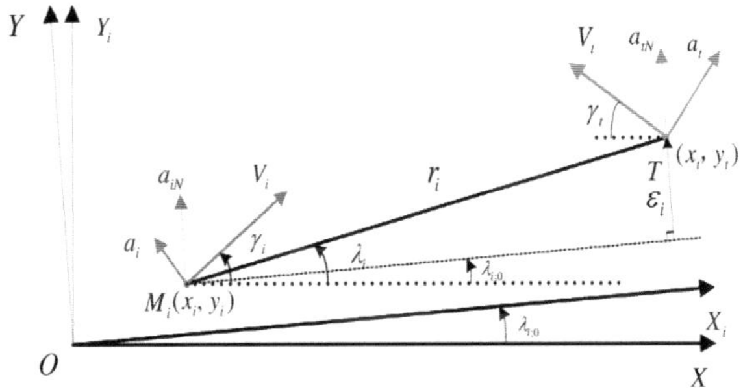

Figure 2. Linearized planar geometric relationship

The accelerations normal to initial LOS $LOS_{i;0}$ is a_{t_iN} and a_N, satisfying

$$a_{t_iN} = k_{ti} a_t, k_{ti} = \cos(\gamma_{t;0} + \lambda_{1;0}) \tag{7}$$

$$a_{iN} = k_i a_i, k_i = \cos(\gamma_{i;0} - \lambda_{1;0}) \tag{8}$$

The subscript 0 indicates the initial value at the start of linearization, and k_i and k_{ti} indicate the k th linearization parameter.

The LOS separation angle $\Delta\lambda$ is defined by the LOS angle in the nonlinear model (1). If $\left|\lambda_{2;0} - \lambda_{1;0}\right|$ is small enough, then $\Delta\lambda$ can be expressed approximately as

$$\Delta\lambda \approx \sin(\lambda_2 - \lambda_{2;0}) - \sin(\lambda_1 - \lambda_{1;0}) \triangleq x_\lambda \tag{9}$$

From Figure 2, we can see that x_λ can be represented by the linearization variables ε_1 and ε_2

$$x_\lambda = \varepsilon_2 / r_2 - \varepsilon_1 / r_1 \tag{10}$$

The distance r_i can be expressed approximately as

$$r_i \approx V_{ci} t_{goi} \tag{11}$$

Where V_{ci} is a constant approach speed

On substituting Eq. (12) in the derivative of Eq. (11), we get

$$\dot{x}_\lambda = \frac{\dot{\varepsilon}_2}{V_{c2} t_{go2}} + \frac{\varepsilon_2}{V_{c2}(t_{go2})^2} - \frac{\dot{\varepsilon}_1}{V_{c1} t_{go1}} - \frac{\varepsilon_1}{V_{c1}(t_{go1})^2} \tag{12}$$

Define the state vector as

$$\boldsymbol{x} \triangleq \left[\boldsymbol{x}_\varepsilon^T \ \dot{\boldsymbol{x}}_\varepsilon^T \ \boldsymbol{X}_m^T \ \boldsymbol{X}_t^T \ x_\lambda\right] \tag{13}$$

Where

$$\boldsymbol{x}_\varepsilon = [\varepsilon_1 \ \varepsilon_2]^T \triangleq [x_1 \ x_2]^T \tag{14}$$

$\dot{\boldsymbol{x}}_\varepsilon$ is the relative speed vector normal to the respective initial LOS. Then the equation of motion can be expressed as

$$\begin{cases} \dot{x}_1 = x_3 \\ \dot{x}_2 = x_4 \\ \dot{x}_3 = k_{t1}a_t - k_1 a_1 \\ \dot{x}_4 = k_{t2}a_t - k_2 a_2 \\ \dot{\boldsymbol{X}}_m = \boldsymbol{A}_m \boldsymbol{X}_m + \boldsymbol{B}_m \boldsymbol{u}_m \\ \dot{\boldsymbol{X}}_t = \boldsymbol{A}_t \boldsymbol{X}_t + \boldsymbol{B}_t \boldsymbol{u}_t \\ \dot{x}_\lambda = -\dfrac{x_1}{V_{c1}(t_{go1})^2} + \dfrac{x_2}{V_{c2}(t_{go2})^2} - \dfrac{x_3}{V_{c1}t_{go1}} + \dfrac{x_4}{V_{c2}t_{go2}} \end{cases} \tag{15}$$

Where $\boldsymbol{A}_m$ and $\boldsymbol{B}_m$ are diagonal matrices

$$\boldsymbol{A}_m \triangleq diag\{\boldsymbol{A}_1, \boldsymbol{A}_2\} \tag{16}$$

$$\boldsymbol{B}_m \triangleq diag\{\boldsymbol{B}_1, \boldsymbol{B}_2\} \tag{17}$$

Assume that the i th missile's guidance law can be expressed as the following linear form

$$u_i = K_{1i}\varepsilon_i + K_{2i}\dot{\varepsilon}_i + \boldsymbol{K}_{mi}\boldsymbol{y}_i + \boldsymbol{K}_{ti}\boldsymbol{y}_t + K_{ui}u_t \tag{18}$$

Equations (16) and (19) can be written in matrix form

$$\dot{\boldsymbol{x}}(t) = \boldsymbol{A}(t)\boldsymbol{x}(t) + \boldsymbol{B}(t)u_t(t) \qquad \boldsymbol{x}(0) = \boldsymbol{x}_0 \tag{19}$$

Where

$$A(\mathrm{t}) = \begin{bmatrix} \boldsymbol{0} & \boldsymbol{I} & \boldsymbol{0} & \boldsymbol{0} & 0 \\ \boldsymbol{A}_{21} & \boldsymbol{A}_{22} & \boldsymbol{A}_{23} & \boldsymbol{A}_{24} & 0 \\ \boldsymbol{A}_{31} & \boldsymbol{A}_{32} & \boldsymbol{A}_{33} & \boldsymbol{A}_{34} & 0 \\ \boldsymbol{0} & \boldsymbol{0} & \boldsymbol{0} & \boldsymbol{A}_t & 0 \\ \boldsymbol{A}_{51} & \boldsymbol{A}_{52} & \boldsymbol{0} & \boldsymbol{0} & 0 \end{bmatrix} B(\mathrm{t}) = \begin{bmatrix} \boldsymbol{0} \\ \boldsymbol{B}_2 \\ \boldsymbol{B}_3 \\ \boldsymbol{B}_t \\ 0 \end{bmatrix} \tag{20}$$

3. Penetration escape strategy design

Lemma 1 [12]: The distance $r^+_{i;k}$ between the target and the i missile can be calculated based on the measurement value $z_{i;k}$, $z_{j;k}$ of the i th missile and the j th missile, which contain noise, and its relative position (r_{ij}, λ_{ij}). And $r^+_{i;k}$ follows the distribution $r^+_{i;k} \sim \mathrm{N}(r_{i;k}, \sigma^2_{\rho i;k})$, $r_{i;k}$ is the real distance at time t_k, $\sigma^2_{\rho i;k}$ is the non-stationary standard deviation of $r^+_{i;k}$.

$$r^+_{i;k} = r_{ij;k}\frac{\sin(\lambda_{ij;k} - z_{j;k})}{\sin(z_{i;k} - z_{j;k})}, i \neq j \tag{21}$$

$$\sigma_{\rho i} = r_{ij}\frac{\sqrt{\sigma_j^2 \sin^2(\lambda_{ij} - \lambda_i) + \sigma_i^2 \sin^2(\lambda_{ij} - \lambda_i)\cos^2(\lambda_j - \lambda_i)}}{\sin^2(\lambda_j - \lambda_i)} \tag{22}$$

Where $r_{ij;k} = \sqrt{x^2_{ij;k} + y^2_{ij;k}}$, $\lambda_{ij;k} = \tan^{-1}(y_{ij;k} / x_{ij;k})$.

If we make $|\lambda_j - \lambda_i| \to 0$, we get

$$\lim_{|\lambda_j - \lambda_i| \to 0} \sigma_{\rho i} = \infty, \qquad i \neq j,\ r_{ij} \neq 0 \tag{23}$$

When the LOS separation angle $\Delta\lambda$ is close to zero, the two LOS angle λ_1 and λ_2 are equal. The variance $r^+_{i;k}$ increase, which in turn makes the distance of the i th missile inestimable and the measured values shared with each other will be equal. Deterioration of the missile's estimation accuracy, especially the estimation of the distance, the speed of the target, and the time-to-go, will lead to a decrease in the guidance accuracy of the missile. Motived by this observation, this section will design the guidance law by minimizing $\Delta\lambda$.

3.1 *First Stage-Optimal Guidance Law*

Defining the intercept time as:

$$t_f \triangleq \min_{i\in\{1,2\}} t_{fi} \tag{24}$$

Selecting the quadratic cost function J to minimize the LOS separation angle and the target's control effort as performance indicators

$$J = \frac{1}{2}\int_{t_0}^{t_s}[Q(\tau)x_\lambda^2(\tau) + Ru_t^2(\tau)]d\tau \tag{25}$$

Where $t_s \le t_f$ is the switching time from first stage to second stage and R is the weight of the target control effort. $Q(\tau)$ is the time-varying weight of the LOS separation angle.

$$Q(\tau) \triangleq \frac{\eta}{t_s - \tau} \tag{26}$$

For a linear time-varying system (20), in order to minimize the performance index (25), the optimal control is determined to be

$$u_t^1(t) = -R^{-1}\boldsymbol{B}^T(t)\boldsymbol{P}(t)\boldsymbol{x}(t), \quad t\in[0, t_s] \tag{27}$$

The nonnegative matrix P satisfies the differential matrix Ricatti equation

$$\dot{\boldsymbol{P}} = -\boldsymbol{A}^T\boldsymbol{P} - \boldsymbol{P}\boldsymbol{A} + \boldsymbol{P}\boldsymbol{B}R^{-1}\boldsymbol{B}^T\boldsymbol{P} - \boldsymbol{C}^T Q\boldsymbol{C} \tag{28}$$

Where $\boldsymbol{P}(t_s) = \boldsymbol{0}, \boldsymbol{C} = [\boldsymbol{0\,0\,0\,0}\,1]$.

3.2 *Second Stage-Escape Strategy*

At the end of the first stage, this paper will make the target do the maximum maneuver escape in the opposite direction of the acceleration command.

Defining the switching time as

$$t_s = t_f - \Delta t \quad 0 < \Delta t < t_f \tag{29}$$

Where the value of Δt determine the time of escape stage. And the acceleration command is

$$u_t^2(\mathrm{t}) = \begin{cases} +u_{t\max} & if\ u_t^1(\mathrm{t}_s) \le 0 \\ -u_{t\max} & else \end{cases} \tag{30}$$

Where $u_t^1(\mathrm{t}_s)$ is the acceleration command of the target at t_s. The final Penetration escape strategy is summarized as:

$$u_t(\mathrm{t}) = \begin{cases} u_t^1 & if\ 0 \le t \le t_s \\ u_t^2 & else \end{cases} \tag{31}$$

4. Numerical simulation

This section demonstrates the effectiveness of the penetration escape strategy by numerical simulation. Facing the two homing missiles launched by the enemy's defense system, the penetration-escape strategy was employed to reduce the missile's estimated performance and guidance performance and achieve the purpose of penetration. The simulation scenario was set as follows: At the beginning of the simulation, two missiles were launched simultaneously. After the first missile passed the target, the simulation was continued until the end of the second missile. The initial conditions of the simulation and the relevant parameters in the guidance law are shown in Table 1 and Table 2, respectively.

Table 1. The values of initial conditions

initial conditions	values
$x_{1;0}, y_{1;0}; x_{2;0}, y_{2;0}$	(0,150); (0,50)
$x_{t;0}, y_{t;0}$	(15000,100)
$u_{1\max}, u_{2\max}, u_{t\max}$	45g,45g,10g
$V_1 = V_2 = V_t$	$2500\,m/s$
$\lambda_{1;0}, \lambda_{2;0}, \lambda_{t;0}$	0rad, 0rad, 0rad,

Table 2. The design parameters

parameters	values
R, η	1,10^{10}
τ_1, τ_2, τ_t	0.01,0.01,0.02
N_{PN}, ε'	4, 0

The target employs the guidance law proposed in this paper. The switching time Δt is set to 0.5s. Its trajectory are shown in Figure 3. It can be seen from the change of the LOS angle (Figure 4) that the guidance law is ideal for controlling the LOS separation angle.

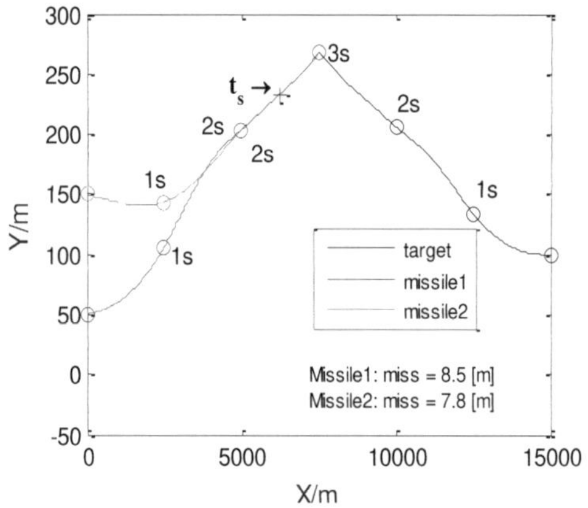

Figure 3. Planar trajectories. Δt=0.5

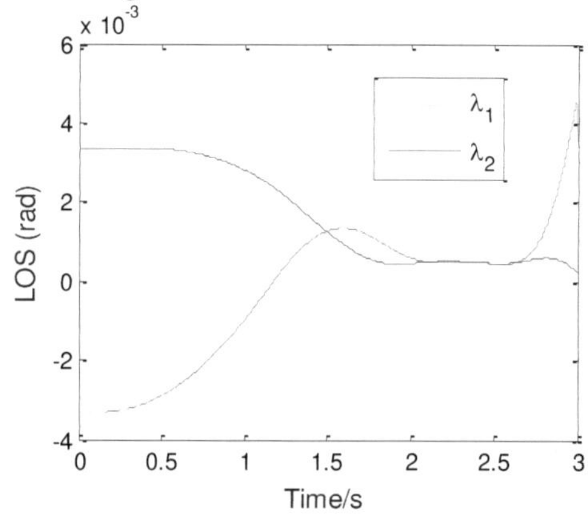

Figure 4. LOS of missiles

In order to analyse the two missiles' estimated performance against a target employing penetration-escape guidance law, the estimated performance is measured by calculating the standard deviation of the error between true and estimated values. Simulation results as shown in Figure 5. As Δt decreases, the duration of the first stage increases, and the accuracy of the missile's estimation of distance, speed, and time-to-go significantly decreases. At the end of the phase, the estimates of all variables deteriorate.

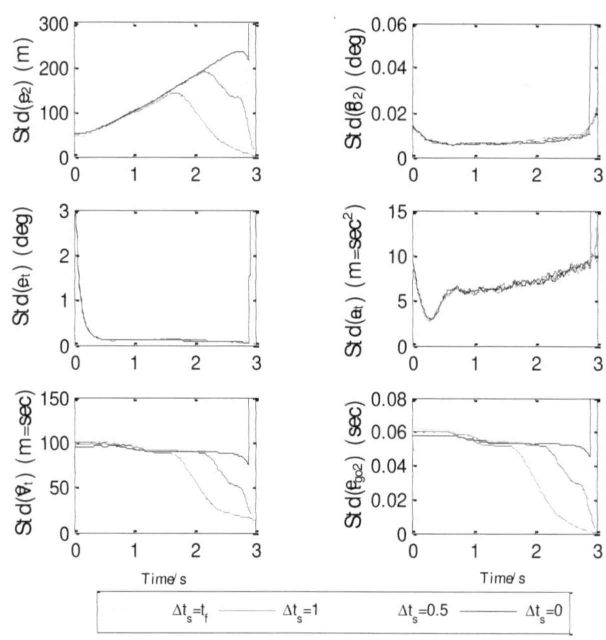

Figure 5. observability performance

5. Conclusion

This paper designs a penetration prevention strategy for the penetration defense missiles. In order to improve the accuracy of the estimation by intercepting the interceptors of the pure azimuth measurement to improve the estimation accuracy, a control strategy is proposed. The antimissile guided the law of equalization of the LOS angles of the two interceptors, thereby exacerbating the estimation accuracy of the interceptor and further reducing the guidance accuracy. Then choose the appropriate switching time to do the maximum escape maneuver to achieve the purpose of penetration. From the Monte Carlo simulation results, if the available overload of the penetration missile is determined, there is an optimal switching time to perform the maximum escape maneuver. Selecting the appropriate switching time can greatly enhance the chance of survival of the penetration missile.

References

[1] Balakrishnan S N, Tsourdos A, White B A. Advances in Missile Guidance, Control, and Estimation[M]. Boca Raton,US: Crc Press, 2012.
[2] Shima T. Optimal Cooperative Pursuit and Evasion Strategies Against a Homing Missile[J]. Journal of Guidance Control Dynamics, 2012, 34(2):414-425.
[3] Fonod R, Shima T. Multiple Model Adaptive Evasion Against a Homing Missile[J]. Journal of Guidance Control & Dynamics, 2016, 39(7):1-15.
[4] Turetsky V, Shima T. Target Evasion from a Missile Performing Multiple Switches in Guidance Law[J]. Journal of Guidance Control & Dynamics, 2016, 39(10):1-10.
[5] LEI H M, WANG H J, ZHOU J, et al. Three-dimensional dynamic surface guidance law based on extended state observer[J]. Systems Engineering and Electronics, 2017, 39(1):138-146.
[6] SONG S M, SONG J H XU S L. A Cooperative Guidance Law for Multiple Missiles to Intercept Maneuvering Target[J]. Journal of Astronautics, 2016, 37(12): 1432-1440.
[7] LV T, LV Y Y, LI C J, et al. Finite Time Cooperative Guidance Law for Multiple Missiles with Line-of-sight Angle Constraint[J]. ACTA ARMAMENTARII. 2018, 39(02):305-314.
[8] He S, Wang W, Lin D, et al. Consensus-Based Two-Stage Salvo Attack Guidance[J]. IEEE Transactions on Aerospace & Electronic Systems, 2017,12(02):31-45.
[9] Seo M G, Tahk M J. Observability analysis and enhancement of radome aberration estimation with line-of-sight angle-only measurement[J]. Aerospace & Electronic Systems IEEE Transactions on, 2015, 51(4):3321-3331.
[10] Kim T H, Lee C H, Tahk M J. Time-to-go Polynomial Guidance with Trajectory Modulation for Observability Enhancement[J]. Aerospace & Electronic Systems IEEE Transactions on, 2013, 49(1):55-73.
[11] Battistini S, Shima T. Differential games missile guidance with bearings-only measurements[J]. IEEE Transactions on Aerospace and Electronic Systems, 2014, 50(4): 2906-2915.
[12] Yanfang Liu, Naiming Qi, and Jinjun Shan. Cooperative Interception with Double-Line-of-Sight-Measuring[C]// AIAA Guidance, Navigation, and Control (GNC) Conference, Boston, MA: 2013. (AIAA 2013-5112)
[13] Shaferman V, Oshman Y. Cooperative Interception in a Multi-Missile Engagement[C]// AIAA Guidance, Navigation, and Control Conference. Boston, MA: 2013.(AIAA 2013-5235)
[14] Shaferman V, Oshman Y. Stochastic Cooperative Interception Using Information Sharing Based on Engagement Staggering[J]. Journal of Guidance Control & Dynamics, 2016, 39(9):1-15.
[15] Fonod R, Shima T. Estimation Enhancement by Cooperatively Imposing Relative Intercept Angles[J]. Journal of Guidance Control & Dynamics, 2017, 40(7):1-16.
[16] Shaoming He, Chang-Hun Lee. Optimality of error Dynamic in missile guidance problem [J/OL]. Journal of Guidance Control & Dynamics, 2018:1-10 doi:10.2514/1.G003343

IOP Conf. Series: Materials Science and Engineering **408** (2018) 012031 doi:10.1088/1757-899X/408/1/012031

An Experimental Framework for Validation of Thermal Modeling for Breast Cancer Detection

Dastan Igali[1], Olzhas Mukhmetov[1], Yong Zhao[1,*], Sai Cheong Fok[1] and Soo Lee Teh[1]

[1] Department of Mechanical Engineering, School of Engineering, Nazarbayev University, Astana 010000, Republic of Kazakhstan

Email: yong.zhao@nu.edu.kz

Abstract. In recent years, breast cancer is considered as one of the most predominant causes of death among women in the world. This also accounts for the 11.7 % of the total cancers diagnosed among ladies in Kazakhstan. The survival chances of the patients can significantly be improved if the disease is detected at its earliest stage. At the present time, mammography is found to be the most common technique for the purpose of detecting the breast cancer. However, this method is very expensive, and the diagnostic ability of this imaging technique fails when it comes to the detection of tumours of small sizes, in addition to its harmful side effect. The IR thermography is inexpensive without any harmful side effect, and promises to be more precise than conventional techniques of detecting breast cancer at its earlier stage. In this study we propose a comprehensive method that includes FEM modelling based on heat transfer principles and requires 3D scanning and IR imaging for disease diagnosis. Here we report the first part of our study, which develops an experimental procedure to generate experimental results for the validation of FEM models for the simulation and detection of breast cancer.

1.Introduction

According to Francis et al [1], breast cancer is currently considered to be one of the most predominant causes of death among young women in the world. The annual mortality rate per 100,000 women from breast cancer in Kazakhstan has also increased by 9 % since 1990, an average of 0.39 % a year [2]. Sejtkazina et al. [3] claimed that breast cancer is the most commonly diagnosed cancer among ladies, representing about 11.7 % of the total cancers. As it has been reported, risk factors causing breast cancer include unfavourable living conditions, chronic stress which has become epidemic in our society, unilateral breastfeeding, breastfeeding less than 3 months or more than 2 years, abortions, and genetic predisposition[4]. Moreover, Bilyalova et al. [5] stated that there is a convincing evidence showing that breast cancer incidence in Kazakhstan varies with respect to ecological regions, thereby demonstrating higher mortality rates from this disease in areas with greater levels of contaminants in the atmosphere.

There are many other techniques of breast cancer detection, such as mammography, ultrasound and MRI. They mostly rely on finding the tumor physically and cannot identify pre-cancerous stage of the breast. X-rays scan readings are strongly dependent on the breast tissue density and hormonal changes and cannot give accurate results if these two markers are not at normal state. Presently mammography is found to be the most common technique for detecting breast cancer. However, the diagnostic ability of this imaging technique fails when it comes to the detection of tumors of small sizes. Thus, the disease may not be brought under control at its earliest stage, leading to further growth. In addition, repeated mammography procedures expose the patient's body to very high levels of ionizing radiation [6].

Patients' chances of survival might be improved significantly if the disease is detected at its early stage, followed by proper prognosis and therapy [1]. Continuous efforts have been made in order to discover additional and safer diagnosis techniques for early tumor detection. One of such technique is thermography, which is applied to detect the abnormalities of thermal profiles in a breast [7]. It was reported by Gautherie et al. [8] that the tumor can be discovered 8-10 years ahead compared to mammography. IR imaging allows the identification of the presence of an early stage tumor by using the thermal signs of variation in blood vessels. Gautherie et al. [8] carried out a study, in which 1527 patients with abnormalities in their breast were under observation for 12 years. As these studies have shown, almost a half of the patients involved developed cancer within 5 years. It was concluded by the researchers that the risk for future development of breast cancer can only be recognized after examination of the thermal images of the breast for abnormalities. However, there are several factors which can affect the accuracy of the IR imaging procedure, such as asymmetrical breast temperature patterns, temperature stability, mental state, and menstrual cycle [9].

It was discovered that the surface temperature over the cancerous region is significantly higher than that of the region surrounding it. Accordingly, a variety of breast abnormalities can be determined by analyzing breast thermal images and the differences of absolute values of temperatures in a region of interest. Furthermore if there is temperature difference between the right and left breasts of more than 0.5 °C, it can be indicative enough to establish an abnormal condition of the breast. Thus, the study of the asymmetrical temperature patterns between contra lateral breasts is widely employed for the detection of breast cancer. The use of thermal imaging technique demonstrates that the metabolic activity and blood circulation in cancerous tissues and their surroundings are at increased rate compared to clinically healthy breast tissues. The cancerous tumors need increased blood circulation in order to provide nutrients to their cells by from existing blood vessels, inactive vessels, and other ones. Thereafter, increase in the surface temperature of the breast can be observed after employing thermal imaging. High precision IR cameras and powerful computers are required in order to obtain very accurate and high-quality images of temperature changes. Early stages of breast cancer can also be detected, if the camera has enhanced thermal sensitivity and infrared resolution of less than 0.1 °C [9].

There are specific thermal features according to which thermographers analyze the thermogram for the presence of cancer. Almost in all cases the contralateral breast is considered as reference to which study refers when detecting cancer. As reported by Hoekstra [10], the following thermographic signs characterize the breast disease:

- Asymmetric and hyperthermic vascular features;
- Focal patterns with +2.5°C temperature difference;
- Asymmetric and unusual vascular patterns with complex features;
- Asymmetric and diffuse hyperthermia with +2°C temperature difference patterns which involves a whole breast or peri-areolar area;
- Concentrated heat along abnormal breast region [11].

These features can also be observed from Figure 1, where the thermographic breast images are provided with distinguishable thermal patterns of healthy and abnormal breasts.

We propose an automated intelligent system that uses accurate 3D scanning, IR thermography and FEM thermal modeling for early detection of breast cancer. This can be achieved through reverse thermal modeling, during which the 3D scanning provides high precision breast geometry, IR thermography generates the breast surface temperature profiles and reverse FEM thermal modeling automatically searches for the location and size of the tumor if there is any. As a first step towards achieving this goal, we need to develop a standard experimental procedure whereby we can validate the FEM model and related simulation results.

The main purpose of this project is to establish such a procedure for FEM model validation for

IOP Conf. Series: Materials Science and Engineering **408** (2018) 012031 doi:10.1088/1757-899X/408/1/012031

future breast cancer detection, wherein silicone breasts with precision geometry are produced with simulated tumor and their temperature profiles are studied by using an infrared camera.

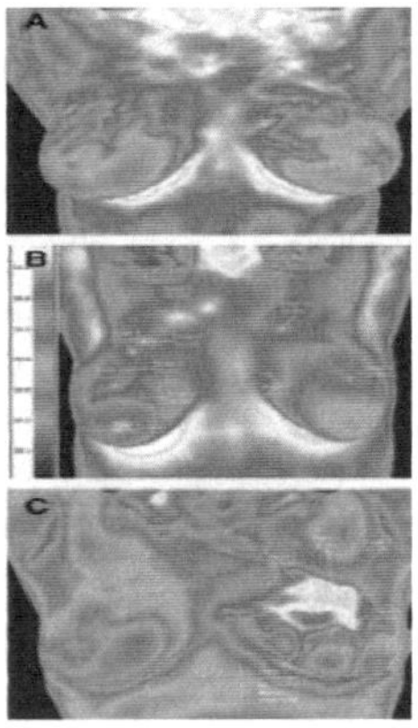

Figure 1. Thermographic breast images: (A) Clinically healthy breast; (B) Pre-cancerous stage of the breast; (C) Developed cancer in the left breast[11]

2. Methodology

2.1. 3D scanning

A 3D scanner, called ZScanner 700, is employed in this study. It is a unique self-positioning handheld device which makes 3D scanning process simple and efficient. It do not require any fixed-position tripods, mechanical arms and external positioning devices which complicates the scanning process of intricate or complex surfaces.

A female mannequin is used with accurate shape and size of a human body (see Figure 2). The mannequin has a height of 800 mm, chest of 838 mm, waist of 635 mm, which is scanned by the ZScanner 700. The breast region of the female mannequin was cut out for the study.

During scanning, positioning targets should be randomly affixed on the dummy breast with a distance of 30 mm between them (as seen in Figure 3). Those targets are required to define the position of the system in space. After the 3D image is obtained, the point cloud data is converted automatically into an STL file or polygon file or triangulated mesh. If further modifications are required (i.e. removal of unnecessary regions), the data can be further converted into CAD data (SLDPRT format).

Figure 2. Female mannequin

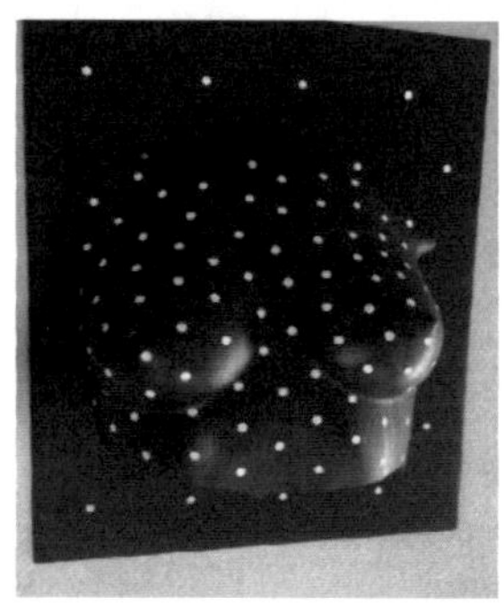

Figure 3. Breast region with targets

Figure 4 illustrates the scanned model with unnecessary regions removed. This image is now ready to be converted into CAD format and will be very used in future analysis. The dimensions of the 3D images are illustrated in the Figure 5 (in mm).

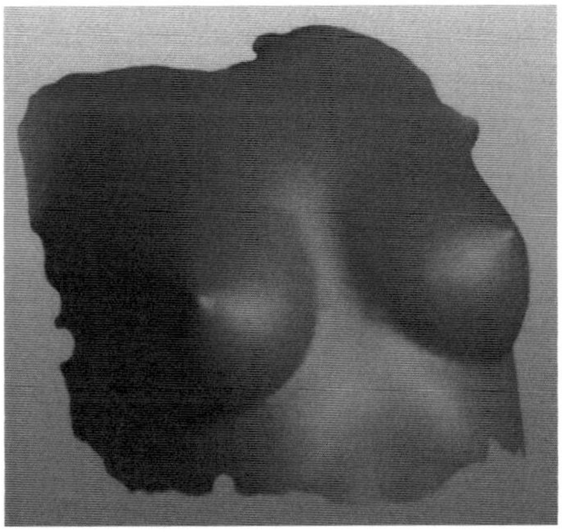

Figure 4. *Scan data with removed unnecessary parts*

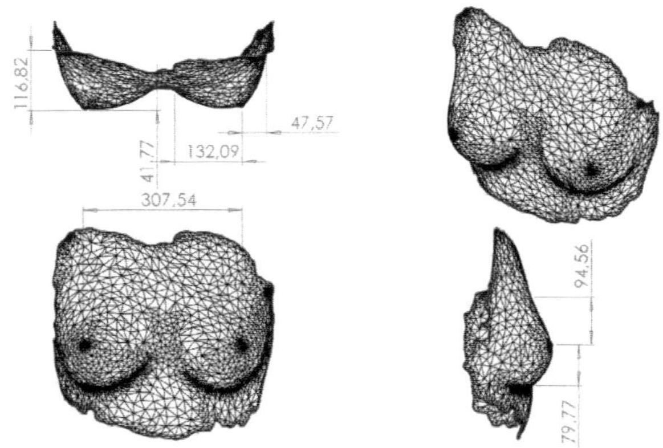

Figure 5. Orthographic projections of the breast

An infrared camera also known as a thermal-imaging camera is a device that measures the amount of infrared radiation falling on its infrared detector. Infrared camera operates at wavelengths roughly from 900 to 14,000 nm, while a simple camera can only operate in a range of 400-700 nm. Infrared image, after being transformed into radiometric one, can be viewed on a LCD screen where the user can observe temperature values. In this experiment, a thermographic camera called Fluke Ti29 is used to observe the temperature variations on the surface of the silicone breast. Fluke Ti29 cameras are able to detect small-scale temperature differences down to 0.1 degree. In addition, Fluke Ti29 has IT-fusion technology and blend modes which allow to display and store full visual images (640X480) and thermal images simultaneously. Thermal and visual images can be saved to a removable SD memory card and then transferred to a PC.

2.2. Resistors as heating elements for the simulation of tumors

Heat source for the experiment is made from several resistors connected very tightly in parallel and wires soldered from both sides to those resistors. Resistors produce heat as electric power is dissipated by them in a working circuit. Breast cancer is customarily associated with increases in the skin temperature around the tumor. Hence, heat sources consisting of resistors and wires are used to imitate the tumors' behaviors inside the real breast. Resistors used for this experiment are of the cylindrical shape with the length of 6 mm and diameter of 2 mm.

The volume of the resistor can be calculated as follows:

$$V = \pi r^2 h = \pi \times 1^2 \times 6 = 18.9\, mm^3$$

The volume of the heat source is determined by multiplying the volume of one resistor by the total number of resistors the heat source contains. Thus, the total volume of a heat source for each breast containing different number of resistors is provided in Table 1.

Table 1. The total volumes of heat sources for all the breasts

№	The total number of resistors in one heat source	Total volume, mm^3
1	4	75.4
2	8	150.8
3	6	113.1
4	4	75.4

In addition, the heat source of each breast is located differently, and their coordinates can be found in Table 2.

Table 2. Coordinates of heat sources

№	x, cm	y, cm	z, cm
1	9.2	8.9	3.5
2	9.5	10	5.4
3	14.5	8	4
4	6	10	4.5

The location of the origin of the coordinate system can be observed from the 3D image in Figure 6.

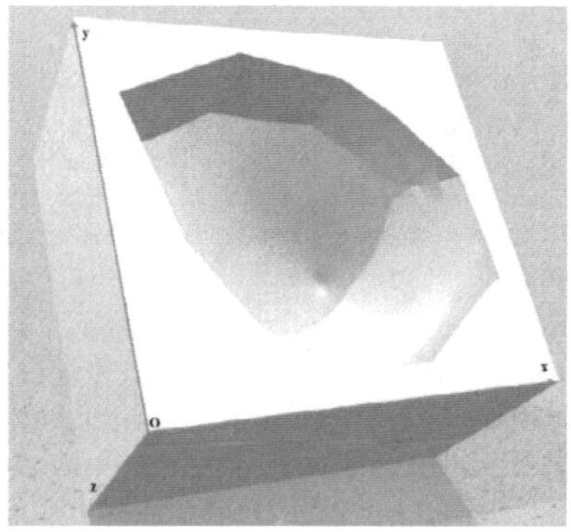

Figure 6. *3D image of the Mold with reference point*

2.3. Molding of silicone breasts

In this experiment, a 3D printer, called ZPrinter 450, is used to print the mold for casting silicone breasts (see Figure 7). This model has very high printing speed and allows printing objects of intricate shapes.

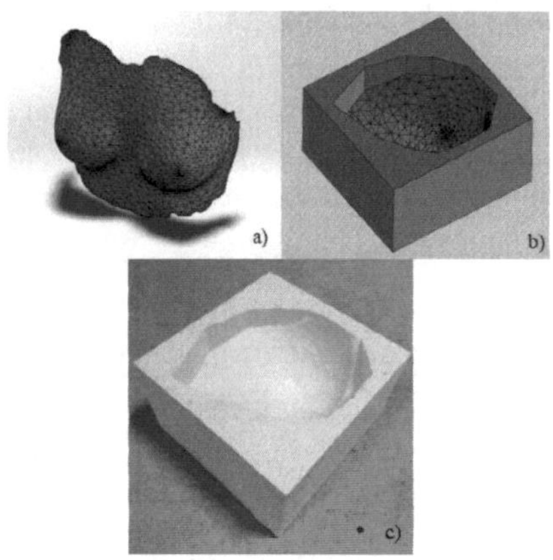

Figure 7. Mold printing: (a) scanned image by ZScanner 700; (b) 3D image of the mold created in Solidworks; (c) Mold manufactured by the 3D printer

The Dragon Skin 10 MEDIUM Set Silicone Rubber is used for casting breasts. One package of silicone rubber contains two ingredients, which should be properly mixed with a ratio of 1:1. The heat source made from resistors and wires is placed inside the mold at proper position, whose X, Y and Z coordinates should be recorded before the mixture is poured into the mold. Afterwards, the liquid silicone inside the mold is left for curing at room temperature for 4-6 hours. Finally, the prepared silicone breast is released from the mold. No releasing agent is required while using Dragon Skin 10 MEDIUM Set Silicone Rubber. The final breasts can be seen in Figure 8.

IOP Conf. Series: Materials Science and Engineering **408** (2018) 012031 doi:10.1088/1757-899X/408/1/012031

Figure 8. Silicone Breasts: (a) with a tumor size of 75.4 mm^3; (b) with a tumor size of 150.8 mm^3; (c) with a tumor size of 113.1 mm^3; (d) with a tumor size of 75.4 mm^3

2.4. Experimental procedure

The aim of the experiment was to take IR images of the silicone breast with the heat source inside by using the infrared camera and record the highest temperature value on the surface of the breast. At the beginning of the experiment, all the necessary equipment should be tested to ensure that they are in good working order before use. In addition, safety measures should be taken in order to prevent accidents or injuries during the experiment. Afterwards, the breast's initial temperature, room temperature and the temperature of the desk on which the silicone breast is placed should be measured by using the infrared camera. The room temperature should be held approximately at 24-26°C. Moreover, the initial temperature of the breast has to be very close to the room temperature and to the temperature of the desk to be sure that ambient temperature will not significantly affect the temperature of the breast. Otherwise significant temperature differences can lead to significant convection heat transfer, causing the temperature of the breast to become unstable over time. Thermographic images are taken by the infrared camera and maximum temperature values are recorded after the power supply is turned on. Measurements will be taken every fifteen minutes.

3. Results and discussion

3.1. Silicone breast with a 75.4 mm^3 tumor size

The first experiment is done on the silicone breast with a *75.4* mm^3 tumor size. The heat source imitating a tumor inside the breast is located at the following coordinates: $x=9.2$ *cm*, $y=8.9$ *cm*, $z=3.5$ *cm*. The thermograms of the breast are shown below in Figure 9, where the presence of the heat source inside the breast can be clearly noticed from Figure 9 (d). The region with relatively hot temperature is located exactly on the top of the heat source and indicated by green color.

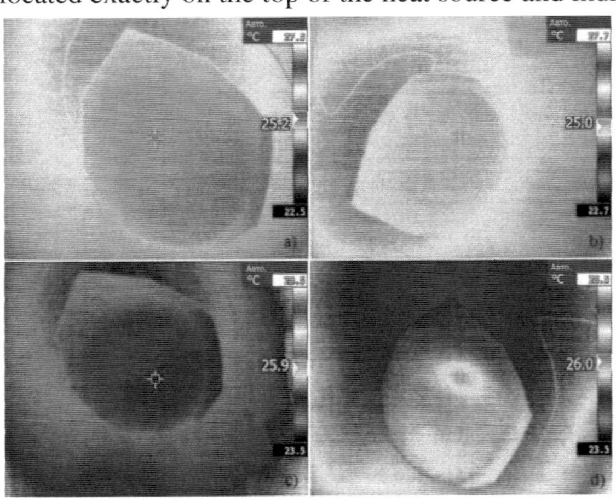

Figure 9 (a) infrared imaging for P=0.0027 W; (b) infrared imaging for P=0.01608 W; (c) infrared imaging for P=0.064 W; (d) infrared imaging for P=0.25408 W.

The current and voltage values for the resistors are given in Table 3. Maximum temperature values with respect to power output values given are also provided in the table.

Table 3. Data recorded with increase in power output in every 15 minutes

№	Voltage, V	Current, A	Power Output, W	Temperature, ℃
1	0	0	0	25.1
2	0.90	0.003	0.0027	25.2
3	2.01	0.008	0.01608	25.0
4	4.00	0.016	0.064	25.9
5	7.94	0.032	0.25408	26.0

It is very important to record the room temperature and the temperature of the desk on which the breast is placed (T_{room}=24.5 °C, T_{desk}=24.4 °C). If the ambient temperature is very low, it becomes very difficult to increase the temperature of the breast, thus leading to inaccurate results. For instance, the maximum temperature dropped from 25.2 °C to 25 °C despite a significant increase in the power output from 0.0027 W to 0.01608 W. The effect of the heat source on the temperature of the breast surface can be viewed from Figure 10. Although there is an slight initial drop in the temperature with a power value of 0.01608 W, the overall trend goes up.

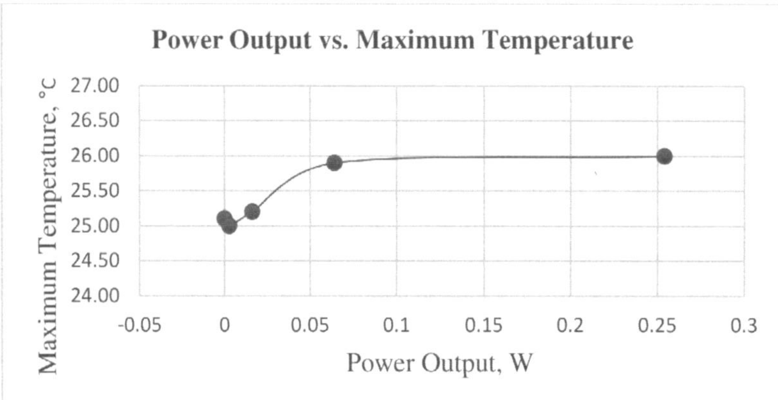

Figure 10. Power Output vs. Maximum Temperature

3.2. Silicone breast with a150.8 mm³ tumor size

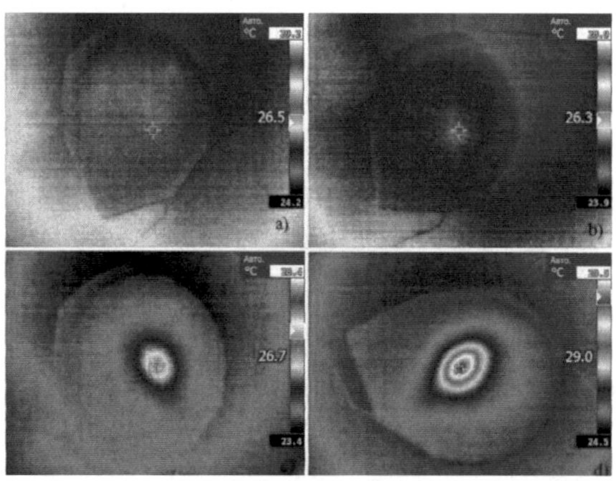

Figure 11. (a) infrared imaging for P=0.0081 W; (b) infrared imaging for P=0.02 W; (c) infrared imaging for P=0.09918 W; (d) infrared imaging for P=0.23085 W

The second experiment is conducted on the silicone breast with a *150.8* mm^3 tumor size. The heat source is located at the following location: *x=9.5 cm, y=10 cm, z=5.4 cm*. The thermograms of the breast are provided below in Figure 11. In Figure 11 (a), there is no observable presence of hotspot, but the general temperature of the surface is higher than its initial value. As the power output increases, the hot spot starts to form in Figure 11 (b). The presence of the heat source inside the breast can be clearly identified in Figure 11 (c) and (d). The region with relatively hot temperature is located exactly on the top of the heat source and indicated by red color. Temperature drop as it goes far from the hotspot can be observed from the different color layers around the hotspot. In this case, the presence of the heat source is very clear, since it is located close to the surface of the silicone breast.

The values of the voltage, current and measured maximum temperature are provided in Table 4.

Table 4. Values recorded with increase in the power output in every 15 minutes

№	Voltage, V	Current, A	Power Output, W	Temperature, °C
1	0	0	0	25.6
2	1.62	0.005	0.0081	26.5
3	2.50	0.008	0.02	26.3
4	5.51	0.018	0.09918	26.7
5	8.55	0.027	0.23085	29.0

The room temperature and the temperature of the desk on which the breast is placed are T_{room}=24.8 °C and T_{desk}=25.6 °C, respectively. Under these circumstances, the maximum temperature dropped from 26.5 °C to 26.3 °C despite the increase in the power output value from 0.0081 W to 0.02 W. The effect of the heat source on the surface temperature of the breast can be observed in Figure 12. Although, there is a drop in the temperature at the power value of 0.02 W initially, the overall temperature increases with power output.

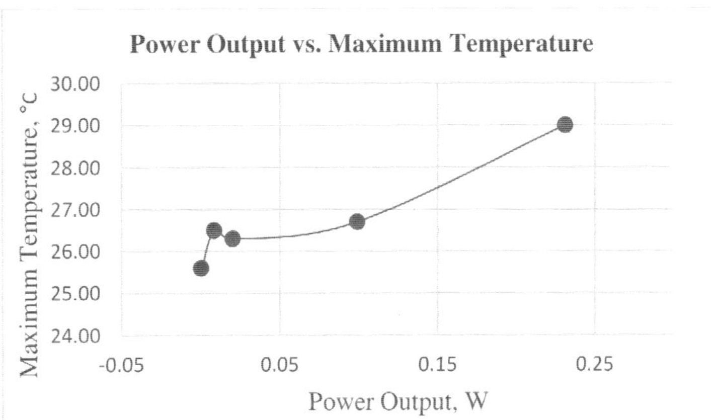

Figure 12. Power Output vs. Maximum Temperature

3.3. Silicone breast with a 113.1 mm^3 *tumor size*

The third experiment is undertaken on the silicone breast with a *113.1* mm^3 tumor size. The heat source imitating the tumor inside the breast is located at the following coordinates: *x=14.5 cm, y=8 cm, z=4 cm*. The thermograms of the breast are presented in Figure 13. In Figure 13 (a), there is no observable presence of hotspot, but the general temperature of the surface higher than on initial. From Figures 13 (b) and (c), it can be seen that the surface temperature increases with increment of power output. The presence of the heat source inside the breast can be clearly seen in Figure 13 (d), when the power output is high. The region with relatively hot temperature is located exactly on the top of heat source and indicated by light green and yellowish green color.

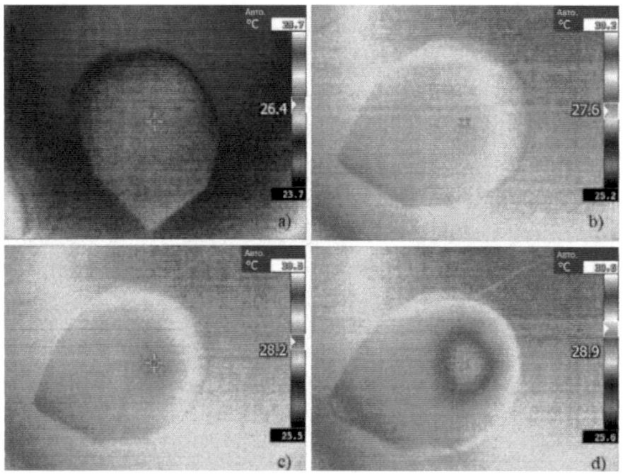

Figure 13. (a) infrared imaging for P=0.00486 W; (b) infrared imaging for P=0.0459 W; (c) infrared imaging for P=0.18685 W; (d) infrared imaging for P=0.42224 W

The voltage, current and measured maximum temperature values are given in Table 5.

Table 5. Values recorded with increase in power output in every 15 minutes

№	Voltage, V	Current, A	Power Output, W	Temperature, °C
1	0	0	0	26.3
2	0.81	0.006	0.00486	26.4
3	2.55	0.018	0.0459	27.6
4	5.05	0.037	0.18685	28.2
5	7.54	0.056	0.42224	28.9

The room temperature and the temperature of the desk are T_{room}=26.3 °C and T_{desk}=26.3 °C, respectively. The effect of the heat source on the maximum temperature of the breast surface can be examined in Figure 14, where it can be noticed that there is only a rising trend without any temperature drop.

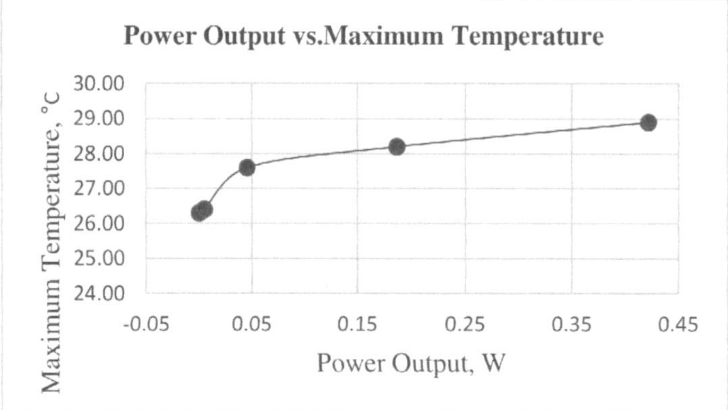

Figure 14. Power Output vs. Maximum Temperature

3.4. Silicone breast with a 75.4 mm³ tumor size

The fourth experiment is done on the silicone breast with a *75.4* mm^3 tumor size. The heat source inside the breast is located at the following coordinates: *x=6 cm, y=10 cm, z=4.5 cm*. The thermograms of the breast are shown in Figure 15, where the presence of the heat source inside the breast can be noticed from Figures 15 (a) and (b). In Figures (c) and (d) the hotspot is clearly seen with red color. Again the region with relatively hot temperature is located exactly on the top of heat source and indicated by red color.

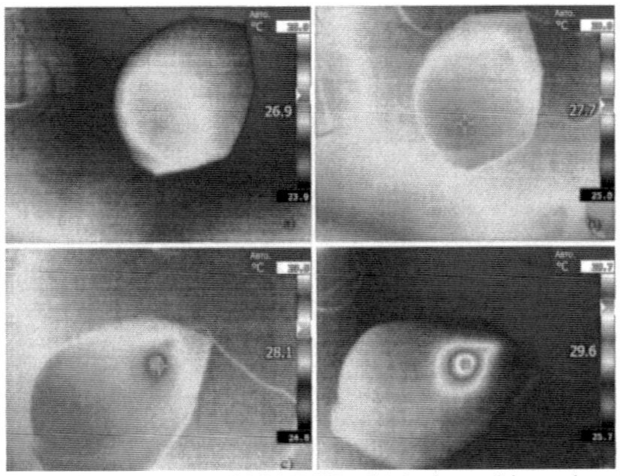

Figure 15. (a) infrared imaging for P=0.00297 W; (b) infrared imaging for P=0.01592 W; (c) infrared imaging for P=0.05012 W; (d) infrared imaging for P=0.16848 W

The recorded power output and maximum temperature values can be found from in Table 6 below.

Table 6. Values recorded with increase in power output in every 15 minutes

№	Voltage, V	Current, A	Power Output, W	Temperature, °C
1	0	0	0	26.1
2	0.99	0.003	0.00297	26.9
3	1.99	0.008	0.01592	27.7
4	3.58	0.014	0.05012	28.1
5	6.48	0.026	0.16848	29.6

The room temperature and the temperature of the desk are T_{room}=25.8 °C and T_{desk}=26.3 °C. The effect of the heat source on the temperature of the breast surface can analysed from Figure 16. It can be noticed from the graph that there is only an upward trend without any temperature drop. Significant temperature rises are observed for power output values of 0.00297 W, 0.01592 W and 0.05012 W.

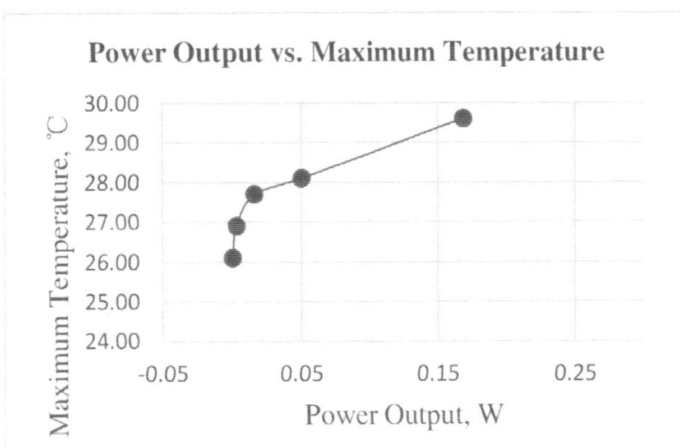

Figure 16. Power Output vs. Maximum Temperature

3.5. Comparison of temperature profiles

Combining all of the graphs in Figure 17, representing the effect of the increase of the power output on the temperature profile, it becomes clear that the maximum temperature values of the third and fourth breasts are going exclusively up as the given value of the power output increases. However, the graphs showing the temperature profiles of another two breasts have negligible temperature drops after increasing the power to the second values, but then temperature goes up again. This is due to the reason that the temperature inside the room where the experiment is conducted is lower compared to the initial surface temperature of the breast. Accordingly, convection heat transfer takes place. But when sufficient amount of power is given to the circuit temperature will start to grow. It can also be noticed that for the second case where the tumor has a size of $150.8\ mm^3$, the magnitude of the temperature is sharply going up, while for the first case where the tumor size is $75.4\ mm^3$ the temperature is increasing gradually. Overall, the temperature increases sharply for the second and fourth cases, since for both cases the tumor is located very close to the surface of the breast and thus showing clear temperature difference on the thermograms provided. In the other two cases, the tumors are located deep inside the breasts, therefore temperature increases on the surfaces of the breasts are at moderate rate.

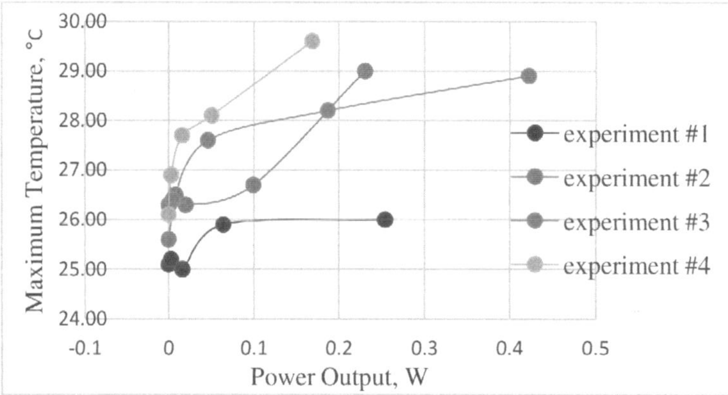

Figure 17. Power Output vs. Maximum temperature

4. Conclusion and future work

Overall the main purpose of this study is to develop an experimental procedure in order to generate experimental results for FEM model validation. The breast of a female mannequin is used as a breast model, which is scanned and its replicas are produced by 3D printing the mold and molding it with silicon. The molded breasts are fitted with resistor heaters as models of tumors. IR imaging and temperature measurements are then taken in the experiment with an accuracy of 0.1 degree in temperature. The results from the experiment are successfully generated and can be used for the validation of FEM simulations.

For future work, FEM thermal models and simulations will be validated with the data obtained in this study. Furthermore, breast cancer patients' data will be obtained using the same procedure with 3D scanning and IR imaging. A toolbox will be developed incorporating a FEM code for inverse thermal modeling based on a search scheme and the patients' data to search for the sizes and locations of tumors in the patients' breasts.

References

[1] Francis, S. V., Sasikala, M., & Jaipurkar, S. D. (2017). Detection of Breast Abnormality Using Rotational Thermography. In *Application of Infrared to Biomedical Sciences* (pp. 133-158). Springer Singapore.

[2] HealthGrove. (2013). Breast Cancer in Kazakhstan. Retrieved November 15, 2017, from http://global-disease-burden.healthgrove.com/l/33097/Breast-Cancer-in-Kazakhstan.

[3] Sejtkazina, G. D., Bajpeisov, D. M., Sejsenbaeva, G. T., & onkologicheskoj sluzhby Respubliki, A. A. P. (2012). Kazahstan za 2011 god (statisticheskie materialy) (Indicators Oncology Service of the Republic of Kazakhstan for 2013 (statistical material)).

[4] Toleutay, U., Reznik, V., Kalmatayeva, Z., & Smigelskas, K. (2013). Risk factors of breast cancer in Kyzylorda oblast of Kazakhstan: a case-control study. *Asian Pacific Journal of Cancer Prevention*, *14*(10), 5961-5964.

[5] Bilyalova, Z., Igissinov, N., Moore, M., Igissinov, S., Sarsenova, S., & Khassenova, Z. (2012). Epidemiological evaluation of breast cancer in ecological areas of Kazakhstan-association with pollution emissions. *Asian Pacific Journal of Cancer Prevention, 13*(5), 2341-2344.

[6] Keyserlingk, M. D. (1998). *Infrared Imaging of the Breast: Initial Reappraisal Using High-Resolution Digital Technology in 100 successive cases of Stage I and II Breast Cancer*(Vol. 4, pp. 245-251, Publication). The Breast Journal.

[7] Jiang, L. J., Ng, E. Y. K., Yeo, A. C. B., Wu, S., Pan, F., Yau, W. Y., ... & Yang, Y. (2005). A perspective on medical infrared imaging. *Journal of medical engineering & technology*, *29*(6), 257-267.

[8] Gautherie, M. (1989). Atlas of breast thermography with specific guidelines for examination and interpretation. *Milan, Italy: PAPUSA*, *256*.

[9] Acharya, U. R., Ng, E. Y. K., Tan, J. H., & Sree, S. V. (2012). Thermography based breast cancer detection using texture features and support vector machine. *Journal of medical systems*, *36*(3), 1503-1510.

[10] Hoekstra P, The autonomic challenge and analytic breast thermology. Thermology International, 2004(14);3:106.

[11] Kennedy, D. M., Lee, T., & Seely, D. (2009). *A Comparative Review of Thermography as a Breast Cancer Screening Technique*(1st ed., Vol. 8, pp. 9-16, Publication). Toronto, Ontario, Canada: The Canadian College of Naturopathic Medicine. doi:10.1177/1534735408326171

The Study of Reducer Modeling Language Design Based on Symmetry

Xiaoli Fu*, Ting Li and Shufeng Yang

Zhongyuan University of Technology, Zhengzhou 450007, China

*fxlzzti@163.com

Abstract. Symmetry design is a kind of common design method in mechanical product design. The rational use of symmetry is helpful to realize the function of the product better, embody the symmetry of shape and meet the design requirements. This paper systematically analyzed the concept and type of symmetry, and established the symmetry system of reducer. From the perspective of material function and art form, the paper decomposed the structure of reducer and expatiated the relationship between global symmetric, component symmetry, and symmetry base elements. This paper by analyzing the symmetry of foreign brand reducer, present the whole shape determination of reducer box and the expression of symmetrical datum. Furthermore, it is used in the modeling language design of reducer, which is beneficial to highlight brand characteristics, optimize the modeling of reducer and upgrade the level of reducer design.

1. Introduction

Symmetry is widely used in mechanical product design. The earliest study of mechanical symmetry was Barrenscheen, a German scholar, that carried out a preliminary exposition of the conceptual system of mechanical symmetry[1]. The team led by Professor Feng Peien from Zhejiang University proposed the symmetry ontology of natural science to engineering science, they clarified the function, principle and time of the symmetry in the symmetry system[2]. Qi Yuxuan and others put forward the concept system of mechanical structure symmetry, and constructed the knowledge and design platform of mechanical product symmetry design[3]. Zhong Kangmin pointed out the basic idea of guiding the innovation of mechanical design with the idea of symmetrical aesthetics[4]. From the viewpoint of aesthetics and mechanics, Qiao Jizhong analyzed the application of symmetry in industrial modeling design[5]. And in the field of the design of the industrial general reducer (hereinafter referred to as "reducer"), Fu Xiaoli and her group studied the modeling of industrial general reducer from the aspects of brand culture and concept characteristics[6]. However, there is no research on modeling language design for product symmetry in previous studies.

In the form design of mechanical products, the form and language of the mechanical product should be given in the form of "deliberate" expression , following the principle of "shape and random energy", which reflects the applicability of the function, the symmetry and beauty of the shape[7]. So symmetry plays an important role in the design of reducer. On the basis of the symmetry analysis of the brand reducer, the symmetry theory and the expression method are put forward to guide the scientific application of symmetry in the design of the reducer's modeling language, which is to improve the design level of the reducer and the general product.

2. The type of symmetry

Symmetry is based on point C/ axis L^n/ surface P. The up and down, the left and right, the ring and the oblique side appear in the form of same and repeated symmetry elements, and the symmetric sum of the set of symmetric elements is also called the symmetric space group. The symmetric space group consists of two parts, one is rotation and the other is the translation group of translational axisymmetric elements.

According to the symmetry benchmark requirements, symmetry graphs are represented by two dimensional and three dimensional views. Three kinds of symmetry are summarized below, as shown in Table.1:

- The mirror symmetry is also called the symmetric mirror Mi. Suppose an axis L^n or plane P, the same part of the graph is repeated by means of the reflection operation of this line or plane.
- Rotational symmetry L^n. Suppose that the number of times that a L^n axis is rotated for one week is n, and the minimum angle of rotation is α. The relationship between them is $\alpha = 360°/n$.
- Translation symmetry Tr. When a symmetry element moves along a certain direction, it will coincide with its symmetrical elements.

Table 1. Symmetry type and space expression.

Space Expression	Mirror symmetry	Rotational symmetry	Translational symmetry
2-dimension	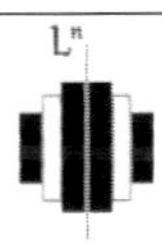	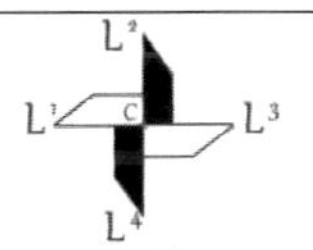	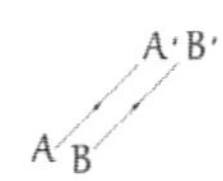
3-dimension	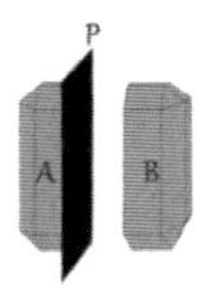	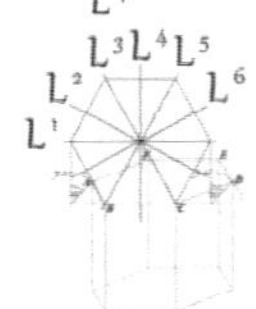	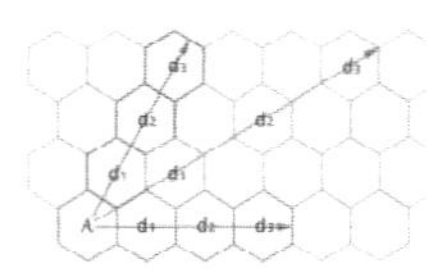

3. The Type of reducer

The reducer is widely used in different equipments. It can be divided into four main parts according to the input and output mode:

- The input and output mode is coaxial R series helical gear Hardened reducer.
- S series helical gear worm gear reducer with input and output mode as direct axis.
- K series spiral bevel gear reducer with input and output mode is direct axis.
- F series parallel shaft helical gear reducer with parallel input and output mode.

4. The symmetry system of the reducer

The reducer is an independent closed driving device between the prime mover and the working machine. It is the high-speed running dynamic generated by the motor or the inter combustion engine, which is through the gear transmission of the input shaft, the layshaft and the output shaft that achieve the purpose of deceleration[8]. The structure diagram of the R series reducer is shown below: schematic diagram is shown in Figure.1.

The reducer mainly composes of transmission parts (gears, shafts, etc.) and supporting parts (boxes, cover plates, flanges, etc.). Its symmetry mainly composes of material function and artistic form. It includes three design elements, which refer to the whole box, structural parts (the box position of M1-M6, schematic diagram is shown in Figure.2), and symmetrical benchmark (the repeatability of symmetry elements).

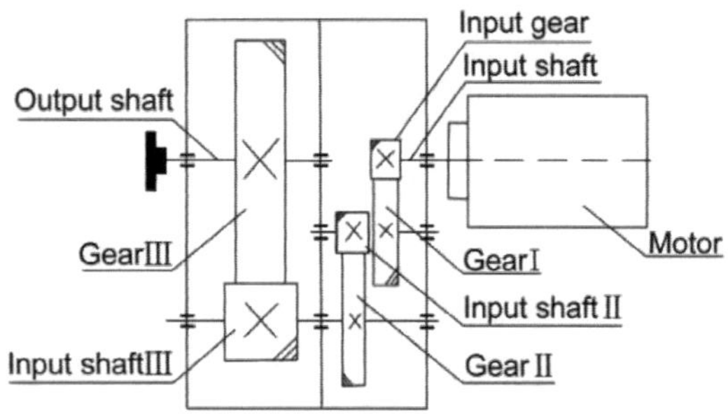

Figure 1. Structure diagram of R series reducer.

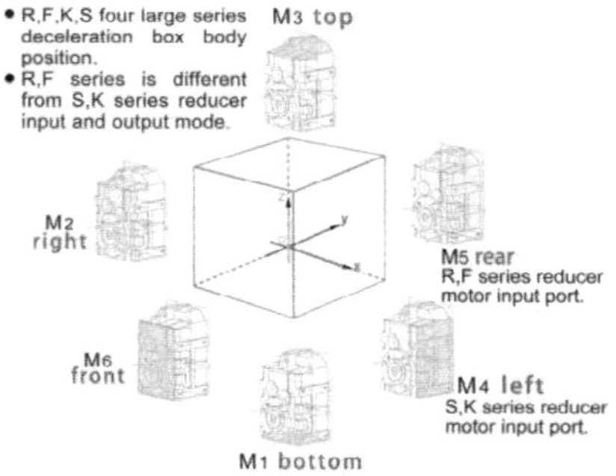

Figure 2. Deceleration box body position.

4.1. Symmetry of material function

On the premise of the function structure of the reducer meeting the requirements of symmetry of material function, taking the R series helical gear reducer as an example, the material function symmetry Sf was divided into functional holistic symmetry Sw, functional component symmetry Sp, functional symmetric reference element Sbe, and the material function symmetry Sf = (Sw, Sp, Sbe) of the reduce, schematic diagram is shown in Figure.3:

- Functional holistic symmetry Sw = (Sem≤ Sre).The weight relationship between reducer Sre and motor Sem is symmetrical.
- Functional component symmetry Sp = (Str, Sbr, Sdis). The transmission part Str is composed of gear, input shaft, intermediate shaft and output shaft, and the support part Sbr is composed of box, flange, ground board, hanger and cover plate; and the radiator Sdis is composed of heat dissipation lines on one side or two sides of the box to complete the symmetry of function.
- The function symmetry benchmark element Sbe = (Sde, Sdes, Sstr). Functional requirements Sde, function design Sdes and function structure Sstr are symmetrical.

The main material function symmetry is functional global symmetry Sw = (Sem = Sre), transmission equilibrium symmetry Str and equilibrium heat sink Sdis.

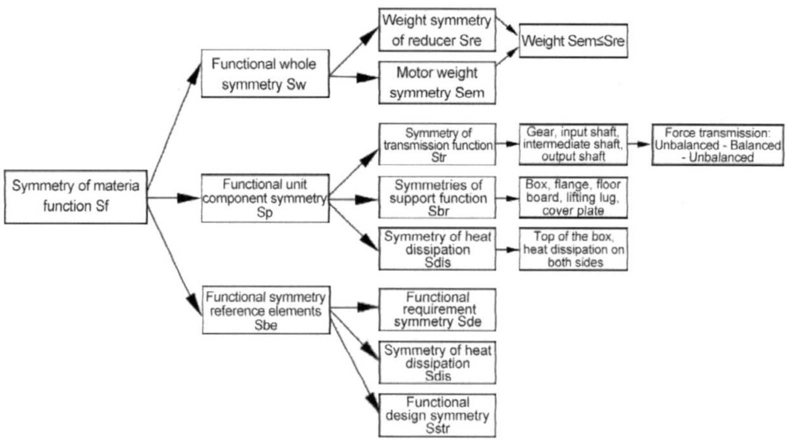

Figure 3. Material function symmetry system of reducer.

4.2. Symmetry of art form

Under the premise of satisfying the symmetry of art form, the art form symmetry of reducer Sa was divided into the overall symmetry of box form Sab , the form symmetry of component Constituent Sap and the form symmetry of base element Sabe, so the art form symmetry of reducer Sa = (Sab, Sap, Sabe), schematic diagram is shown in Figure.4:

- The overall form symmetry of box Sab. It is composed of the symmetry of the overall M_6 shape of the reducer box Sdc and the symmetry of the shape details Sda. Among them, the form symmetry of the reducer box M_6 Sdc is determined by the symmetry square of 4 axis. And the symmetry of line segment a, a '~h, h' evolves into the symmetric form of 1 axis, 2 axis and 6 axis. After the initial shape is determined, the symmetry of the shape details Sda adjust the symmetry in detail through the surface Pa, Pa '~Ph, Ph', final shape is obtained. Refer to the overall shape determination of the box and the symmetry of detail adjustment expression method, schematic diagram is shown in Figure.5.
- The form symmetry of component Constituent Sap = (Spar, Sdh). The support Spar is composed of a box, a flange, a floor board, a hanging ear, and a cover plate. The heat dissipation component, Sdh, is composed of the heat dissipation lines on one side or both sides of the box to complete the symmetry of the shape.
- Form symmetry base element Sabe. The symmetrical base element is based on the symmetrical point C/ axis L^n/ surface P/ angle alpha as the symmetry reference. The form of the reducer is expressed by symmetrically, and the symmetry datum expression method is referenced, as shown in Table.2. To checks the form symmetry of reducer for correctness, refer the symmetry theorem to judge as follows[9]:
- Definition 1: if there is a secondary axis L^2 perpendicular to the n subaxis L^2, then there are nL^2 vertical n times L^n ——$L^n nL^2$;
- Definition 2: if there is A symmetric surface P Contained the n subaxis L^n, then there are n P Contained n times L^n ——$L^n nL^2$;
- Definition 3: if there is an even time axis vertical symmetry plane P, then a symmetric center C must be generated at the intersection point between the symmetry axis and the symmetric plane——$L^2 nPC$.

The main artistic form symmetry Sa has the form symmetry Sdc, flange form symmetry Spar and heat sink shape symmetry Sdh.

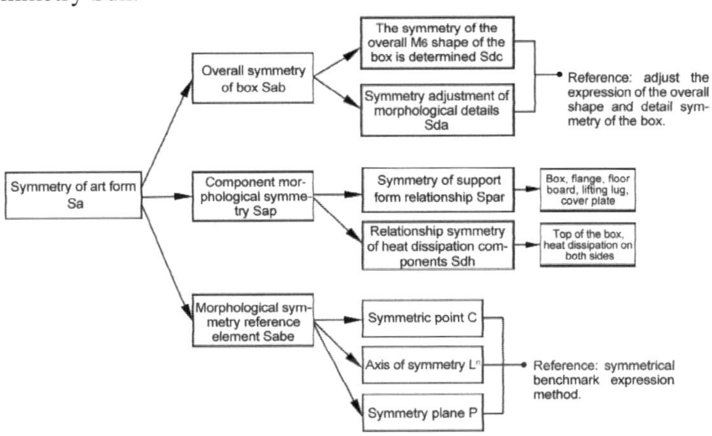

Figure 4. The symmetry system of the art form of the reducer.

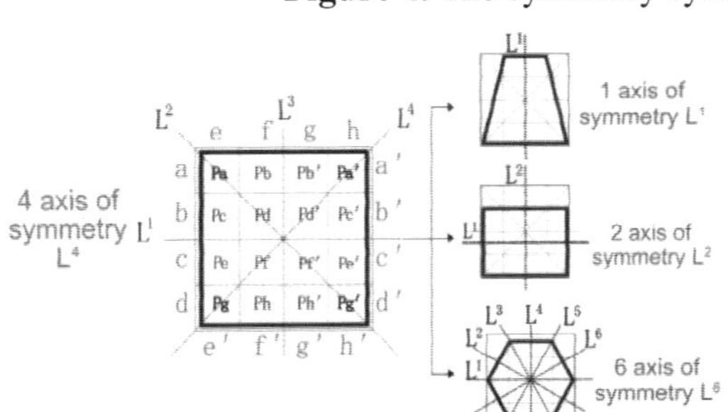

Figure 5. The overall shape determination of reducer box and the symmetry adjustment expression method of detail.

IOP Conf. Series: Materials Science and Engineering **408** (2018) 012032 doi:10.1088/1757-899X/408/1/012032

Table 2. Symmetric benchmark expression method.

Axis of symmetry L^n	1	2	4	6
Center of symmetry C	0	1	1	1
Symmetry plane P	1	2	4	6
Base angle α	360°	180°	90°	60°
Morphological symmetry	L^1	L^2	L^4	L^6

5. The analysis of the symmetry modeling language for the brand reducer

In the aspects of the reducer's form and design,The design principle of "shape random energy"must followed by us, which can reflect the practicability of function and it should emphasize on "form follows emotion", which means that express the emotion expressed by the reducer through morphological language. Because most of the domestic reducer products imitate the foreign brand reducer, we select the three main functions and forms of the four major series of reducer, BAUER, SEW and BONFIGLIOLI, to carry out symmetry extraction, and carry out the symmetrical elements analysis of the selected reducer through the material function, the symmetry system of art form and the symmetry datum expression method, as shown in Table.3. The modeling language is summarized and the symmetry rules of the reducer series and the brand are explored.

Under the established condition that the main material function symmetry of the reducer is balanced, the modeling language of the symmetry for the art form can be divided into two parts. The first part is the main body modeling language M_6, which mainly refers to the overall symmetry modeling language of the reducer box; the second part is the auxiliary body modeling language M_1-M_5, mainly includes the flange and the heat dissipation symmetry modeling language, as shown in Table.4.

Table 3. Symmetry element analysis of brand reducer (a).

Series	BAUER						
	Product form	Function Sf			Form Sa		
		Sw	Str	Sdis	Mi	L^n	Tr
R		Weight balance	Balance of transmission force	Heat dissipation balance	Heat dissipating pattern of box	flange	—
		—	—	—	Curve	Octagon	—
S	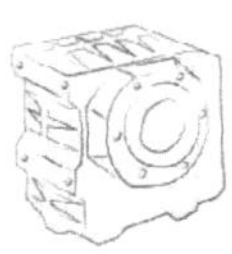	Weight balance	Balance of transmission force	Heat dissipation balance	Heat dissipating pattern of box	Flange	—
		—	—	—	Curve	Octagon	—
K	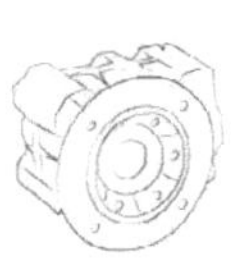	Weight balance	Balance of transmission force	Heat dissipation balance	—	Box Flange	Heat dissipation
		—	—	—	—	Round	Curve
F		Weight balance	Balance of transmission force	Heat dissipation balance	Heat dissipating pattern of box	Flange	—
		—	—	—	Curve	Round	—

IOP Conf. Series: Materials Science and Engineering **408** (2018) 012032 doi:10.1088/1757-899X/408/1/012032

Table3. Symmetry element analysis of brand reducer (b).

Series	SEW						
	Product form	Function Sf			Form Sa		
		Sw	Str	Sdis	Mi	L^n	Tr
R	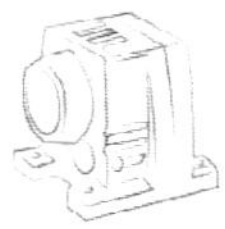	Weight balance	Balance of transmission force	Heat dissipation balance	Box	Flange	Heat dissipation
		—	—	—	—	Octagon	linear
S	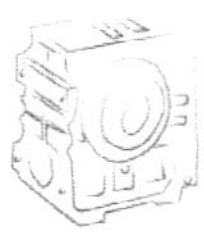	Weight balance	Balance of transmission force	Heat dissipation balance	Box	Flange	Heat dissipation
		—	—	—	—	Round	Linear
K		Weight balance	Balance of transmission force	Heat dissipation balance	—	Flange	Heat dissipation
		—	—	—	—	Round	Linear
F		Weight balance	Balance of transmission force	Heat dissipation balance	Box	Flange	Heat dissipation
		—	—	—	—	Round	Linear

Table3. Symmetry element analysis of brand reducer (c).

Series	BONFIGLIOLI						
	Product form	Function Sf			Form Sa		
		Sw	Str	Sdis	Mi	L^n	Tr
R	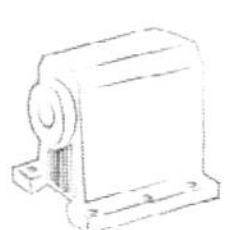	Weight balance	Balance of transmission force	Heat dissipation balance	Box	Flange	Heat dissipation
		—	—	—	—	Round	Linear
S		Weight balance	Balance of transmission force	Heat dissipation balance	Box	Flange	Heat dissipation
		—	—	—	—	—	Linear
K	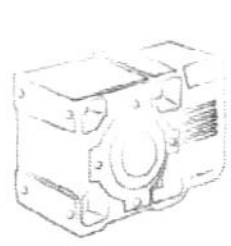	Weight balance	Balance of transmission force	Heat dissipation balance	Box	Flange	Heat dissipation
		—	—	—	—	Round	Linear
F		Weight balance	Balance of transmission force	Heat dissipation balance	Box	Flange	Heat dissipation
		—	—	—	—	Round	Linear

Table4. Modeling language analysis of the symmetrical law of brand reducer.

Series	BAUER	SEW	BONFIGLIOLI	Modeling Language BAUER box	BAUER Flange	BAUER Heat dissipation	SEW box	SEW Flange	SEW Heat dissipation	BONFIGLIOLI box	BONFIGLIOLI Flange	BONFIGLIOLI Heat dissipation
R	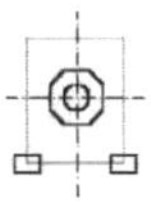	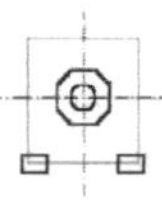	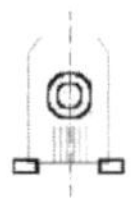	Biaxial cube	Octagon	Curve	Biaxial cube	Octagon	Linear	Uniaxial cube	Round	Curve
				Rule	Symmetrically	Fluency	Rule	Symmetrically	Precise	Light	Full	Fluency
	$L^2$2PC	$L^2$2PC	$L^1$1P	Stability	Full		Stability	Full	Order	Stable	Rigorous	
S	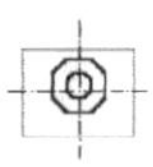	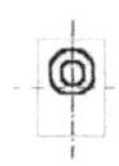	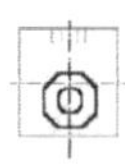	Biaxial cube	Octagon	Curve	Biaxial cube	Round	Linear	Biaxial cube	Octagon	Linear
				Rule	Symmetrically	Fluency	Rule	Full	Precise	Rule	Symmetrically	Precise
	$L^2$2PC	$L^2$2PC	$L^2$2PC	Stability	Full		Stability	Rigorous	Order	Stability	Full	Order
K	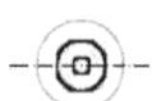	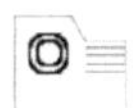	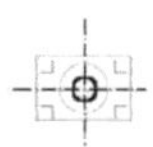	Round	Round	Curve	Irregular body	Round	Linear	Biaxial cube	Round	Linear
				Full	Full	Fluency	Bias	Full	Precise	Rule	Full	Precise
	$L^1$1P	—	$L^2$2PC	Rigorous	Rigorous		Variable	Rigorous	Order	Stability	Rigorous	Order
F	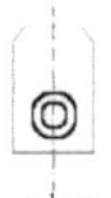	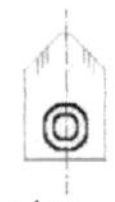	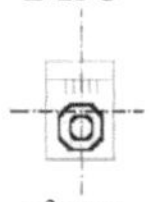	Uniaxial cube	Round	Curve	Uniaxial cube	Round	Linear	Biaxial cube	Round	Linear
				Light	Full	Fluency	Light	Full	Precise	Rule	Full	Precise
	$L^1$1P	$L^1$1P	$L^2$2PC	Stable	Rigorous		Stable	Rigorous	Order	Stability	Rigorous	Order

5.1. Analysis of subject symmetry modeling language

The whole box of brand BAUER, SEW R, S series, BONFIGLIOLI S, K, F series reducer is dual axis symmetrical body, box M_6 has L^2 sub symmetric axis, 2 symmetric plane P, a symmetric center point C, symmetrical rule combination for $L^2$2PC can be upward / down / left / right mirror, rotating symmetry; The main body of the speed machine is a single axis symmetrical body, and the box M_6 has L^1 sub symmetric axis, 1 symmetric surface P, no symmetric center point C, and the symmetry rule is composed of $L^1$1P to the left and right mirrors, while the K series of the brand SEW is irregular and the box M_6 has no symmetrical C/ axis L^n/ P.It is concluded that the subject symmetry law of the reducer box subject of the brand BAUER reducer is 2 times $L^2$2PC=2 times $L^1$1P, and the subject symmetry law of the reducer box subject of the brand SEW is composed of 2 times $L^2$2PC>1 times $L^1$1P, and the subject symmetry law of the reducer box subject of the brand BONFIGLIOLI is composed of 3 times $L^2$2PC>1 times $L^1$1P.

It is concluded that the symmetry law of the M_6 body modeling language of the deceleration chassis is $L^2$2PC>$L^1$1P, and the form is 2 axisymmetric $L^2$2PC, followed by the uniaxial symmetry $L^1$1P. The 2 axisymmetric shape is regular and stable, while the uniaxial symmetry shape modeling language not only shows symmetry and stability, but also presents a lightweight and flexible visual effect.

5.2. Analysis of auxiliary symmetry modeling language

In the R, S, F series of the brand BAUER, the flange of the R series of SEW and the S series of BONFIGLIOLI is the rotation symmetry of the multi symmetry axis L^8, showing the positive eight sides, and the symmetry rule combination is $L^8$8PC;the K series of brand BAUER, SEW's S, K, F series and BONFIGLIOLI's R, K and F series are flanged symmetrical circular flanges, and the symmetry rule combination is $L^1$1P.It is found that the shape of flanges is mostly circular rotationally symmetric $L^1$1P, while the shape of some flanges is L^8 with a symmetrical eight axis $L^8$8PC. Most of the flanges of the four Chinese-brand reducers are also circular symmetrical $L^1$1P. Therefore, the external shape of the flange is a design principle that follows the function of the internal structure, which conforms to the functional requirements, functional structure and function design of the torsion and transmission of the bearing, and determines the form of the speed reducer.The form of symmetrical pattern of flanges is $L^1$1P>$L^8$8PC. The shape of the flange is symmetric and round, followed by regular octagon of axis, the shape of regular octagon is close to round, and the circular shape language is plumpness and rigorous, and the eight sides are plumpness and symmetry.

There are two forms of heat dissipation symmetry in the four series of reducer of brand BAUER, SEW and BONFIGLIOLI: one is the side of the BAUER heat dissipation line with a curve in the upper and lower unit mirror symmetry, while the SEW and BONFIGLIOLI heat dissipation lines are symmetrical in line. The other because of BAUER, SEW, BONFIGLIOLI deceleration chassis most is 2 axisymmetric bodies, the symmetry rule is $L^2$2PC, and the heat dissipation lines are distributed on the symmetric surface M_2, M_4, and then the other side heat dissipation lines are shot through the center symmetric axis mirror of the box. From the above analysis, the most symmetrical form used in the design language design is the straight line, followed by the curve. The modeling language of the straight line is accurate and order, but the curve is rich and fluent.

To sum up, the design of the brand reducer should reflect the "stability" from the perspective of the language of the main subject modeling or the auxiliary modeling language. And the expression language is conveyed in a stable, accurate and all term.

6. The example of the reducer symmetry modelling language design

Based on the above knowledge, in the case of the symmetry of the material function, taking the R series reducer as an example, the knowledge of symmetry rules is applied to design a stable and symmetrical reducer with a precise and symmetrical shape. And its form drawing will be given.

6.1. The design of the symmetry modeling language for the whole shape of the box

In order to meet the factors such as the overall material function and the stability of the art form of the reducer box, the frequency of the symmetry axis n must be set to n<5 and cannot be equal to 3, so that the whole shape of the box is more stable, The whole shape of the reducer is designed as a single axisymmetric square $L^1$1P by means of the Sdc method for determining the symmetry of the whole M_6 shape of the box body. Design of morphological details using symmetric Sda method. Some symmetrical detail adjustments are made by the line segment a, e mirror symmetric out of a 'h, Pa (P1~P4) mirror (P1~P4) mirror and Pa' (P1' ~P4'), and the final shape is determined, schematic diagram is shown in Figure.6. The design language of the reducer embodies the stability that must be embodied in the function form, and after the adjustments of the symmetry of the shape details, the modeling language is lighter and more flexible.

6.2. Design of flange shape symmetry modeling language

The shape of flange must conform to the demands of the functional connection of bearing torsion and transfer force. Its function determines that most of the flanges should be set to round and rotated symmetry, distributed on the surface Pb, Pc, Pd, Pf, Pb', Pc', Pd ', Pf '. The modeling language used for the reducer flange is called strong hexagon, and the symmetry combination rule is $L^6$6PC, schematic diagram is shown in Figure.7.

6.3. Shape symmetry modeling language design of heat dissipating pattern

The design of heat dissipating pattern is a symmetrical hexagonal arrangement with good symmetry and stability, schematic diagram is shown in Figure.8. The combination of symmetry rules is $L^6$6PC, which is divided into translation and mirror symmetry. The heat dissipation lines are distributed on the M_2 and M_4 surfaces of the box uniformly, and play the role of balanced heat dissipation.In the form presentation, the curve is arranged with the regular hexagon, and the curve in the stable box can make the modeling language more varied and fluent.

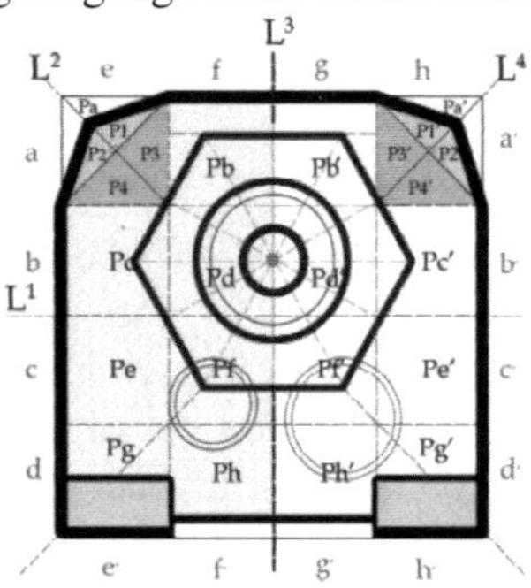

Figure 6. Overall shape of reducer box.

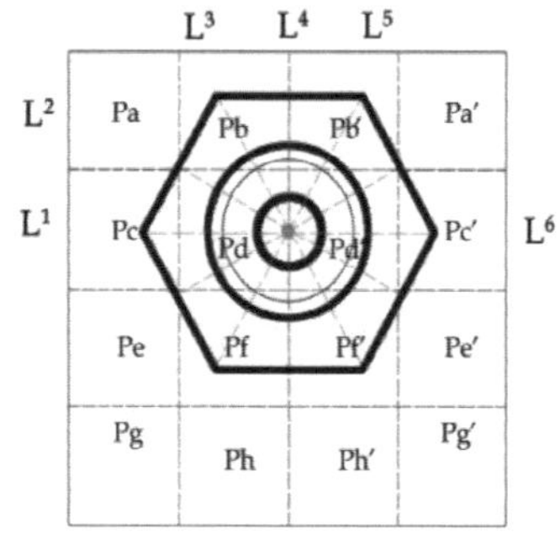

Figure 7. Flange shape of reducer.

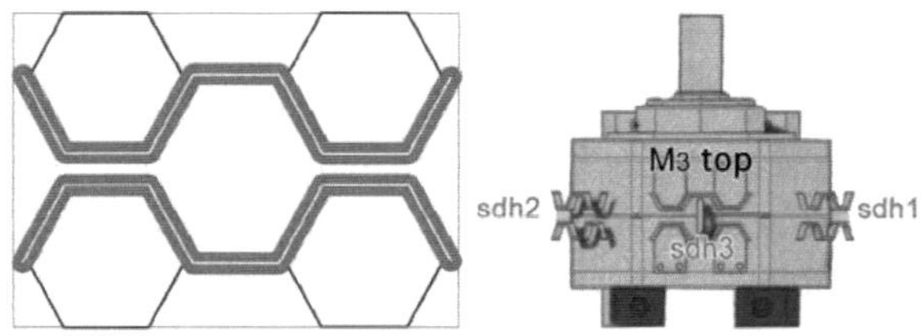

Figure 8. Heat dissipation pattern of reducer.

7. conclusion

Symmetry is ubiquitous in the field of mechanical product design and contributes to meeting the requirements of mechanical products. Not only in the field of the function of the reducer, but also in the field of the art form, the mechanical is reflected the symmetry. In order to meet the demands of reducer design, the symmetry system of reducer should be set up, and the relationship between components symmetry and symmetry datum elements should be clarified. Based on the analysis of the symmetry of the foreign brand reducer, the symmetry law and symmetry method of the reducer's form datum are put forward. It is used in the design of the reducer's modeling language to highlight the brand characteristics, optimize the modeling of the reducer and improve the design level of the reducer and its similar products.

References:

[1] J Barrenscheen 1990 *Die Systematische Ausnutzung von Symmetrieeigenschaften beim Konstruieren* (Tec hnische Universität Braunschweig Fakultät für Maschinenbau und Elektrotechnik)
[2] Feng Peien, Ma Zhiyong, Qiu Qing Ying, Shen Menghong and Ceng Lingbin 2008 Research on symmetr y ontology from natural science to engineering science (natural science exhibition) p 1441-1450
[3] Qi Yuxuan, Qiu Qing Ying, Feng Peien and Li Lixin 2016 *Conceptual structure and application of combi ned symmetry of mechanical structure* (Zhejiang:Journal of Zhejiang University Engineering Edition) p 1 889-1901
[4] Zhong Kangmin and Wang Mingdi 2004 *Symmetry beauty and mechanical design innovation* （Suzhou:J ournal of Soochow University Engineering Edition) p 36-37
[5] Qiao Jizhong 1988 *Application of symmetry in industrial design* (Xi'an:Journal of Xi'an Technological Un iversity) p 64-68
[6] Fu Xiaoli, sun Xu Fang and Chen Fang 2014 *Research on Modeling Design of industrial general reducer* (Tianjin:mechanical design) p 108-110
[7] Zuo Tiefeng, Shu Wei, Zhang Shuai and Geng Dian Mei 2015 *product form language and design* (Shenyan g: Liaoning Art Publishing House) p 12-13
[8] Yang Zhi and Chen Xuedong 1991 *Vibration analysis of the axisymmetric reducer's shaft system* (Wuhan: Wuhan Polytechnic University) p 84-89.
[9] Shi Nicheng and Li Guowu 2008 *Symmetry and crystallography* (Shanghai:natural journal) p 44-49

IOP Conf. Series: Materials Science and Engineering **408** (2018) 012033 doi:10.1088/1757-899X/408/1/012033

Usability Evaluation of Adaptable Urban Park Furniture Product with Cellular Light-weight Concrete as Material

Diana Thamrin[1,*], Grace Mulyono[1] and Antoni[2]

[1] Interior Design Department, Petra Christian University, Siwalankerto 121-131, Surabaya, Indonesia

[2] Civil Engineering Department, Petra Christian University, Siwalankerto 121-131, Surabaya, Indonesia

[*]dianath@petra.ac.id

Abstract. In this research, cellular light-weight concrete has been developed as a material for a modular design of a multifunction and adaptable park furniture, as it is durable and has excellent resistance to extreme weather. Inspired by Froebel play blocks, the design attempts to accommodate variable activities, increase the value of limited urban parks and to change their settings from time to time in order to enhance their recreational values. This paper aims to evaluate the usability of the furniture product based on human-centred design during the design process and post-production. Formative usability test was conducted through conceptual exploration of multiple activities that could be accommodated through the product design by a range of users and different availability of park spaces. Summative usability test was then conducted post-production through lead users in order to expand the findings from the explorative usability tests. Findings reveal that multiple activities could be generated through the adaptable compositions and material durability of furniture design, even in the smallest areas of parks. It also grants opportunities for users of different age groups to perform multiple activities together, hence supporting urban parks to achieve their essential purpose of accommodating social interaction and healthy motoric activities.

1. Introduction

Urban parks are vital in maintaining the natural environment as an essential part of a sustainable city. However, they also have the potential in improving the social and physical health of a city's inhabitants [1]. Hence, besides preserving green landscape, the design of urban parks should also support the surrounding non-green spaces to become recreational habitual areas that encourage social interactions and healthy motoric activities [2]. Consequently, the design of park facilities such as furniture and other outdoor furnishings should also be a primary concern in the planning of urban parks [3]. In most urban cities there are very few large parks with a preserved natural environment, while many other parks are pocket parks that consist of smaller non-green spaces, located on street edges and road corners, sometimes appearing lifeless and inhabited. Many park furniture are made of conventional materials such as iron or wood that cannot tolerate the intense sunlight radiation, snow and heavy rainfall. Apart from that, the conventional singly-functioned designs of the furniture are insufficient in accommodating the variety of social and motoric activities that can take place in public parks besides sitting. This research develops the idea of using cellular light-weight concrete (CLC) as material for a modular design of park furniture, inspired by Froebel play blocks, in order to provide an innovative way of accommodating various activities in public parks. CLC is a material made of concrete mortar mixture that has been injected with air bubbles using a foam generator before solidification, resulting in a light-weight concrete block [4] that is resistant to sunlight radiation, rain, rust and termite. The use of CLC compared to conventional concrete enables mobility, supporting changes in park settings from time to time in order to keep the parks attractive enough to draw in visitors. In order to ensure its usability to park users, this research paper focusses on performing usability tests based on Bailey's human performance model in which a design product ought to have equivalent considerations to aspects of human (somebody), activity (something) and context (someplace) [5]. Two types of tests were conducted: formative usability tests during the explorative design process, followed by summative usability test that is conducted post production. Through the usability tests, it is hoped that the modular CLC furniture product yielded can truly accommodate various social and motoric activities in public parks and support the recreational as well as social function of urban parks in contemporary urban cities.

2. Research Method

In a successful human-centered design product, usability tests form a crucial part of the design process. Two methods of usability tests consisting of formative usability test and summative usability test were conducted in this research. Formative usability test is an explorative study conducted early in the design process when a product is still in the preliminary stages of being defined and designed [5]. The main objective of this study is to examine the effectiveness of the design concepts. In this research, formative usability study was conducted based on Bailey's human performance model on performance usability, through conceptual testing of activities that can be conducted with the modular light-weight concrete furniture in variable spaces of public parks by a range of users. This test was done to determine the appropriate final form, dimensions and approximate concrete material mixture of the final product. After production of prototypes with real CLC material, summative usability tests were performed through the lead user method. The purpose of the test is to expand the findings of the exploratory test by direct testing of the product with potential users. In this research, a total of 5 children, 5 teenagers and 5 adults were invited as lead users. The results of the test were then documented and compared with the findings expected from the formative usability tests.

3. Findings and Discussion

From four design alternatives produced prior to this research paper [6], the fourth design, that can accommodate the largest range of activities, has been chosen for development. It also has the smallest size of module that can adapt to variable park spaces and has a massive form that suits light-weight concrete production (figure 1). The form of this module was inspired by Froebel geometric blocks in children toys and modern Froebel architecture. It is a simple cube with cavities to support the interlocking system. The voluminous form is suitable to the CLC concrete material, as the instalment of metal structures as generally found within conventional concrete would be unnecessary.

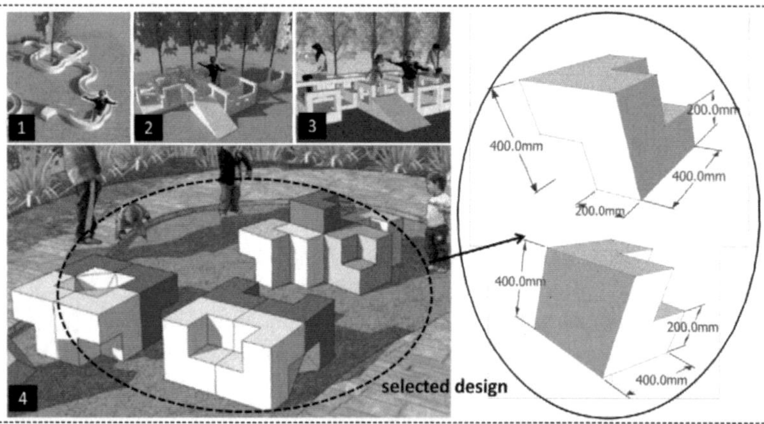

Figure 1. Modular design of cellular lightweight concrete park furniture inspired by Froebel blocks.

IOP Conf. Series: Materials Science and Engineering **408** (2018) 012033 doi:10.1088/1757-899X/408/1/012033

The simple geometric block grants opportunities for multiple compositions according to creativity and context, as such is the adaptable purposive function of toy blocks and modern Froebel architecture. The function of Froebel-inspired products would be similar to its pedagogic purposes that garners high considerations on both physical and psychological experiences in a natural environment that facilitates children play [7]. From the selected module, we further explored a range of possible compositions, using three dimensional animation software. Variable compositions could be produced even with a few blocks (less than or equal to 4), medium number of blocks (more than 4 and less than 10) and a large number of module blocks (more than 10) as seen in figure 2.

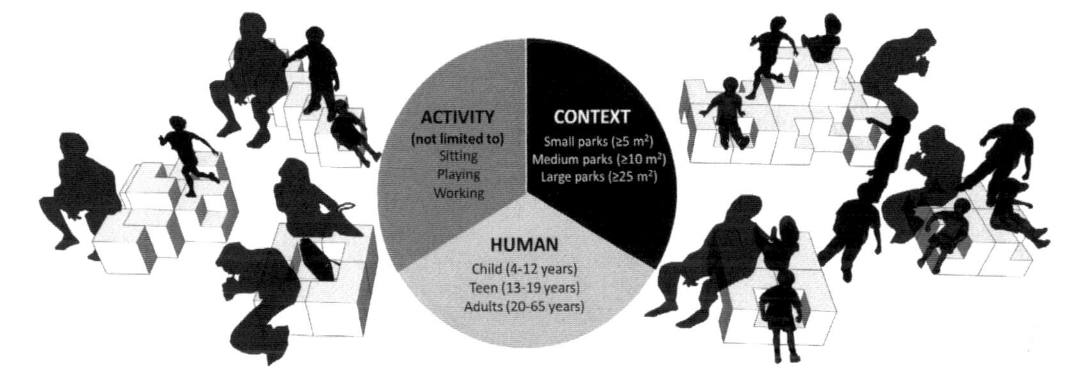

Figure 2. Varying formations of CLC park furniture module accommodating varying activities and spaces based on Bailey's Human Performance Model.

In order to test its usability before further development, the researchers used the triangulation method of considering three aspects in the design based on Bailey's human performance model in testing design products [5]. The three aspects are human, activity and context (place). In regards to the human aspect in this research, the targeted users are children users (4-12 years), teenagers (13-19 years) and adults (20-65 years). The activities targeted for accommodation revolves around sitting, playing and working. Meanwhile the aspect of place considered are parks that has an area of more than or equal to 25m^2, more than or equal to 10m^2 but less than 25m^2, and more than or equal to 5m^2 but less than 10 m^2, like in the case of pocket parks and street corners. The results of the formative or explorative usability test are articulated and summed up in table 1.

Through the formative usability tests conducted, the authors explored that at least three activities that are not limited to sitting but also climbing and balancing could be conducted with only 2-4 blocks of furniture modules that could be placed on the smallest areas of parks (5m^2). The explorative tests, using computer software, showed that they could be used by all target users. Meanwhile, the use of 4-10 module blocks can accommodate more physical or healthy playing activities such as climbing, balancing and jumping. Formations of more than 10 module blocks can combine smaller compositions of 2-4 and 4-10 modules that garner even more variable activities. The explorative possibility of activities that could be accommodated enabled the authors to determine the appropriate dimensions of the blocks. We formulated the dimensions based on standard popliteal and hip-breadth measurements of Indonesian users [8], as majority of the activities would revolve around sitting in the case of both adults and children. This resulted in a 400 x 400 x 400 mm of cubic form with a 200 x 200 200 mm of cavity (figure 2). In terms of module weight, literature studies reveal that the standard maximum load for a single male person is 21.3 kg [9]. With the determined dimensions, the explorative usability test assisted us to adjust the suitable volume of foam in the concrete mixture in order to achieve a weight of about 30-35 kg per module, as it has to be heavy enough to avoid theft, yet light enough to be moved by at least two park workers. The volume of air bubbles should be about 60-70% to achieve targeted density between 750-800 kg/m^3.

IOP Conf. Series: Materials Science and Engineering **408** (2018) 012033 doi:10.1088/1757-899X/408/1/012033

Table 1. Formative (explorative) Usability Test of CLC Adaptable Park Furniture Design

Modules	Possible Modular Arrangement	Activity	Human			Context (Park Space)		
			Child	Teen	Adult	$\geq 5\ m^2$ $\leq 10\ m^2$	$\geq 10\ m^2$ $\leq 25\ m^2$	$\geq 25\ m^2$
≥2 ≤4		• Casual Sitting • Chatting • Drinking	√	√	√	√	√	√
≥2 ≤4	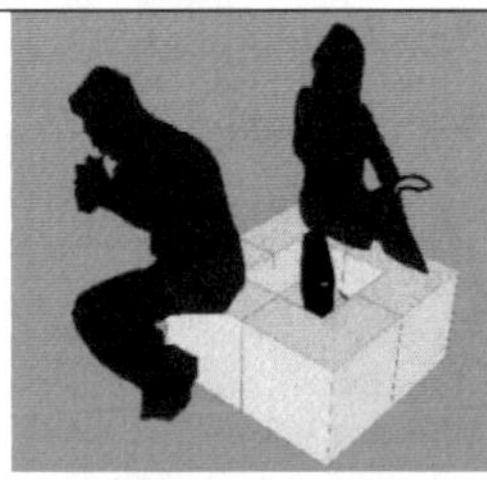	• Casual Sitting • Chatting • Drinking • Storing bags		√	√	√	√	√
≥2 ≤4		• Casual Sitting • Chatting • Drinking • Climbing • Balancing	√	√	√	√	√	√
≥4 ≤10		• Casual Sitting • Chatting • Operating mobile devices • Climbing • Balancing	√	√	√	√	√	√
≥4 ≤10		• Casual Sitting • Chatting • Operating mobile devices • Climbing • Balancing • Jumping	√	√	√	-	√	√
≥ 10	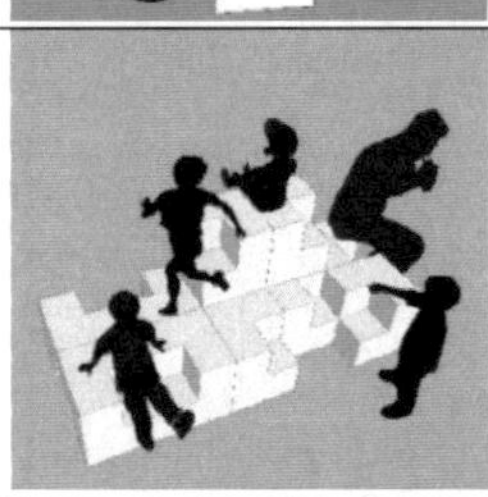	• Casual Sitting • Chatting • Operating mobile devices • Climbing • Balancing • Jumping	√	√	√	-	√	√
Conclusion		Variable activities not that are not limited to just sitting and playing could be accommodated even in the smallest area of parks (10 m^2) by all target users.						

IOP Conf. Series: Materials Science and Engineering **408** (2018) 012033 doi:10.1088/1757-899X/408/1/012033

Figure 3. Summative usability study of the load weight of CLC module

To validate and expand the results of the explorative usability test above, we performed a summative usability test after the production of the experimental CLC module blocks, using the lead user method of evaluation. We invited 5 teenagers, 5 adults and 5 children as lead users in the usability test. The first summative test was performed on behalf of park workers as users and managers of the furniture. With the calculated concrete-air bubble composition as a result of the explorative study, each module block produced weighed between 35-42 kg. Each concrete block could be lifted by two workers but difficult to be moved by one user, as expected in the research (figure 3). In terms of park visitors as users, we compared the results of the formative usability test with the summative usability. The summative test validated and expanded some of our findings in the formative tests as summed up in table 2.

Table 2. Summative usability test of CLC Adaptable Park Furniture Design

Modules	Explorative Usability	Summative Usability	Conclusion
≥2 ≤4			This module formation took up only 0.8 m^2 of space + 30% for circulation and hence can be used in the smallest area of parks. The formation was able to be used as a sitting facility for adults, teenagers and children. The sequential leveling between children and adult seats also benefits social interactions between parents and children.
≥2 ≤4			The module formation took up only 0.8 m^2 of space + 30% for circulation and hence can be used in the smallest area of parks. The formation was able to be used as a sitting facility for adults. Storage areas in the middle proved to be quite functional for college and high school students.
≥2 ≤4			The module formation took up only 0.8 m^2 of space + 30% for circulation and can be used in the smallest area of parks. The formation was able to be used for sitting and climbing for adults, teenagers and children. The sequential leveling between children and adult seats also benefits social interactions between parents and children.

≥4 ≤10			The module formation took up only 1.5 m^2 of space + 30% for circulation and hence can be used in the smallest area of parks. The formation was able to be used as a bench for adults/teenagers, sitting, balancing and climbing facilities for children.
≥4 ≤10			The module formation took up only 2.1 m^2 of space + 30% for circulation and hence would be more suitable for medium to large parks The formation was able to be used for sitting, balancing, jumping and climbing facilities for children, and a sitting facility for teenagers and adults. The varying seat heights also benefit social interactions among parents and children.
≥ 10			The module formation took up 3 m^2 of space + 30% for circulation and hence would be more suitable for medium to large parks. The formation was able to be used for sitting, balancing, jumping and climbing facilities for children, and a sitting facility for teenagers and adults. The varying seat heights also benefit social interactions among parents and children. It is also found that when not used for playing, the adult seat could be used as a table for children to read and write. A large park space can also be used for multiple formations at one spot that can be changed from time to time. This multiple formation can accommodate different activities in different ways, creating multi-function spaces in parks.

Through the summative usability test conducted in the case of visitors as users, it was found that multiple activities could be generated through the Froebel concept of furniture design, even in the smallest areas of parks, just as what has been expected in the explorative usability test. This makes limited park spaces on street edges and pocket parks usable and multifunctional. Expanding results reveal that the varying heights of seats and multi-function nature of the furniture encourages social interaction between adults and children that are not generally accommodated in conventional park furniture. Hence, the design may provide opportunities for parents to play together with their children rather than sit and watch them play in conventional singly-functioned parks. This supports the role of urban parks in enhancing social cohesion between different people of different ages and hobbies, contributing to social equality and social awareness in neighbourhood communities [10].

Other findings from the summative usability test also include limitations in the design product that could be addressed in future design development. First, on a few occasions, the light-weight concrete surfaces were chipped off during the lifting process. This could be countered by re-adjusting the concrete mortar mixture, for instance, by adding fibre or resin to increase the cohesive characteristic of the material. Other limitations include the existence of gaps between the modules when stacked or placed together. This problem could also be addressed in future design development by using thicker steel casts for the moulding process in order to increase their precision when joined perpendicularly. The weight of the material could also be re-adjusted to become lighter than the blocks that have been produced in this research, as some modules weighed as heavy as 40 kg, which was deemed too heavy for some park workers. Nevertheless, the concept of a modular park furniture with a material that is resistant to extreme weather, theft and vandalism that is usable even in limited park spaces, can serve as an initial idea for future developments of urban parks that should encourage outdoor activities in the indoor-attached and gadget-engrossed society.

4. Conclusion and recommendations

From the usability tests conducted, it can be concluded that modular designs of park furniture using materials that suit outdoor exposure such as CLC have the potential in making parks fulfil their primary purpose of supporting social interactions and physical health in urban cities. It grants opportunities for users of different age groups to perform multiple activities together rather than individually, hence, encouraging family and community bonds within neighbourhoods. Park managers can also play a certain role in creative settings of urban parks. Despite a few limitations found in this research, it can serve as an initial concept of developing future urban parks into novel spaces. Future urban planners and product designers can develop the concept of multifunction products or spaces for urban parks with other modular forms and newly developed materials that can counter typical problematic issues in public facilities such as exposure to extreme weather conditions, lack of space, theft or vandalism. Through this research-design project, the authors hope that urban parks may evolve to become habitual multifunctional spaces for social engagements and healthy recreation that can compete against people's attachment to indoor spaces and gadget devices in the contemporary society.

Acknowledgments

The authors of this research would like to thank the Ministry of Research, Technology and Higher Education, Indonesia for their financial support in this research.

References

[1] Chiesura A 2004 The role of urban parks for the sustainable city *Landscape and Urban Planning* Vol 68 (Amsterdam: Elsevier) p 129-38

[2] Larson L R Jennings V and Cloutier S A 2016 Public Parks and Wellbeing in Urban Areas of the United States *Plos One* Vol 11

[3] Kargar R and Ghassemi M 2016 Urban Furniture and Leisure of Citizens (Case Study of Sabzevar) *Journal of Civil Engineering and Urbanism* Vol 6 (Turkey: Scienceline publication) p 40-7.

[4] Antoni, Jos R and Lukito MM 2011 Influence of Water Absorption on Properties of AAC and CLC Lightweight Brick *Proceedings of the 4th ASEAN Civil Engineering Conference* (Yogyakarta)

[5] Rubin J and Chisnell D 2008 *Handbook of Usability Testing* Second Edition (Indianapolis: Wiley Publishing Inc)

[6] Mulyono G Thamrin D and Antoni 2017 Development of Modular Outdoor Furniture Product Using Lightweight Concrete for Public Parks in Surabaya *IOP Conference Series: Materials Science and Engineering* Vol 234 (IOP Publishing) p 012022

[7] Provenzo Jr E F 2009 Friedrich Froebel's Gifts: Connecting the Spiritual and Aesthetic to the Real World of Play and Learning *American Journal of Play Vol 2* (Board of Trustees of the University of Illinois) p 85-99

[8] Hartono M 2018 Indonesian anthropometry update for special populations incorporating Drillis and Contini revisited *International Journal of Industrial Ergonomics* Vol 64 (Amsterdam: Elsevier) p 89-101

[9] A S M Hamid and S B M Tamrin 2016 Recommended Standard For Acceptable Weight For Lifting Activities Among Young Asian Adults *Malaysian Journal of Human Factor and Ergonomics* Vol 1 (Selangor: Human factors and Ergonomics Society Malaysia) p 62-7

[10] Peters K, Elands B and Buijs A 2010 Social interactions in urban parks: Stimulating social cohesion? *Urban Forestry & Urban Greening* Vol 9 (Amsterdam: Elsevier) p 93-100

Participatory Design of Portable and Adaptable Furniture Product for Village Traditional Pastry Vendors in Surabaya

Diana Thamrin and Grace Mulyono

Interior Design Department, Petra Christian University, Siwalankerto 121-131, Surabaya, Indonesia

dianath@petra.ac.id

Abstract. Traditional pastry making is an authentic skill and source of livelihood for many village vendors in Indonesia. Despite their potentials in improving regional economy and promoting traditional cuisine, village pastry vendors face marketing problems due to their poor product display system, lack of retail space, and uncertainty of available time to make and sell their pastries. This research develops the idea of designing a portable and adaptable furniture product that can help pastry makers display and sell pastries efficiently and independently. Using the participatory method of human-centred design, the designers became participants of the pastry-making community in order to empathize with their problems and aspirations. Meanwhile, the pastry makers became participants in the following design processes of defining problems, ideating concepts and testing of furniture design prototypes. The result of this research is a furniture product for displaying and selling traditional pastries that is movable, adaptable to space and place, visually representable to the natural authentic image of Indonesian traditional cuisine, could be assembled and dissembled easily, and practical for storage and mobility. From the participatory design method, product designers also gain a reflexive awareness of their role in the society as agents of social change and economic innovations.

1. Introduction

In culturally rich countries like Indonesia, local communities, with their authentic skills and culture, have the potential in promoting a city's social and economic values [1]. Since 2010, the government of Surabaya city has issued a program called *Kampung Unggulan Surabaya* (Surabaya leading villages) as an effort to improve the regional economy by training and facilitating village communities as part of Indonesian small to medium enterprise (SME). Up to ten villages have been selected in this program and were chosen based on their existing local skills. One of them is *Kampung Unggulan Kue*, a village where a community of traditional pastry makers routinely work together to sell and distribute pastries in small markets. Despite their potentials in improving the regional economy and promoting traditional cuisine to the modern society, these pastry makers face marketing problems due to their poor product display system, lack of retail space, and uncertainty of available time to sell them to the public. Majority of the community members are housewives who daily tend to their families and do not have the knowledge or skill in managing their business, especially in terms of visual merchandising. They mainly depend on external traditional markets or non-pastry food stalls to entrust the selling of their products. Such conventional display of traditional pastries observed do not have any unique visual merchandizing that could represent the authenticity and cultural values of these heritage delicacies. Sometimes, the women get offers to sell their products in village food courts; other times they sell their products via manual carriers or mass orders for special occasions. Yet, the role of these skilful women in improving social and economic power in the family has been proven to be truly significant [2].

On the basis of this background, this research develops the idea of designing a portable and adaptable furniture product that can assist these village vendors to improve the visual merchandizing of their pastries and help them sell their pastries independently and efficiently, regardless of the availability and uncertainty of time and space. Using the participatory method of approach in design, the designers became participants of the pastry community and engaged with their pastry making activities in order to empathize with their problems and needs. At the same time, the village vendors became participants in the following design processes of defining problems, ideating concepts and testing of furniture design prototypes. Through the design of furniture product yielded from this research, it is hoped that village vendors can enhance the practice of their authentic skills, promote them to the society and improve socio-economic conditions through the existing local genius of their community.

2. Research Method

This research uses the method of participatory design, a human-centered design approach that attempts to involve users and experts to assist designers in the design process in order to ensure the usability of the product design [3]. This research was conducted in five stages of design process consisting of Empathize, Define, Ideate, Prototype and Test. In this research, not only the vendors became participants in the design process, but the designers also participate in the pastry making, packaging and selling activities as part of the Empathize stage, in order to gain a deep understanding of their difficulties, aspirations and goals in pastry making. From the Define-Ideate-Prototype to the Test stage, the vendors and members of the village community were invited and visited frequently to provide ideas and feedbacks. After the production of the furniture prototype, the designers also performed a usability test of the product by participating in a sales trial using the furniture product to ensure that it would be truly usable. The failure to engage with users that could be part of a progressively impactful market force would cause a loss in a commercial design opportunity [4]. Hence, these participatory involvements form a crucial part of the design process in the development of the furniture product.

3. Results and Discussion

3.1. Empathize-Define

The designers visited the pastry making community and used ethnographic methods of data collection, a qualitative method of observation in which the researchers or designers would stay in a community for a certain amount of time and participate in the community's activities in order the grasp a deep understanding of their daily lives at their natural settings [5]. The design team conducted informal discussions that emphasize more on dynamic interactions rather than structured interviews, analysed daily behaviours of the community and also participated in their pastry making activities. The activities included preparing raw materials, baking, cooking, packaging and eating them together (figure 1). Through being participants of the pastry community, the designers obtained valuable information on the community that could not have been grasped through conventional structured interviews. The designers gained a deep understanding of the community's aspirations in marketing the authentic Indonesian taste of their pastries, the natural ingredients used, the type of pastries being made, the product packaging techniques and difficulties in marketing and selling. Being mostly housewives, the pastry makers usually have to produce pastries as early as 2 am in the morning, entrust them to vendors in traditional markets or common food stalls while tending to their families in the morning. However they have free time late during the day till evening, a span of time they could use for selling their products independently. Unfortunately, they do not have the channel and space to sell their products directly to the public.

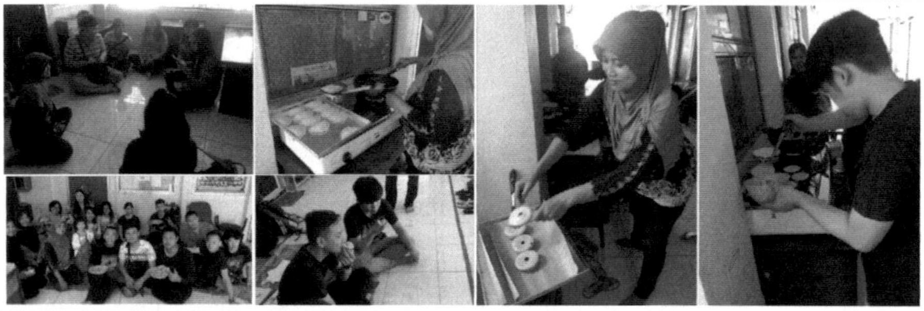

Figure 1. Designers learn to empathize and participate in the activities of the pastry community

Hence, together with the community, the design team discovered that an adaptable, practical and movable furniture product would truly be beneficial for them such that they could independently sell their products at any time and space available. Through empathy maps, discussions and affinity diagrams, the design team defined the problems and factors to be considered in the product design in terms of form, function, performance, producer's interest, marketing and production (figure 2).

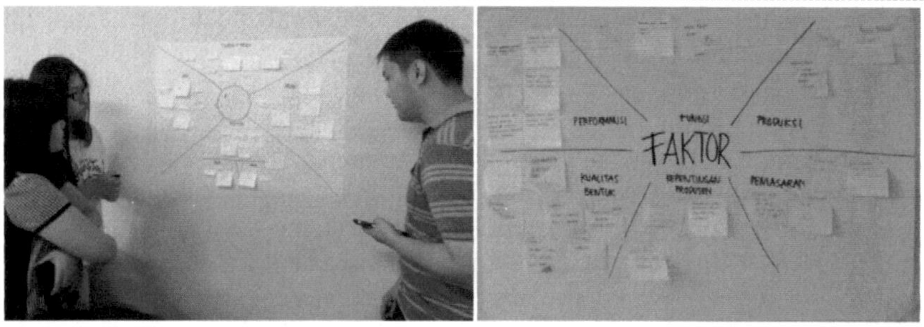

Figure 2. Designers define problems to be solved and consider factors related to product design needed by the community

The product defined in this research must have a modular form to ease its assembly when in use, and its storage when not in use. In terms of function and performance, it should accommodate the basic need of displaying at least ten different Indonesian pastries in a visible and attractive way such that it can encourage sales. Thus, it must support the visual merchandising of the pastry products. The furniture should also accommodate their use both indoors and outdoors and hence should be adaptable to the type of space available. It must easily be transported from one place to another. For assisting in branding and marketing, the furniture should also support the product image: natural, traditional and authentically Indonesian. Finally, in terms of production, the furniture ought to use materials that are readily available and are easy to be built in order to ensure the possibility of independent repairs or development in future.

3.2. Ideate-Prototype-Test

After defining the factors of the product design, the researchers then began the process of ideation. Various design alternatives were produced that displayed modern systems of furniture product yet incorporating traditional Indonesian modes of display such as the use of bamboo display plates known as *tampah*, banana leaves, stems, *alang-alang* (straw) roof, *goni* fabric and wooden materials (figure 3). Most of the design alternatives aimed at producing a variety of display methods with varying heights to accommodate the display of different types of Indonesian pastry. Some are placed on a surface and others could be hung. As a crucial part of the participatory design method, the designers then involved the community members in the review and selection of design through presentation and dot voting (figure 4). The designers themselves provided feedbacks to each of their team members in order to further develop the design. From both participatory processes, many useful feedbacks were acquired. The pastry makers favoured the traditional display modes of using woven bamboo plates and display levelling system as they enhance the image of authentic Indonesian culture. They also favoured a display system that can adapt to exterior and interior spaces, and that could be kept easily when not in use.

Figure 3. Designers develop modern furniture products for pastry display that incorporates traditional Indonesian modes of display (Mock-up by Claudia Filiana and Eko Prasetyo).

Figure 4. Participatory method of design evaluation with pastry makers and design peers

Based on the participatory method of obtaining feedbacks and developing concepts, the designers further established a concept of an adaptable portable furniture with an authentic traditional image abbreviated as "MANTAP," which stands for Movable, Adaptable, Natural, Traditional, Assembling, and Practical. The word *mantap* is also an Indonesian term for "awesome", especially spoken after the tasting of good food. The design was redeveloped with 8 different modules that are detached from one another, each having a variety of functions (figure 5). The first module is a large table-storage module, having dimensions of 1100 x 600 x 820 mm, formulated based on the standard dimensions of a dining table that has a slightly lower height than food counters or stalls. This height enables a much broader view of displayed pastries. The table-like form of the module, made of pallet wood, also enables flexibility of use for both display and pastry packaging, as needed when products are being sold outdoors to street buyers. On the front of the table is a designed logo containing of an image of a woman with a typical traditional Javanese hair bund, placed on a woven bamboo plate to enhance the traditional and natural branding. The second module consisted of traditional bamboo plates of 500 mm, 300 mm and 200 mm in diameters, mounted on bamboo stems that could easily be detached. The third module is a rack made of bamboo stems. Pastries could be displayed by placing on top of each row or hanging them with threads tied on the stems as often used in the display of traditional nuts and crackers. The fourth module is a slanted rack with a combined dimensions of 400mm x 400 mm 250 mm, in which each wooden plain could be detached from one another in order to ease its assembling and dissembling. When not in use, the boxes: the eighth and sixth modules, on which the rack is stacked on, can function as a packaging table. The fifth module is a simple wooden box with holes that support the vertical position of bamboo stems mounting the bamboo plates. The seventh module is a simple stair-like box that helps in dividing the display between different types of pastries. Finally, the sixth and eight modules are simple storage boxes needed by the pastry vendors to store excess pastries and other belongings when they are selling on the streets.

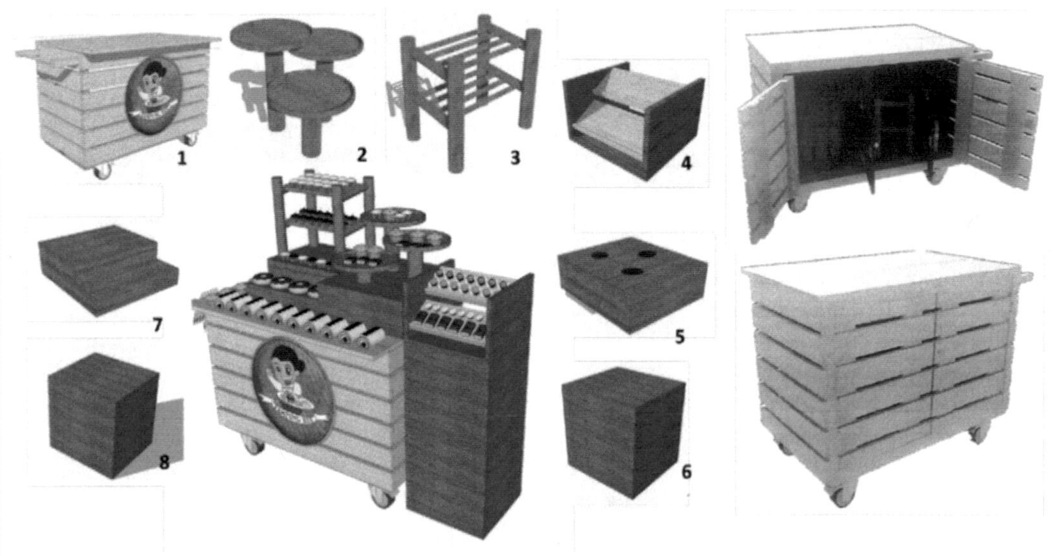

Figure 5. Designers develop a multi-function furniture product for pastry display that incorporates traditional Indonesian modes of display (image by Diana Thamrin and Charles Hartanto)

As a whole, the furniture product design display the features of "MANTAP" as introduced previously. The Movable concept was made possible by the modular design, in which every single module could be kept inside the first module and transported with the help of wheels installed beneath the table. Using modules that are stacked upon one another supports the concept of Adaptability, as each module can be placed at different directions and positions according to the availability and characteristics of the space available. The modules themselves can also be placed separately on retail spaces and hence can adapt as both indoor and outdoor furniture. Furthermore, the various modules enable a larger variety of display compared to conventional display stalls and hence can increase the visual merchandizing of the pastry products. The concept of Natural and Traditional have been enhanced by the use of Indonesian customs in the method of display. Displaying pastries and other food on large woven bamboo plates, often with banana leaves, is an authentically Indonesian way of displaying traditional cuisine in Javanese traditional markets [6]. The use of natural materials such as bamboo and pallet wood, with its distinct textural pattern, is a suitable visual representation of the natural ingredients in these pastries. Lastly, the concept of Practicality is evident as the assembling and dissembling of each module can be done at ease due to the temporary stacking system of construction. The design was produced with the help of both manual force and wood workshop machines such as handheld and stationary electric planner (figure 6). The use of teak pallet wood for majority of the furniture product is favourable compared to multiplex or solid wood as it is artistic, largely available, light-weight, durable, stackable, versatile, easy for construction or repair and is an environmental-friendly material [7]. The pallet sheets, having a thickness of 20 mm, were joined together using glue for parallel joints, and added with nails for perpendicular joints. The resulting form was then finished with a clear coat to expose and enhance the natural artistic wood texture. Modules made of bamboo were also joined using the mortise-tenon construction. The surfaces of the bamboo were also finished with a clear coat to maintain its natural texture. Bamboo plates are available in the market and the stacking system of construction enables them to be changed from time to time. The readily-available and simple materials used yielded a relatively economic product in terms of production. This would truly benefit the pastry vendors if they need to repair or develop the product in the future.

Figure 6. Production of furniture product using naturally artistic yet economical materials such as teak pallet wood, bamboo stems and traditional woven plates.

Figure 7. Testing of final furniture product by both designers and pastry makers

In order to ensure its usability to end users, the designers performed a summative usability test of the furniture product using the participatory method [8]. The designers themselves became lead users of the product by opening a bazaar in a public place and selling pastries made by the village community for a duration of three days. The pastry makers were also invited to assist in the sales. There were many positive results obtained from the test (figure 7). First, the different types of display, with different height levels has made a dynamic display of different pastries, supporting visual merchandizing. This is evident through the high number of sales in three consecutive days such that the products were sold out by the afternoon during the three days of trial. The assembling and dissembling of the furniture product was also done with little effort. The whole process of dissembling took up a duration of less than ten minutes. All modules could fit into the table module as expected in the design. Transporting the furniture product and the pastry goods was easily accommodated by the lightweight modules, cubic storage boxes, a large bamboo handle on one side of the table for pulling and pushing, and wheels beneath the table. In terms of visual appearance, the natural and artistic look of teak pallet wood and bamboo, along with the authentic Indonesian mode of food display using bamboo woven plates also yield a mixture of modern and traditional look, enhancing the authentic branding image of the traditional pastry products, as hoped by the pastry makers. One limitation discovered through the summative test is that the background surfaces on which the pastries are placed is too similar in terms of tone and colour to that of the traditional pastries. A contrasting yet natural colour like green or yellow would make the pastries more visible and attractive. This could be easily addressed by using green banana leaves or traditional coloured cloth beneath the items displayed.

4. Conclusion

The result of this research is a novel form of furniture product for displaying and selling traditional pastries that is movable, adaptable to space and place, visually representable to the natural authentic image of Indonesian traditional pastries, could be assembled and dissembled easily, and practical for storage and mobility. As displayed by the results of this research, participatory design could be an essential key to both product and social innovation [9]. Through the participation of designers in the Empathize and the Test stage, as well as the participation of target users (the village community) in the whole design process, the furniture product design yielded can support the practice of local culture in traditional pastry-making and increase the social and economic values of a community. Product designers also gain a reflexive awareness of their important role in the society as agents of social change.

5. Acknowledgments

The authors would like to thank Mrs. Anik Pudjiati from the Kampung Kue Penjaringansari community, Charles Hartanto, Claudia Filiana, Eko Prasetyo, Meinasari, Monica Santoso, Alexander Ferdinand, Alfonsus Reynaldo, Yulianto Kurniawan, Jonathan Andrianto, and Nico Maximillion for their participation in this research.

References

[1] Adi I R 2008 *Intervensi Komunitas Pengembangan Masyarakat sebagai Upaya Pemberdayaan Masyarakat* (Jakarta: PT Raja Grafindo Persada)

[2] Setyati A W 2016 Peran Perempuan Anggota Kampung Unggulan Kue Penjaringansari dalam Meningkatkan Ekonomi Keluarga di Kecamatan Rungkut, Kota Surabaya *AntroUnairdotNet* Vol 5 (Surabaya: Airlangga University) p 461-70

[3] Bijl-Brouwer M d and Dorst K 2017 Advancing the strategic impact of human-centred design *Design Studies* Vol 53 (Amsterdam: Elsevier) p 1-23

[4] Wilkinson C R and Angeli A D 2014 Applying user-centred and participatory design approaches to commercial product development *Design Studies* Vol 35 (Amsterdam: Elsevier) p 614-31

[5] Maginn P J 2007 Towards more effective community participation in urban regeneration: the potential of collaborative planning and applied ethnography *Qualitative Research* Vol 7 (California: Sage Publishing) p 25 - 43

[6] Nurhayati E, Mulyana, Ekowati V I and Meilawati 2014 Inventarisasi makanan tradisional Jawa unsur sesaji di pasar-pasar tradisional kabupaten Bantula *Jurnal Penelitian Humaniora* Vol 19 (Yogyakarta: Universitas Negeri Yogyakarta) p 124-140

[7] Blake P A 2012 *A Process to Apply Hardwood Pallets as Viable Raw Material for Furniture Making* Master's thesis (Georgia Institute of Technology)

[8] Rubin J and Chisnell D 2008 Handbook of Usability Testing Second Edition (Indianapolis: Wiley Publishing Inc)

[9] Freire K, Borba G and Diebold Luisa 2011 Participatory Design as an Approach to Social Innovation *Design Philosophy Papers* Vol 9 p 235-250

IOP Conf. Series: Materials Science and Engineering **408** (2018) 012035 doi:10.1088/1757-899X/408/1/012035

Generative Design Approach for Modeling Creative Designs

Haibing Li[1, a] **and Roland Lachmayer**[1, b]

[1]Institute of Product Development, Welfengarten 1a, Hannover, 30167, Germany

[a]li@ipeg.uni-hannover.de, [b]lachmayer@ipeg.uni-hannover.de

Abstract. Design problems with variable requirements force designers to explore multiple alternatives between problem space and solution space of a design task. Despite several methods have been already proposed, CAD models hardly meet these variable requirements. To close this research gap, this study develops a generative design approach, addressing the automation of designing alternatives by capturing design knowledge into a set of design elements. These elements use essentially knowledge-based parametrics in terms of design requirements. Such design elements are used to assist designers in the creation of required design variants. Thus, the exploration of design alternatives is transformed into configuration of desired design elements. As a result, the invested time in routine work is reduced and designers are able to concentrate on the exploration of required design alternatives. Moreover, the output configuration also offers potentials in computer-aided optimization, since the design elements not only capture the design knowledge, but also incorporate manufacturing restrictions. At last, this approach is illustrated by taking a bicycle peddle crank as an example.

1. Introduction

Product design is a multidisciplinary activity that requires constant exploring, optimizing and testing until an efficient solution is designed and built. Such activities should be agile and well-organized to provide enough flexibility for adapting to constantly evolving markets. Thus, the key steps of these activities are that designers are required to explore multiple alternatives between problem space where all the design requirements are described, and solution space that consists of all possible design solutions for the design task [1]. Besides, ongoing understanding and new knowledge require continuous modification of concepts and constraints throughout the design lifecycle, inevitably impeding the search for alternative designs [2]. Hence, it is needed a robust design method to generate dynamic geometric models that can be easily adapted to satisfy changes and increases the number and diversity of design alternatives. As a result, computer-aided design (CAD) has been long accepted as an essential enabler, providing formal validation methods to create efficient and adaptable models that guarantees the delivery of truly innovative products in a timely and cost effective manner.

However, despite several computer-aided design methods have been proposed, e.g., parametric design [3], feature-based design [4] and knowledge-based design [5], CAD models hardly meet these variable design requirements, because design is an iterative process, which is limited to a reduced number of constraints and simple relations among parameters that are clearly insufficient to explore multiple design alternatives. To close this research gap, this study develops a Generative Design Approach (GDA) to address the automation of design alternatives by capturing design knowledge into a set of design elements, which are later used to assist designers in the creation of required design alternatives. As a consequence, the invested time in routine work is reduced, allowing designers to concentrate on the exploration of required design alternatives.

2. Related works of generative design

Generative design is a design method in which the exploration of a new design is described as a series of transformations from an early initial data to a large number of possible knowledge-integrated designs, which have high qualities that meet the design requirements in the best possible way. Hence, after studying the techniques that support generative design, Cui overviewed the framework of generative design by several examples, a design method is presented using automatic transformation algorithms of AI nature to support the exploration and generation of design alternative solutions [6]. Singh proposed an integrated generative design framework for supporting design exploration and creativity by reviewing five generative design techniques [7]. The potential challenges and building strategies towards developing such an integrated framework are discussed in this research as well. Krish developed a generative design method that is focused on generating multiple variants and alternatives of one proposed product without human intervention [8]. This method is suitable for complex multi-criteria design problems where important performance criteria are un-computable. However, in their works, there is a significant burden of selecting amongst a large quantity of design alternatives due to the lack of an evaluation method.

In the aspect of generative design implementation, Barros implemented a method for exploration of solutions for chair generation in the age of mass customization [9]. This method leads to a better exploration of design alternatives, stimulates multidisciplinary collaboration and finds potential best solutions prior to the manufacturing phase. Bagassi revised the improvements of generative design principles within traditional design procedures in aircraft design [10]. Thus, the output is not only a suitable design solution, but also a family of different results that designers could properly select and modify. Gulanova developed a generative design application within complex CAD systems, which provides an automatic adaptation of detailed design in case of design modification [11]. This is capable to improve data quality and reduce development time through collaborative workflow. However, this method is only applicable with enhanced CAD systems with parametric associative surface modeling and the designer has to follow predefined methodological rules.

3. GDA formulation

Generative design is formulated to support the automatic generation of a large number of design solution alternatives with computer algorithms that can be selected and controlled by designers. Despite the existence of many success stories in the area of generative design, the design procedures are still not well-documented. For that, in this paper the design procedures of GDA are formulated by generating design alternatives in a closed loop, as shown in figure 1.

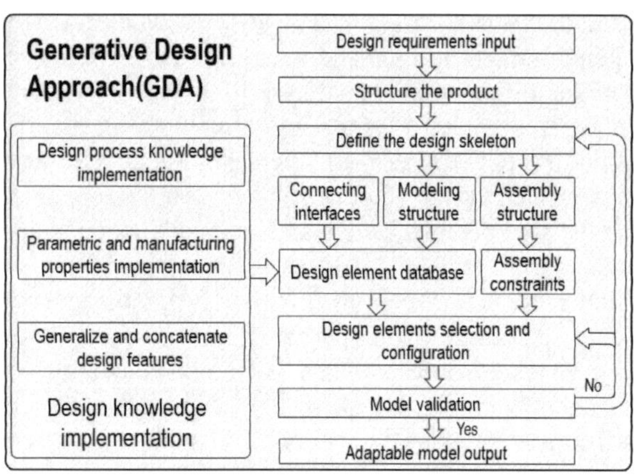

Figure 1. Framework of GDA.

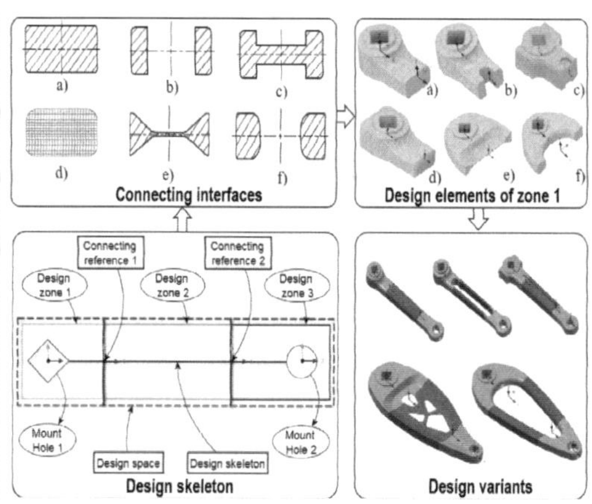

Figure 2. Bicycle peddle crank design by GDA.

According to the design requirements, a product is structured and analyzed by its functionalities from a systematic point of view, dividing the product into several design zones. After researching the product features and functions, a design skeleton is established, where the modeling structures, connecting reference planes and assembly structure are defined. Following the design skeleton, the connecting interfaces and design elements can be created. Desired design elements are linked through the design skeleton and configured via multiple levels of parametrizations. This parametrizations and configurations of design elements enable the generation of a large number of design alternatives that are robust against topological changes, which is beneficial for computer aided optimization. At the end, the exploration of the possible design alternatives is automated to find the best design solution.

4. GDA implementation: Bicycle peddle crank design

4.1. Design element database creation

Design element is the essential building block of the GDA. A design element in GDA is normally a CAD file that can be defined as "the representation of a family of objects that share the same topological constraints but have different geometry" [12]. The design element needs to capture the design intent, which is defined as "that is a set of geometric and functional rules, which the final product has to satisfy" [13]. The building sequence of a design element plays also an important role in determining its robustness and adaptability. Therefore, the creation processes of a design element are in detail clarified later in this section.

4.1.1. Design skeleton. The idea of a design skeleton is based on the observation that a new product is often comparable to preview ones, but with different dimensions and element positions. The result of a design skeleton is defined based on reference elements, such as points, axes, planes and surfaces, which provides either a modeling structure to create a design element or a reference to drive the main dimensions and positions of the design elements, e.g., an assembly structure. A well-constructed design skeleton contains all critical positions and dimensions, which are controlled via multiple levels of parametrization inside the model [14], which offers the following advantages:

- Specification-driven design: Key design information is stored in the design skeleton. Design constraints are defined within the skeleton to help allocate space for the design elements.
- Design changes management: The design skeleton helps to manage high-level design changes and propagates them throughout the assembly. Modifications of the design skeleton propagates the changes to all related design elements and sub-assemblies. This provides designers more control over changes in product development.
- Collaborative design: Key information stored in the design skeleton can be associatively copied into the appropriate design elements used in the product. Different designers can then edit the design element separately. Changes during the design can be made in the design skeleton and all models will update to reflect these modifications, realizing high degree of collaborative design.

To use the design skeleton properly, design elements are constrained using only the skeleton as reference for positioning and parametrization. The modification of the design element dimensions is done by manipulating values of the design skeleton parameters as well. As shown in figure 2, the design skeleton of a bicycle peddle crank contains 2 mount holes, which are the most important criteria that have to be considered. According to its structure and its functions, this crank is divided into 3 design zones, resulting in a skeleton that is defined by 4 local coordinate systems. These coordinate systems address the assembly positions as well as the orientations of all 3 design zones. Likewise, the connecting reference planes are also defined at every local coordinate system to set a reference for modeling and placing different design elements. Based on this skeleton, each design zone of the crank has a unique modeling structure, which is already linked to neighbor design zones.

4.1.2. Connecting interface parametric design. The modeling structure coupled with connecting interface explores the design boundaries of every design zone, so that every design element represents an inherent fraction of a product. Thus, parametric design of the connecting interfaces of a design element plays a vital role in design element creation. Further, the connecting interface has to incorporate design restrictions, e.g., manufacturing restrictions [15], so as to increase the robustness and the diversity of design elements. Going into details, parametric design of a connecting interface is identified by a shared set of parameters. On one hand, there are some fundamental parameters characterizing the shape information of a design element. On the other hand, there are special parameters characterizing the individual design information of the connecting interface. Besides, some parameters that have a limited number of values corresponding to design standards, e.g., DIN EN 10210-2 [16] for a hollow interface, should be defined as well. Therefore, a connecting interface is fully constrained by a set of parameters and integrated necessary design knowledge and restrictions.

4.1.3. knowledge-integrated design element creation. Design knowledge enables designers to setup a design, where the product is described in detail for the purpose of covering the allowable design space. Knowledge integration translates the design requirements to fully defined outputs via various methods, e.g., configuration concepts, formulas, macros and design tables. Thus, the knowledge related to a product is translated to a set of formulas and rules that describe its behaviors, so that the design requirements are formulated into a language that can be understood by the CAD system.

In figure 3, a creation example about the design elements of design zone 1 is presented. Here the design problem is how to balance the weight and structural stiffness of the bicycle peddle crank. The evaluation of the weight and stiffness requires geometrical models of the design. Normally, the evaluations are done with only a few design alternatives, since traditional parametric CAD modeling techniques hardly cover the whole design space. However, the evaluation here is carried out through different connecting interfaces by integrating different manufacturing techniques, e.g., lattice interfaces for additive manufacturing. With the help of a modeling structure, a design element addresses both design knowledge and manufacturing knowledge. When a design element is established, a check is made by design requirements and design guidelines to verify if the current set of parameters is allowed or conflicting. A design element is finished if all parameters are solved and no conflicts with the requirements on the weight and stiffness of the crank are found. Through detailed calculation, the designed elements have the same function but different topologies, which are saved together as a design element database for design zone 1 (figure 2).

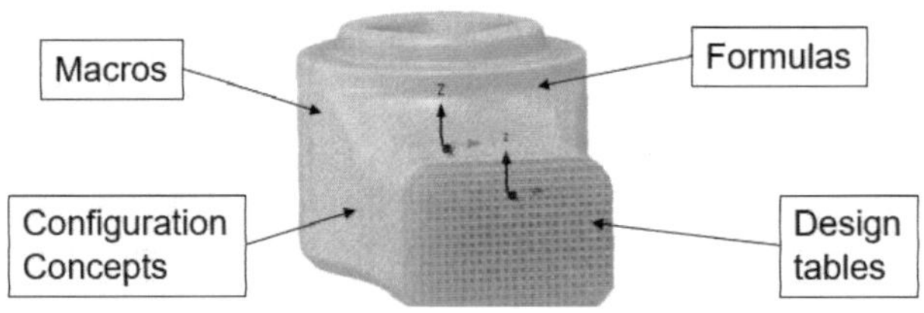

Figure 3. Knowledge implementation via various methods.

4.2. Model configuration and optimization

Based on different design requirements, design elements have to be assembled together. The possible design solutions are created by a combination of design skeleton, connecting interface and design element database. A computer-aided design tool that supports parametric design in parts and assembly is necessary to facilitate the implementation of GDA. During the design process, the user selects the desired design elements from the database and a generative model is built up step-by-step by applying automated placing and assembling procedures according to the design skeleton and connecting interfaces. A configurator, e.g., a spreadsheet or a macro, is also programmed being responsible for selecting desired design elements from the design element database for feeding model generation. In order to ensure that the selected design elements are compatible with each other, a rule is defined with the knowledge integrated in logical expressions to determine whether they are compatible or not.

Figure 4 shows the generation processes of the crank. For that, Autodesk Inventor CAD system is extended with a CAD plug-in application that supports the generation of such a crank. In this application, by defining the connecting interfaces, the design rules behind will retrieve the corresponding design elements and configure those at the right place in Inventor, which is basically done by inheriting parameters of the design skeleton to the design elements. Here these parameters are the positions and orientations of design elements. The local coordinate systems in the design skeleton aid for positioning design elements one after another and orient of them, so that a desired peddle crank is generated. In this way, different models can be created by changing the connecting interfaces. At last, the most relevant solution alternative is identified.

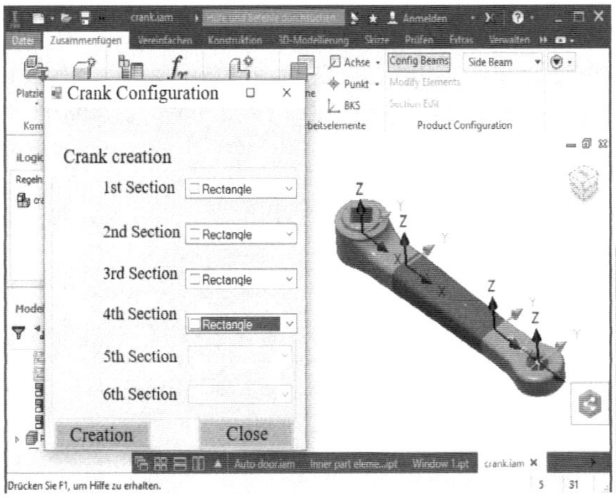

Figure 4. Model configuration.

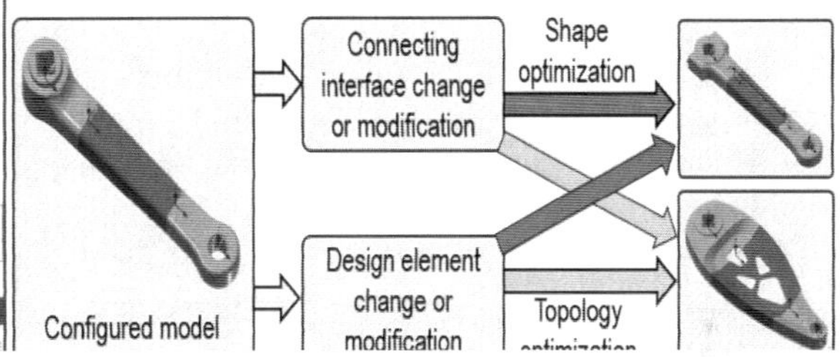

Figure 5. Computer-aided optimization.

The presented approach offers also possibilities for computer-aided optimization, since the design elements not only capture the design knowledge, but also incorporate knowledge of manufacturing restrictions (figure 5). On receiving initial results, designers will usually study the various options to refine the design skeleton and the connecting interfaces more accurately. With the help of the design element database, the designed model allows adding, deleting and modification of design elements or connecting interfaces throughout the product development process.

Basically, the parametrizations of connecting interfaces and design skeleton support parameter based shape optimization, i.e., the positions and orientations of design element as well as the parameters of connecting interface can be refined, realizing a parameter based shape optimization. Further, the design solution space is explored by topological design elements of every design zone. Therefore, the exchange of design elements leads to a topology optimization problem. The topology optimization by GDA is structured as the search for the design variation ranges within the design element database and the design knowledge by variable design rules, according to the design requirements.

5. Discussion

Design problems in GDA are represented as a design element configuration problem, where design solution alternatives are explored based on the design elements creation and configuration throughout the product development process. As discussed, GDA has the following advantages:

- Reduction of development time and cost: Through the use of limited computing, a large number of design alternatives can be investigated at the same time it might take to create one design using a more traditional approach, which is a timely and cost effective manner.
- Customize product development: With GDA, complex geometries specifically designed and optimized easily to suit an individual need are much more accessible than ever before.
- Provide optimization ability: Based on the parametrizations of design elements and design skeleton, computer aided optimization is easily implemented by either exchanging the design elements or manipulating the parametrizations of connecting interfaces or design skeleton.

However, there are also some negative aspects of GDA. It is clearly observed that GDA has an increased demand on knowledge of design engineers in terms of design element creation and knowledge integration. Besides, in GDA, the focus is supporting knowledge for engineers to capture and formalize the necessary design knowledge rather than preparing the knowledge for end users.

6. Conclusion and outlook

The contribution of this research is the development of GDA, which supports the automatic generation of a large number of design solution alternatives. It does not only save time by exploring and storing all the necessary knowledge of the whole lifecycle through design skeleton and design element database, but also integrates CAD system through friendly user-interfaces that provide reliable technical supports. The output model is not only a design solution, but also a concept that the designers could manipulate that facilitates design optimization, because the design elements capture the geometric knowledge and incorporate design restrictions as well. Therefore, the optimization is structured as the search of the variation ranges within the design elements and the design knowledge by variable design configuration rules according to the design requirements.

After verifying GDA with the mentioned case study, the evaluation of the generated models still have to be investigated. Rather, in the application and example presented in this paper, design restrictions have been integrated in the design elements, so that the evaluation was left for the user to decide. In this way, the user can understand how to configure a satisfied outcome. Otherwise, the application will be exhausted when all the design rules have been applied. Nevertheless, an evaluation method will be developed in the future to suggest which design alternative is better than others. Another important issue will concentrate on the development of design methods and tools for a further integration of design knowledge, in which the manufacturing restrictions has to be implemented as well.

References

[1] Gembarski P, Li H and Lachmayer R 2017 Template-Based Modelling of Structural Components *Int. J. of Mech. Eng. and Robotics Res.* **6** pp 336-342

[2] MacLellan C, Langley P, Shah J and Dinar M 2013 A Computational Aid for Problem Formulation in Early Conceptual Design J. Computer Inf. Sci. Eng. 13 pp 0310051-10

[3] Wawre S and Kabade D 2016 An overview of recent trends in parametric modelling Int. J. of Modern Trends in Eng. and Res. 3 pp 381-389

[4] Shahin T 2008 Feature-based design - an overview Computer-Aided Des. a. Appl. 5 pp 639-653

[5] Hirz M, Gfrerrer A and Lang J 2013 Knowledge-Based Design Integrated Computer-Aided Design in Automotive Development ed Hirz M (Berlin: Springer) chapter 5 pp 309-330.

[6] Cui J and Tang M 2017 Towards generative systems for supporting product design, Int. J. of Des. Eng. 7 pp 1-16

[7] Singh V and Gu N 2011 Towards an integrated generative design framework Des. Stud. 33 pp 185-207

[8] Krish S 2011 A practical generative design method Computer-Aided Des. 43 pp 88-100

[9] Barros M and Chaparro B 2014 Integrated generative design tools for the mass customization of furniture Design Computing and Cognition '12 ed Gero J (Dordrecht: Springer) pp 285-300.

[10] Bagassi S, Lucchi F, De Crescenzio F and Persiani F 2016 Generative design: advanced design optimization process for aeronautical application Proc. of 30th Int. Council of Aeronautical Sci., Daejeon, South Korea, September 25-30

[11] Gulanová J, Gulan L and Hirz M 2017 Generative engineering design methodology used for the development of surface-based components Computer-Aided Des. and Appl. 14 pp 642-649

[12] Solano L and Brunet P 1994 Constructive constraint-based model for parametric CAD systems Computer-Aided Des. 26 pp 614-621

[13] Mun D, Han S, and Kim J 2003 A set of standard modeling commands for the history-based parametric approach Computer-Aided Des., 35 pp 1171-1179

[14] Sauthoff B 2017 Generative Parametrische Modellierung von Strukturkomponenten für die Technische Vererbung, PhD thesis, Technische Universität Hannover.

[15] Gembarski P, Sauthoff B, Brockmöller T and Lachmayer R 2016 Operationalization of Manufacturing Restrictions for CAD and KBE-Systems Proc. of 14th Int. Des. Conf., Dubrovnik, Croatia, May 16 - 19

[16] DIN EN 10210-2 2016 Hot finished structural steel hollow sections (Berlin: Beuth Verlag)

IOP Conf. Series: Materials Science and Engineering **408** (2018) 012036 doi:10.1088/1757-899X/408/1/012036

Design of Low-power Ultrasonic Anemometer Based on STM32L476

H B Fan, J L Liu[*] and G Z Sun

School of Electric and Electronic Engineering, North China Electric Power University, Baoding 071003, China

[*]E-mail:940175139@qq.com

Abstract. The power consumption of ultrasonic anemometers in engineering application has always been too high, which could limit their application in the long-time field measurement for wind speed. To solve the above problem, an ultrasonic anemometer with low power consumption, which employs the time-difference method to measure wind speed, has been designed in this paper. The ultra-low-power microprocessor STM32L476 is adopted as the core processor and low-power chips are used extensively in peripheral circuits. The whole circuit consists of power circuit, ultrasonic wave transmitter, signal receiver and memory chip. Besides, it can work in both online mode and self-contained mode. Simultaneously, Kalman filter algorithms have been employed to reduce data errors. It has been indicated in many experiments that the anemometer has lower power consumption, as well as higher precision.

1. Introduction

In recent years, with the development of new ultrasonic transducers and the popularization of high-speed processors, ultrasonic has been extensively applied in measurement technology. Compared with the traditional mechanical anemometer, the ultrasonic anemometer possesses a higher precision and a larger measuring range, and does not suffer from the mechanical wear [1]. Therefore, it has been widely used in engineering applications currently.

However, in some wind speed measurement experiments, carried out in the field, the ultrasonic anemometer can only be powered by the battery, which gives a limitation to its working hours. To meet the requirement of the long-time field experiments, it is of great significance to reduce the power consumption of ultrasonic anemometers. This paper introduces a low-power ultrasonic anemometer with the self-contained working mode. This low-power anemometer can not only output real-time wind speed, but also work in the self-contained working mode, which can realize the sleep-measurement, storage-sleep work. This newly designed ultrasonic anemometer has a great application value in engineering practice.

2. Wind speed measurement method

The ultrasonic anemometer designed in this paper uses time difference method to measure wind speed and direction, as shown in figure 1. Two groups of ultrasonic transducers are placed orthogonally, and the distance between two transducers in each group is chosen as the same value [2], with four transducers driven in turn. The wind speed v can be orthogonally decompositioned into v_x and v_y. L stands for the distance between two transducers in the same group, and c represents the ultrasonic speed in the current environment. For the x direction, the time difference between ultrasonic waves sent from transducer A and received by transducer B in the downwind condition is T_{AB}. It can be expressed by the following equation.

$$L = (c + v_x) \cdot T_{AB} \tag{1}$$

And, in the headwind condition, the time difference between ultrasonic waves sent from transducer B and received by transducer A is T_{BA}.

$$L = (c - v_x) \cdot T_{BA} \tag{2}$$

Combine the two above equations and $\boldsymbol{v}_x$ can be given as

$$v_x = \frac{L}{2} \cdot \left(\frac{1}{T_{AB}} - \frac{1}{T_{BA}}\right) \tag{3}$$

Similarly, the wind speed in y direction $\boldsymbol{v}_y$ can also be obtained.

$$v_y = \frac{L}{2} \cdot \left(\frac{1}{T_{CD}} - \frac{1}{T_{DC}}\right) \tag{4}$$

So, the wind speed can be expressed as:

$$v = \frac{L}{2} \cdot \sqrt{\left(\frac{1}{T_{AB}} - \frac{1}{T_{BA}}\right)^2 + \left(\frac{1}{T_{CD}} - \frac{1}{T_{DC}}\right)^2} \tag{5}$$

Moreover, the wind direction angle α can be calculated with arc tangent equation.

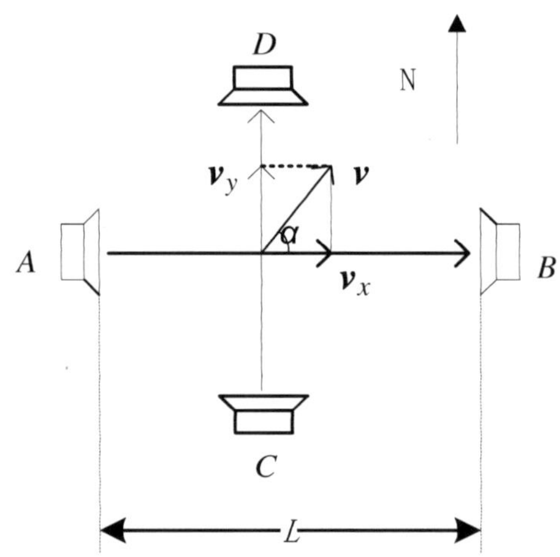

Figure 1. Time difference method.

3. System design

3.1. Overall design

With STM32L476 employed as the core processor, the system is mainly composed of power unit, signal control and transmitting circuit, signal receiving and processing circuit, memory chip and communication unit, as shown in figure 2.

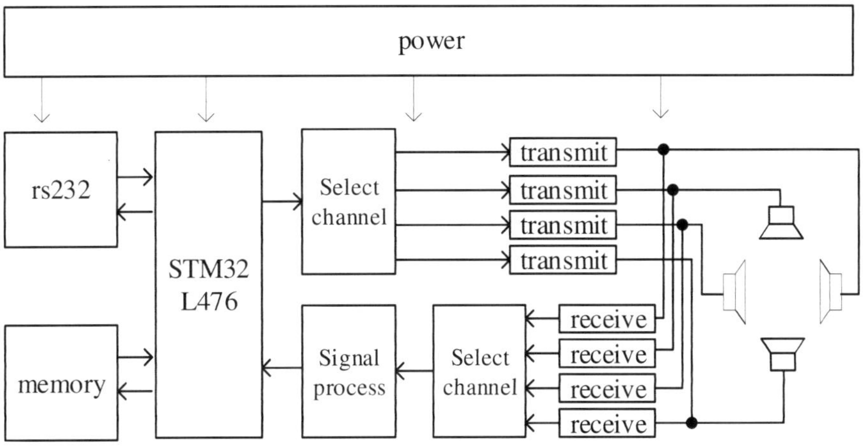

Figure 2. The Anemometer System.

The power unit supplies power to the circuit. And the processor can control the power supply of the peripheral circuit by controlling the power chip, which can subsequently decide whether the system sleeps. MCU generates two complementary PWM waves to drive transducer to transmit ultrasonic wave, which can be received by the corresponding transducer to generate an electrical signal. After further amplification and envelope detection, the signal can be converted to a square wave by voltage comparator, which can be easily captured by the MCU. Finally MCU calculate the speed of sound based on the time difference between transmitting and receiving. Different instructions can determine two different working modes of the system. The first mode is to continuously output the real-time wind speed online, and the second is the self-contained mode, which adopts sleep-measurement, storage-sleep work cycle.

3.2. Circuit design

3.2.1. Microprocessor. Based on ultra-low-power ARM Cortex-M4 processor core, the STM32L476 is a high-speed microprocessor of STMicroelectronics. The current in shutdown mode is only 30 nA, and the current in standby mode is only 300 nA. This system employs the stop2 low-power mode of the processor, and the current is only 1.4 uA while RTC is working. However, other types of microprocessors usually operate at more than 20 uA in the same operating mode. Therefore, choosing STM32L476 as the core processor can reduce power consumption of system greatly.

3.2.2. Power unit. The system circuit mainly consists of analog circuits and digital circuits. The digital circuit includes microprocessor, memory chip and RS232 chip, and the analog circuit is mainly composed of ultrasonic wave transmitting circuit and signal receiving circuit. The microprocessor needs the long-term stable power-on mode to control other peripherals. In addition, in order to lower power consumption of system, the analog circuit should be shutdown when the system enter sleep mode. Therefore, this system introduces two different power units that power to digital circuits and analog circuits respectively, with 12 V DC input voltage converted into different output voltages.

3.2.3. Signal control and transmitting unit. As shown in Figure 3, in order to drive four transducers in turn, the system adds a analog switch 74HCT4052D in the signal control part. The microprocessor changes the voltage of the EN pin to control the channel selecting. Also 74HC45 is used here to improve driving ability of the signal, so that the signal can drive transducer to transmit ultrasonic waves more easily.

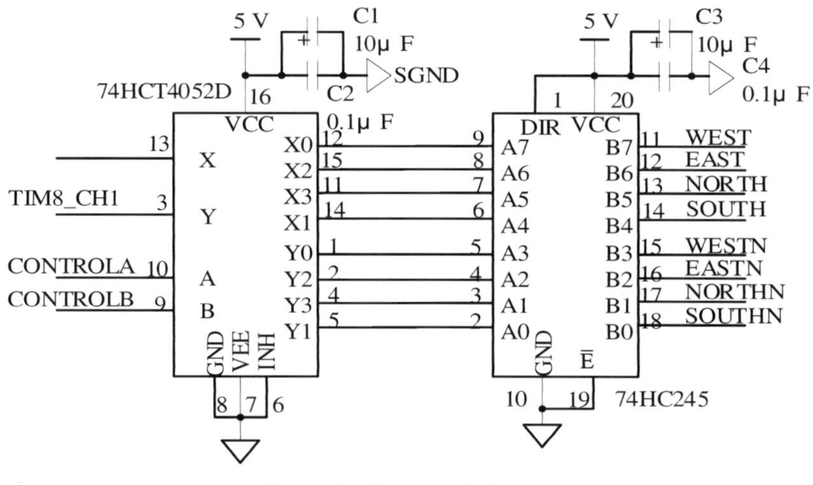

Figure 3. Channel Selecting.

Since transducers connect the ultrasonic transmitting together with receiving circuits, there will be some interference between the transmitted signal and the received signal in the circuit. In order to avoid this interference, two parallel diodes are laid in transmitting circuits to prevent the transmitting signal from disturbing the receiving circuits. Besides, two diodes are incorporated in the receiving circuits to limit voltage so that the circuits will not be damaged by the high voltage of transmitting circuits, as shown in Figure 4 [3].

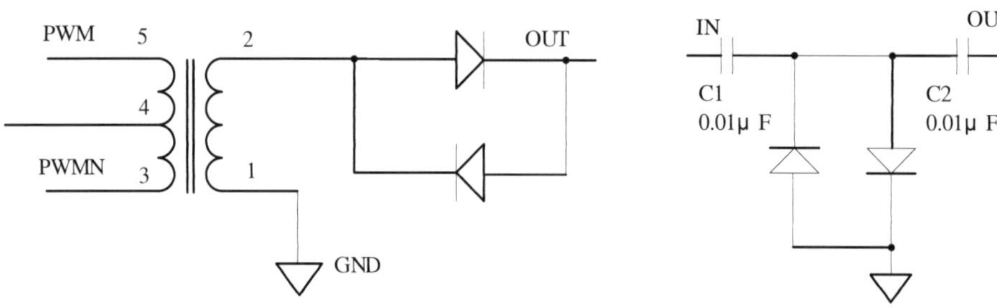

(a) (b)

Figure 4. Anti-interference Circuit.

3.2.4. Signal receiving and detection unit. Ultrasonic waves decay rapidly while it travels in air. Its amplitude attenuation law can be expressed by following equation.

$$A = A_0 \cdot e^{-\alpha x \pm \beta v} \tag{6}$$

Where, A_0 is amplitude of the acoustic wave before travelling. α is attenuation factor, which is proportional to the square of frequency; β is attenuation offset coefficient due to wind speed; v is wind speed. and v is positive when v is coincident with the direction of propagation [4]. Due to the attenuation process, the transmitted signal is quite weak after being received by the transducer.

In order to capture the weak signal accurately, filtering, amplification, and detection circuits are usually employed in the receiving circuit, which would result in high power consumption. In order to simplify the circuit and improve the anti-interference ability of the receiving circuit as well as lower power consumption, SA614A is selected to process signals, which has low power consumption, strong anti-interference and can achieve amplification and envelope detection for the input signal.

After envelope detection, the frequency of the received signal decreases, but its edge rises/falls slowly, which will induce a large error when the processor capture the signal. In order to improve the capture accuracy and reduce the signal rise/fall time, a voltage comparator is adopted here [5]. Figure 5 and figure 6 show the waveforms before and after the voltage comparator, which are observed by the oscilloscope. As shown in the figure, the signals processed by the voltage comparator have much better edge characteristics.

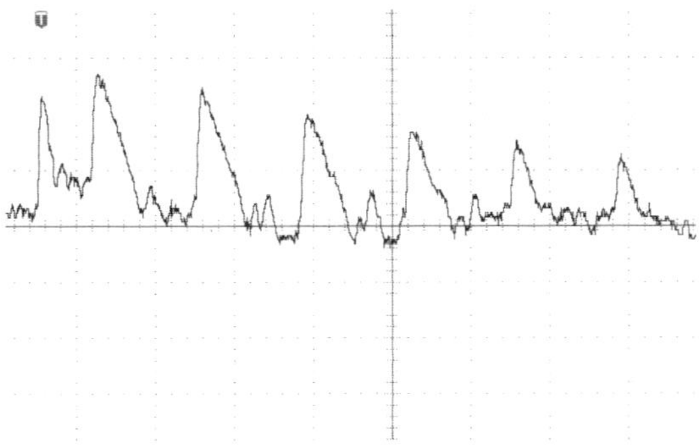

Figure 5. Signals before processing.

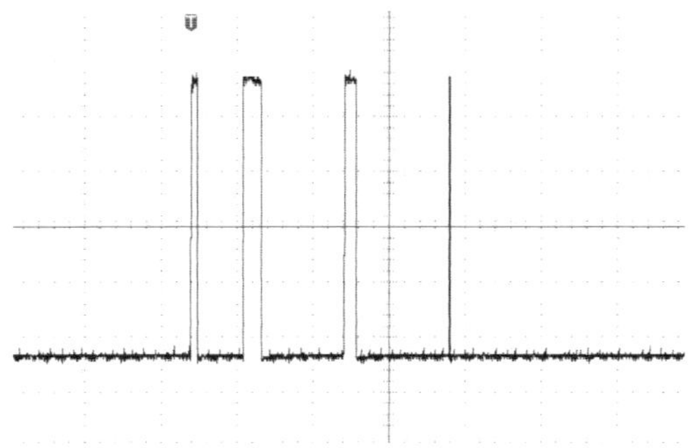

Figure 6. Signals after processing.

3.3. System operation flow

As shown in figure 7 (a), after initialization, the system first selects the working mode by instructions from serial port. In the online output mode, the system continuously calculates the current wind speed and direction, and outputs it via the serial port. If self-contained mode is selected, the system will sleep after measuring and storing a set of wind speeds. When the MCU sleeps, it controls the EN pin of the power chip to power off the external circuit at first, and then the MCU enters STOP2 mode, so that the entire system enters a sleepy state with extremely low power consumption. The system will work again after waking up.

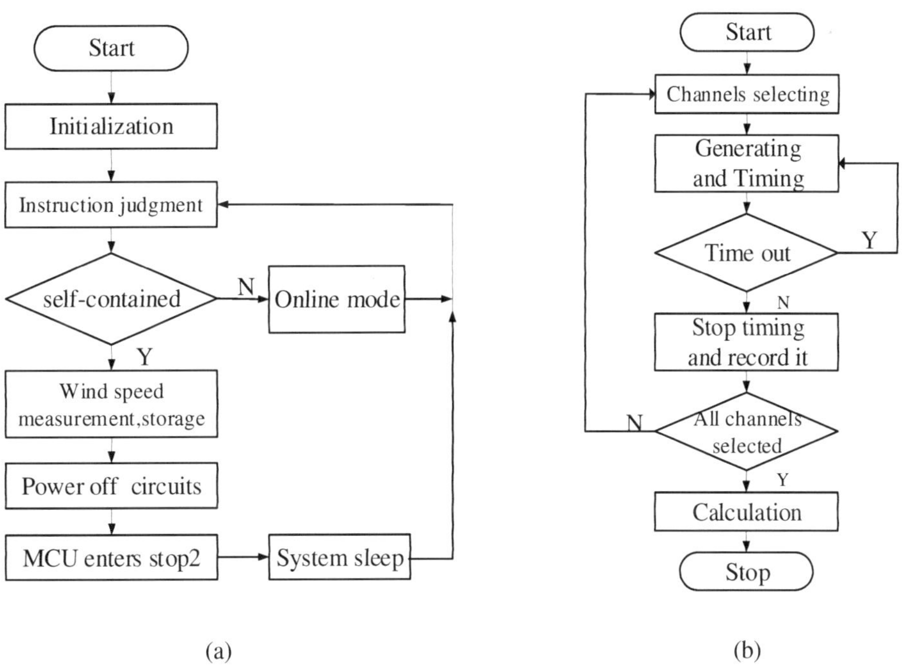

Figure 7. System operation flow and wind speed calculation flow.

IOP Conf. Series: Materials Science and Engineering **408** (2018) 012036 doi:10.1088/1757-899X/408/1/012036

3.4. Wind speed measurement
As shown in Figure 7 (b), the microprocessor first determines the transmitting channel and receiving channel. Then it transmits a series of pulses and starts timing. After capturing signal, the processor stops timing and records the time difference. If the time difference exceeds the preset threshold, the process will restart. Then the remaining 3 channels are selected and the time difference will be recorded respectively. After four transmissions and receptions being completed, the wind speed can be calculated using the equation 5. Finally, the Kalman filter algorithms are employed to correct wind speed.

3.5. Kalman filter algorithms
The errors of wind speed mainly result from interference of airflow at the test site, occasional fluctuations in sensors, noises from circuit components and other random factors. These errors can be considered as mainly composed of Gaussian white noise. In order to reduce the impact of noise on the accuracy, the system decided to use Kalman filter algorithms to process the data. Because the wind speed is the only output of system, all variables of the classical Kalman filter state equation and prediction equation are one-dimensional. The equations are as follows [6].

$$X_k^- = AX_{k-1} + BU_{k-1} + q_{k-1} \tag{7}$$

$$Y_k = HX_k + r_k \tag{8}$$

Where, X_k is the state vector(wind speed) at time k; X_{k-1} is wind speed at time k-1; A is the state transition matrix which applies the effect of each system state parameter at time k-1 on the system state at time k and we assume that the wind speed is unchanged in a short time, so A=1; B is the control input matrix which applies the effect of each control input parameter in the vector U_{k-1} on the state vector; q_{k-1} is the vector containing the process noise, with covariance given by the covariance matrix Q; And r_k is the vector containing the measurement noise with covariance R. By experiments, we determine the value of Q *and* R, with the covariance matrix P_0 =1. Then the following recurrence equations can be obtained [7].

$$X_k^- = X_{k-1} \tag{9}$$

$$P_k^- = P_{k-1} + Q \tag{10}$$

$$K_k = P_k^-(P_k^- + R)^{-1} \tag{11}$$

$$X_k = X_k^- + K_k(Y_k - X_k^-) \tag{12}$$

$$P_k = (I - K_k)P_k^- \tag{13}$$

By substituting the measured wind speed and the previous wind speed into the above equations, the accurate wind speed present can be calculated.

4. Test

4.1. Power consumption
The author investigated a variety of ultrasonic anemometers that are widely used in engineering application and found that their power is up to 1 W, and most of them cannot work on the self-contained mode. We measured the current in different working modes when 12 V DC voltage was used here to power our newly designed ultrasonic anemometer. The current in the on-line working mode is 70 mA, and 40 uA in sleeping mode. It can be indicated that this ultrasonic anemometer has obvious advantages in terms of power consumption.

4.2. Accuracy test
In order to test the accuracy of the ultrasonic anemometer, the author conducted a test in a small wind tunnel. The temperature of the wind tunnel is 23.72°C and the relative humidity is 52.34%. Due to the

wind tunnel conditions, 5 points of 1 to 20 meters per second were selected to measure wind speed in the experiment. We conducted 20 measurements for the wind speed at each point and obtained average value to compare with the standard wind speed. The results are shown in Table 1, which can prove the high accuracy of this newly designed ultrasonic anemometer.

Table 1. Comparison of standard wind speeds and measured wind speeds (m/s)

Standard	1.00	5.00	10.00	15.00	20.00
measured	0.92	4.87	10.17	14.78	20.35

5. Summary

Based on the fact that the ultrasonic anemometer on the market generally consumes too much power and can only work in online mode, a low-power ultrasonic anemometer with a self-contained working mode is newly designed in this paper. Two working modes of the anemometer can meet different experimental requirements, and the power consumption is greatly lowered. In addition, the Kalman filter algorithms are introduced to reduce data errors and improve system stability. This newly designed anemometer has been proved to be very useful in engineering applications by power consumption test and accuracy test.

References

[1] Liu D R, Wang Y, Wang T L and Zhang B Y 2017 Design of Ultrasonic Anemometer Based on STM32F *Measurement & Control Technology* vol 36 no 6 pp 97-100

[2] Wu Z L, Yan L S, Zeng D B, Chen J Z 2012 Design of ultrasonic wave anemometer with high precision based on ARM+CPLD *Transducer and Microsystem Technologies* vol 31 no 6 pp101-3+107

[3] Zhou K and Xin X S 2011 Wireless Ultrasonic Anemometer Based on Direct Time Difference Method *Microcontrollers & Embedded Systems* vol 11 no 12 pp 54-6+60

[4] Cao K J, Cui G H and Zhu Y B 2010 Modeling and analysis of ultrasonic anemometer *Technical Acoustics* vol 29 no 4 pp 388-91

[5] Ren B F, Chen L, Liu B J and Zhang Z W 2010 Application of Voltage Comparator in Test System *Instrumentation and Technology* no10 pp 63-5

[6] Fan H B, Jia S S, Zuo B B and Li R Q 2016 Design of Single Beam Acoustic Doppler Velocity Meter *Instrument Technique and Sensor* no 9 pp 37-40+97

[7] Kim K, Lee J and Park C 2009 Adaptive two-stage extended Kalman filter for a fault-tolerant INS-GPS loosely coupled system *IEEE Transactions on Aerospace and Electronic Systems* vol 45 no 1 pp 125-37